Advancements in Numerical Analysis and Quantitative Decision-Making

Serkan Derici
https://orcid.org/0000-0003-2581-6770
Nevşehir Hacı Bektaş Veli University, Turkey

IGI Global
Scientific Publishing
Publishing Tomorrow's Research Today

Vice President of Editorial Melissa Wagner
Director of Acquisitions Mikaela Felty
Director of Book Development Jocelynn Hessler
Production Manager Mike Brehm
Cover Design Jose Rosado

Published in the United States of America by
IGI Global Scientific Publishing
701 East Chocolate Avenue
Hershey, PA, 17033, USA
Tel: 717-533-8845 | Fax: 717-533-7115
Website: https://www.igi-global.com E-mail: cust@igi-global.com

Library of Congress Cataloging-in-Publication Data

Names: Derici, Serkan, 1991- editor
Title: Advancements in numerical analysis and quantitative decision-making / edited by Serkan Derici.
Description: Hershey, PA : IGI Global Scientific Publishing, [2026] | Includes bibliographical references and index. | Summary: "This book serves as a comprehensive academic resource that brings together current developments and innovative approaches in numerical analysis and quantitative decision-making processes"-- Provided by publisher.
Identifiers: LCCN 2025053798 (print) | LCCN 2025053799 (ebook) | ISBN 9798337367460 hardcover | ISBN 9798337367477 paperback | ISBN 9798337367484 ebook
Subjects: LCSH: Statistical decision | Numerical analysis | Decision making--Mathematics
Classification: LCC QA279.4 .A378 2026 (print) | LCC QA279.4 (ebook)
LC record available at https://lccn.loc.gov/2025053798
LC ebook record available at https://lccn.loc.gov/2025053799

British Cataloguing in Publication Data
A Cataloguing in Publication record for this book is available from the British Library.

Table of Contents

Detailed Table of Contents

Big data has evolved as a disruptive force in the digital age, altering how businesses acquire, store, and analyse data. Big data, defined by the "3Vs" of volume, variety, and velocity, has transformed database management systems (DBMS). Traditional DBMS, which were created for structured and constrained datasets, struggle to handle the massive, unstructured, and rapidly generated data of current applications. Hadoop, MapReduce, and NoSQL databases have addressed these difficulties by providing distributed, scalable, and fault-tolerant systems for big data management. Big data applications are found in a wide range of industries, including healthcare, education, logistics, and finance, and they drive innovation and efficiency. Despite its potential, data security, regulatory compliance, and infrastructure scalability remain major concerns. The future of DBMS is to fulfil the demands of a data-driven world by harnessing new technologies such as quantum computing and AI.

Modern decision-making depends on extracting reliable patterns from noisy, heterogeneous data—tables, time series, text, and graphs. After collecting and unifying these sources, data must be cleaned, transformed, and converted into model-ready features. Modeling techniques span time-series methods, gradient boosting for tabular data, neural models for text, and graph analytics, with probabilistic tools adding uncertainty estimates. Rigorous evaluation ensures models align with business goals. Decisions are then operationalized through transparent rules, and continuous monitoring detects drift and guides retraining. In banking, integrating repayment histories, income data, network links, and text enables credit scoring, cash-flow forecasting, fraud detection, and risk-aligned pricing. This framework provides a general blueprint for turning messy data into evidence-based intelligence.

This chapter explores the shift from traditional statistical analysis to advanced deep learning in quantitative decision-making. It argues for integrating classical paradigms—such as descriptive and inferential statistics conducted in tools like SPSS—with modern computational techniques to address complex, large-scale challenges. The discussion highlights the pivotal role of big data platforms, including Hadoop and Spark, in enabling real-time analytics and distinguishes core machine learning algorithms from deep neural architectures designed for unstructured data. Interdisciplinary case studies in healthcare, finance, and engineering demonstrate the practical synergy of these approaches. By embracing multi-paradigm strategies, the chapter offers insights into building transparent, scalable, and effective decision-support systems across diverse domains. It concludes by addressing challenges of ethics and governance and by pointing to future directions such as explainable AI and federated learning, emphasising that robust decision-making depends on a hybridised analytical framework.

Quantitative decision-making challenges in management, technology, and finance increasingly rely on data-driven forecasts and heterogeneous preferences, which are often only partially observable. This chapter presents an integrated decision-making framework that explicitly incorporates forecast and preference uncertainty to facilitate robust and explainable decisions. The proposed explainable and robust multi-criteria decision-making (xMCDM) approach integrates a predictive layer with uncertainty quantification, employs a Bayesian preference-learning method to model uncertain weights, and uses distribution-robust decision aggregation. Empirical case studies in supplier selection, renewable energy site planning, and credit portfolio allocation demonstrate that xMCDM enhances rank stability, reduces regret, and enables transparent evaluation of robustness trade-offs. Consequently, xMCDM advances the focus from score-based evaluation to robust decision-making under uncertainty.

potential of machine learning as a numerical tool for supporting quantitative financial decision-making.

The present chapter lists a specific literature review from 2010 till 2025 about Educational Data Mining to determine steps in developing effective predictive models for identifying low-performing students, the parameters that influence academic achievements, various data mining techniques that can efficiently predict students' academic outcomes along with the advanced methods that may improve the prediction results. Finally, it may be concluded that students' academic and background features are the most influential features. Further, among the data mining techniques that could be used for predicting students' performance classification (especially Decision Tree) was a widely used data mining technique due to its simplicity, interpretability, and capability to provide significant results consistently. The present analysis also reveals that the use of feature selection, data balancing, hyper-parameter tuning, and ensemble models may further improve the prediction results.

This study applies quantitative decision-making and numerical modeling to analyze interactions between dietary behavior, obesity, and perceived stress among 1,036 students at Ibn Tofail University, Morocco. Using binary and multinomial logistic regression models, key determinants were identified for adopting a balanced diet, developing obesity, and experiencing high stress. Age, home-prepared meals, and health awareness increased the likelihood of a balanced diet, whereas sugary drinks and long engagement time reduced it. Regular physical activity and better nutritional knowledge lowered obesity risk, while prolonged daily engagement increased it. Obesity strongly predicted high perceived stress, supported by an E-value of 6.075, indicating robust associations. These findings highlight the relevance of numerical analysis in understanding student health dynamics and support evidence-based decision-making for targeted university health interventions.

of market efficiency and highlights the critical impact of deviations from a normal distribution on the analysis results. The results demonstrate the heterogeneity of the capital markets of the countries studied and emphasize that the choice of method determines inferences regarding the efficient market hypothesis.

Preface

The accelerating growth of data availability, computational power, and algorithmic sophistication has profoundly reshaped numerical analysis and quantitative decision-making across disciplines. Contemporary decision problems whether in economics, management, health sciences, education, or environmental systems—are increasingly characterized by scale, uncertainty, heterogeneity, and the need for transparency. Traditional analytical methods, while foundational, are no longer sufficient on their own to address these challenges. Instead, robust decision-making today requires an integrated framework that combines classical numerical and statistical reasoning with modern machine learning, optimization, and data-driven methodologies.

Advancements in Numerical Analysis and Quantitative Decision-Making brings together contributions from researchers across diverse geographical regions and academic domains to reflect this methodological evolution. The volume emphasizes not only predictive accuracy, but also interpretability, robustness, and practical relevance, key requirements for translating numerical results into actionable decisions. By spanning topics from big data infrastructures and explainable multi-criteria decision-making to health analytics, educational data mining, financial markets, and sustainable agriculture, the book illustrates how quantitative tools can be adapted to complex real-world contexts.

The primary objective of this book is threefold. First, it provides a conceptual and methodological bridge between classical numerical analysis and emerging computational approaches. Second, it demonstrates applied quantitative decision-making through empirical case studies grounded in real datasets. Third, it highlights future research directions, including explainable AI, uncertainty-aware modeling, and sustainability-oriented decision frameworks. As such, this volume is intended for researchers, graduate students, and practitioners seeking both theoretical insight and applied guidance in modern quantitative analysis.

OVERVIEW OF CHAPTERS

Chapter 1 introduces the foundations of big data and examines its transformative impact on database management systems (DBMS). It discusses the limitations of traditional DBMS in handling high-volume, high-velocity, and high-variety data and presents distributed technologies such as Hadoop, MapReduce, and NoSQL databases as key enablers of scalable data management. The chapter sets the technological context for data-intensive numerical analysis.

Chapter 2 focuses on algorithmic frameworks for extracting reliable structure from complex and noisy data. It presents a full decision pipeline, from data collection and preprocessing to modeling, evaluation, and deployment, emphasizing uncertainty handling and continuous monitoring. Practical examples from banking illustrate how heterogeneous data sources can be transformed into evidence-based decisions.

Chapter 3 explores the transition from traditional statistical analysis to machine learning and deep learning in quantitative decision-making. By advocating a hybrid analytical paradigm, the chapter demonstrates how classical statistics, big data platforms, and neural architectures can be combined to build scalable, transparent, and ethically grounded decision-support systems.

Chapter 4 develops an integrated framework for explainable and robust multi-criteria decision-making (xMCDM). By combining predictive analytics, Bayesian preference learning, and distribution-robust optimization, the chapter addresses uncertainty in both forecasts and stakeholder preferences. Empirical case studies highlight improvements in decision stability, robustness, and interpretability.

Chapter 5 examines strategic decision-making through the integration of qualitative and quantitative tools. By combining SWOT analysis with breakeven analysis and multi-criteria decision-making methods such as AHP and QSPM, the chapter demonstrates how numerical techniques can enhance strategic planning in both public and private sector contexts.

Chapter 6 presents a comparative study of machine learning models for budget prediction. Using multiple real-world datasets, the chapter shows that ensemble methods such as Random Forest and Gradient Boosting outperform traditional linear regression, underscoring the nonlinear nature of financial decision problems.

Chapter 7 reviews educational data mining techniques for predicting students' academic performance in higher education. It synthesizes findings from over a decade of research, identifying influential features and highlighting the effectiveness of interpretable models such as decision trees, alongside advanced ensemble and optimization techniques.

Chapter 8 applies logistic and multinomial regression models to analyze dietary behavior, obesity, and perceived stress among university students. The chapter demonstrates how numerical modeling can uncover complex health-related relationships and support evidence-based interventions in educational settings.

Chapter 9 investigates tobacco-related health risk factors across OECD countries, with particular attention to psychosocial dimensions such as age and gender. Through multivariate analysis, the chapter provides quantitative insights into smoking and vaping patterns and their implications for public health decision-making.

Chapter 10 introduces constrained ordination techniques, specifically RDA–PLSR modeling with permutation-based inference, to study environmental structuring of biochemical and antioxidant traits. The chapter illustrates the power of numerical ordination methods for ecological and biochemical decision analysis.

Chapter 11 explores machine learning approaches for predicting plant water retention under salt stress. Using Random Forest regression and feature importance analysis, the chapter demonstrates how interpretable AI models can support agronomic decision-making and stress management strategies.

Chapter 12 proposes a multi-criteria predictive framework for sustainable tomato irrigation. By combining machine learning models with TOPSIS-based ranking, the chapter establishes a baseline architecture for future integration of treated wastewater data, addressing sustainability and resource optimization challenges.

Chapter 13 concludes the volume with an empirical investigation of the Efficient Markets Hypothesis using the RALS method. By comparing classical and robust unit root tests across multiple capital markets, the chapter highlights how methodological choices influence financial decision-making and inference.

Together, these chapters offer a comprehensive and interdisciplinary perspective on contemporary numerical analysis and quantitative decision-making. The volume underscores that effective decisions in complex systems arise not from isolated methods, but from the thoughtful integration of data, models, uncertainty, and domain knowledge.

Chapter 1
Big Data and Its Impact on DBMS

Tanishka Prashant Madke
Ajeenkya D.Y. Patil University, India

Amisha Allauddin Shaikh
Ajeenkya D.Y. Patil University, India

Susanta Das
https://orcid.org/0000-0002-9314-3988
Ajeenkya D.Y. Patil University, India

ABSTRACT

Big data has evolved as a disruptive force in the digital age, altering how businesses acquire, store, and analyse data. Big data, defined by the "3Vs" of volume, variety, and velocity, has transformed database management systems (DBMS). Traditional DBMS, which were created for structured and constrained datasets, struggle to handle the massive, unstructured, and rapidly generated data of current applications. Hadoop, MapReduce, and NoSQL databases have addressed these difficulties by providing distributed, scalable, and fault-tolerant systems for big data management. Big data applications are found in a wide range of industries, including healthcare, education, logistics, and finance, and they drive innovation and efficiency. Despite its potential, data security, regulatory compliance, and infrastructure scalability remain major concerns. The future of DBMS is to fulfil the demands of a data-driven world by harnessing new technologies such as quantum computing and AI.

DOI: 10.4018/979-8-3373-6746-0.ch001

1. INTRODUCTION

In 2017, The Economist reported that "the world's most valuable resource is no longer oil, but data". The big data market is currently generating over 55 billion dollars in revenue and is predicted to increase to 103 billion dollars by 2027 (Acciarini et al., 2023). The adoption of Internet-connected gadgets contributes to the amount of digital data produced. Thus, cell phones, tablets, and laptops send information about their users. Connected smart items provide insights about consumer use of everyday objects. Aside from linked devices, data is collected from a variety of sources, including demographic data, temperature data, scientific and medical data, energy usage data, and so on. All of these data include information about device users' locations, travel, interests, consumption patterns, leisure activities, projects, and so on (Riahi, 2018). Big data has evolved as a disruptive force in the current digital era, influencing how companies collect, store, and analyze large datasets. Business operations worldwide have made Big Data Analytics (BDA) essential because massive data growth from various sources including autonomous vehicles and social media and sensor networks. Users encounter three core factors they must manage when dealing with Big Data which are volume because of extensive datasets and variety because of diverse data formats together with velocity from high-speed data generation. The Big Data Ecosystem consists of six major tools required for efficient large-scale data handling. Data technologies, such as Apache Hadoop and Apache Spark, analyze and handle large amounts of data in ways that exceed traditional capabilities. Analytics and visualization tools, such as Tableau and SAS, reveal trends, whereas Business Intelligence solutions, such as Cognos, translate raw data into business intelligence. Cloud service providers, such as AWS and GCP, provide core infrastructure (Badshah et al., 2024). The development of big data evaluation followed statistical computing which launched during the 1970s and then included large data sets from the 1980s before advancing to data mining with statistical learning in the 1990s and business analytics in the 2000s until big data analytics appeared in 2010. Businesses possess the potential to adopt the discussed big data analytics system yet they encounter various implementation obstacles. SHMM introduces active and passive data that serve to support multiple healthcare system applications ranging from prediction to diagnostics to forecasting. Big data management functions as a requirement for scaling up data needs to operate multiple applications. Big data functions as an enabler to many different applications. The advancement of scientific research methods and additional evaluation approaches for economic growth receives through expanding methods. The electronic collection of e-commerce data through acquisitions results in multiple assessments of the processed information. Big data functions as a tool for combining different healing sources while also developing an ongoing healthcare data sharing lifecycle. The complete process of big data life cycle includes data generation followed by acquisition and storage before analysis and visualization stages (Rahul et al., 2023). Big Data software comprises parallel and distributed file systems, retrieval software, and data mining software. Applications developed through Big Data analytical technologies find their use in criminal prevention and business management and finance alongside GPS tracking and commercial and travel industries and urban information management and weather forecasting alongside genomic analysis and biological and environmental research and healthcare applications (Hong et al., 2018).

2. FEATURES OF BIG DATA

All huge data sets have following distinguishing qualities:

2.1. Volume

Volume is the amount of data. When dealing with BD (Big Data), this is the most important trait in people's minds. Many firms have massive amounts of data saved in the form of logs but lack the ability to handle it. The ability to manage massive amounts of data to generate insights is the most important advantage of BD analytics (Al-Dulaimi & Ku-Mahamud, 2020). The advent of cloud computing has given businesses access to zettabytes of data! Regardless of the perceived relevance, all data is saved. According to big data experts, answers to business concerns can occasionally be found in unexpected data (Thelin, 2020).

2.2. Velocity

Velocity refers to the speed of data during processing as well as the rate at which data is generated. As a result, this suggests the need for speed when handling, storing, and evaluating data. Hundreds of hours of video are uploaded to YouTube every second, and more than 200 million emails are received over Gmail each day (Thelin, 2020).

2.3. Variety

The term "variety" refers to the diversity of data types and sources. Data analysts should identify the appropriate group for BD since unstructured and disorganized data makes it difficult to store in relational databases. Because 90% of produced data is unstructured, the problem of retaining and interpreting BD will be exacerbated (Thelin, 2020).

2.4. Veracity

Veracity is associated with the truth of data, which is critical for analytical precision. When dealing with a large volume of data at a high velocity and diversity, it is hard to assure that all data is 100% accurate. The data quality will vary, and the precision of the analysis will be determined by the reliability of the data resource (Thelin, 2020).

2.5. Value

The relevance of the data is represented by its value, which is a critical characteristic in business intelligence. BD has significant potential utility. However, it's worth cannot be realized unless appropriate access is obtained (AI- Dulaimi and Ku-Mahamud, 2020).

3. TRADITIONAL DBMS VS. BIG DATA

Conventional data can be compared to a well-maintained filing cabinet that contains well labelled folders with standardized data, such as client names and sales numbers. The wide, wild jungle of big data, on the other hand, is home to a diverse range of information, including tweets, images, videos, and more. The primary distinctions are in the sheer number, variety, and speed with which they operate. Traditional data refers to structured information in digestible volumes, such as a monthly spending report. Big data, on the other hand, faces the issue of dealing with large amounts of unstructured data, such as social media posts, photographs, and sensor data. Traditional data excels in regular corporate processes, which necessitate exact computations and consistent reporting. On the other hand, big data shines when it comes to uncovering hidden trends, making precise predictions, and gaining deep insights into client behavior (Kaur, 2024). There are several variables that you can use to distinguish between Big Data and Traditional Data. There is less emphasis on the volume of data between the two than on its diversity, pace, and management (Roger, 2024).

3.1. Big Data vs Traditional Data: Flexibility

A static relational database provides the foundation for typical data functions. It is more rigid by definition, allowing only organized data to fit into the patterns with ease. Big data, on the other hand, is more unstructured and employs novel approaches to store both structured and unstructured data. Because of its dynamic design, Big Data can hold a broader range of information and is more versatile (Roger, 2024).

3.2. Big Data vs. Traditional Data: Real-Time Analytics

Traditional databases only allow for post-event analysis. This method is advantageous since it reduces the opportunity for error. However, it severely limits the data's usefulness. However, thanks to big data, real-time analytics is now possible. This skill is tremendously beneficial in a variety of areas, including manufacturing, transportation, healthcare, and safety (Roger, 2024).

3.3. Big Data vs. Traditional Data: Distributed Architecture

Big Data and Traditional Database designs also differ. The typical database's centralized architecture may limit its utility. Big data, on the other hand, is designed with a distributed architecture. This feature makes it less expensive and more scalable. Big Data storage is less expensive because to open-source software and cloud-based storage (Roger, 2024).

3.4. Big Data vs. Traditional Data: Multitudes of Sources

Enterprise resource planning (ERP) with customer relationship management (CRM) systems alongside Internet transactions add extra enterprise-level information to traditional data sources. Big data collection processes extract data from enterprise and non-business databases by including information extracted from social media in addition to sensor readings and audiovisual records. The continuous growth of these sources happens together with evolving data structure changes. Big data derives from unstructured data sources which include both text documentation and multimedia files such as image and video and audio. Such data cannot be processed using traditional relational database organization which utilizes tables with rows and columns. Analytical studies become essential because organizations deal with growing unorganized data streams multiple platforms (Roger, 2024).

3.5. Big Data vs. Traditional Data: Analysis

Traditional data analysis proceeds in a sequential manner, beginning with event occurrence and ending with data production and subsequent analysis. Real-time operations are available for undertaking big data analytics. Big data products emergence in real-time makes possible the execution of analysis functions directly upon their creation. Big data analysis provides firms dynamic insights into their entire range of requirements and objectives. A business implements a training program for its staff then needs to evaluate how effective this initiative was (Roger, 2024).

3.6. Big Data vs. Traditional Data: Critical Future Considerations

Big data applications are linked to traditional data operations, despite the fact that they serve distinct goals. Big data provides a broader range of possible benefits, but it lacks the right fit and application for every circumstance. Big data analytics enables businesses to do in-depth market trend analysis as well as through consumer behavior assessments. Big data enables analysts to discover conclusions that traditional analysis methods cannot achieve. It also provides insights faster. Real-time learning becomes possible through using big data by organizations. The increased efficiency enables organizations to establish a superior position within big data analytics. Human activity coupled with business operations both produce extensive data collections that happen continuously or with each passing hour since society transitions towards digital practices (Roger, 2024).

3.6.1 The rise of big data does not imply that traditional data is going away. Traditional data

Secure storage is easier to achieve which makes this method suitable for the protection of sensitive personal and confidential data collections. Standard-sized data sets do not need distributed processing or third-party storage their smaller capacity. Standard data processing tools along with standard system architecture enable the handling of this data. A complicated setup of big data processing systems leads to higher costs and increased system requirements unless traditional data processing techniques fulfil the requirements. The management and comprehension of this data type prove easier to most users. The relational data structure of traditional data is east to handle through standard processing algorithms while non-professionals can manage it using standard functions. The choice between implementing big data and traditional data approaches does not exist. More businesses that create massive unstructured data will require compatible technologies to support their needs. Your big data preparedness requires mastering the appropriate utilization and supportive practices of these two data systems to upgrade your strategic framework (Adiba et al., 2016).

4. BIG DATA LIFE CYCLE AND APPLICATION'S

According to Kimball "Big Data represents a fundamental change in data management strategy because it determines what data we obtain along with approaches for analysis and methods to extract revenue from insights." The examination of the Big Data life cycle holds critical importance because it enables identification of development issues that occur in each phase (Adiba et al., 2016).

4.1. Data Acquisition

Data are gathered from a data-generating source. Large amounts of data can be filtered and compressed to remove irrelevant information. The challenge exists in establishing these filters so they do not eliminate beneficial information. Processing raw data into usable information about small quantities for end-users requires information reduction science studies during data acquisition. Streaming data necessitates online analysis systems because it is not feasible to first store it before reduction operations can take place. Automatic generation of correct metadata became necessary to describe measurements and recorded data. Data provenance represents another important matter that needs immediate resolution. Data is only useful if it can be analyzed and carried through the analysis pipeline after it is collected (Adiba et al., 2016).

4.2. Data Cleaning

Recording and storing data in a repository is insufficient due to its diverse nature. To do this, data structure and semantic distinctions must be stated in computer-readable and "robotically" tractable ways. Research on data integration can offer some answers. However, further work is needed to accomplish automatic error-free difference resolution. To help other professionals, such as domain scientists, create effective database designs, we can create tools to assist them or develop techniques to use databases effectively even without intelligent design (Adiba et al., 2016).

4.3. Big Data Aware Applications

Large numbers of institutions along with academic researchers perceive digital data storage to provide essential wisdom and insights.

(i) Improve marketing and customer service effectiveness;
(ii) Conduct sentiment analysis;
(iii) Monitor epidemic progress;

Social media analytics of tweets and networks should be performed for maximizing the use of information to serve the public. Multiple applications have experienced the collection of data at a level never seen before. Modern decision systems based on data have eliminated both unsystematic guesses and detailed, manual creations of predictive models. Researchers throughout the world have found the Sloan Digital Sky Survey to be an essential astronomical database. Science of astronomy received its defining transformation during that period. Researchers in astronomy dedicated large amounts of their work days to sky photography at one point. Astronomers perform their investigative searches through images stored within a single database. Scientists in biological research maintain the habit of keeping their scientific data within public repositories while developing public databases as search tools for their fellow researchers. Bioinformatics is primarily concerned with analysing such data. The introduction of Next Generation Sequencing has led to an exponential increase in the size and quantity of available experimental data sets (Adiba et al., 2016).

5. BIG DATA AND DBMS TECHNOLOGIES

5.1. MapReduce (Hadoop) system

MapReduce provides big data programmers an approach to parallel computation that combines scalability and fault tolerance features. The desirable features of MapReduce led to its implementation across multiple computer architectures which include multi-core clusters and clouds as well as Cubie boards and GPUs. Cloud companies use MapReduce as widely adopted system to deliver analytics services to their clients. A large number of traditional algorithms together with data processing tasks are transferred to multiple-machine environments by using the MapReduce method with grid indexing which optimizes polygon overlay procedures in GIS spatial time needs immediate attention in numerous Big Data applications that utilize MapReduce. The academic world along with industry organizations actively pursue improved MapReduce job performance with multiple targets which include job scheduling while addressing resource allocation and memory issues and avoiding I/O bottlenecks (Yang et al., 2016).

5.2. Hadoop capabilities

Numerous users adopt Apache Hadoop as a widely recognized technology for Big Data processing. This software addresses obstacles related to speed along with data complexity which occur when businesses use traditional methods for Big Data analysis. Hadoop's ability to analyze enormous data volumes quickly is one of its primary advantages, thanks to its parallel clusters and distributed file system. In reality, unlike older solutions, Hadoop does not copy all remote data into memory in order to run computations. Hadoop, on the other hand, accomplishes operations that include data storage. Hadoop functions a s a powerful framework which depends on two principal elements including the Hadoop Distributed File System and the MapReduce framework. The following sections explain these subcomponents. Users have the flexibility to add modules on top of Hadoop in order to fulfil their specific application requirements. The storage system of Hadoop consists of two elements: the HDFS file system and the non-relational database Apache Hbase (Oussous et al., 2018).

The HDFS system functions as a data storage facility. This system provides clusters that enable hundreds of nodes alongside economical and sustained storage solutions. The system maintains organized data and unstructured data in huge quantity. The storage capabilities of HDFS extend only to certain specific tasks because it functions as a specialized filesystem. The design of HDFS intended it for batch processing that involves operations with high latency requirements. Having to look up records in files through HDFS is not possible because the system does not provide instant file search features. Portable across multiple hardware and software platforms makes HDFS the core advantage of this system. The location of data storage directly improves system performance since HDFS reduces network congestion to bring calculations closer to their relevant storage locations. The system maintains redundant data duplicates for achieving fault tolerance. These features explain why HDFS brings extensive adoption in practice (Oussous et al., 2018).

Hbase operators as an open-source non-relational distributed database system. As an open-source development Hbase operates under the foundations of HDFS. The developers built this system with the main goal of achieving low latency. Hbase follows a data structure based on an intersection of keys and values which utilizes column-based organization. The system enables prompt table updates as well as cluster scaling across distributed networks. Hbase implements a flexible database structure to host enormous tables that operate according to Big Table methods. Data arranged in rows and columns exists as an efficient organizational structure known as tables. These tables enable storage of billions of items across multiple millions of categories. Hbase provides a system where numerous properties belong to column families that maintain their constituents as a unifies group. A Hbase table follows a different storage method than relational databases because it maintains all columns in separate rows. The system architecture of Hbase provides better flexibility than traditional relational databases do. Hbase enables users to supply updates through its system which helps organizations address changing application requirements. The only major limitation of HBase is its inability to handle structured query languages that include SQL (Oussous et al., 2018).

Hbase features a collection of abilities such as dynamic sharding management and table partitioning in addition to real-time querying and natural language search and linear scalability. The messaging platform in Facebook uses Hbase as part of its Big Data solutions and data-driven websites infrastructure. When using Hbase users get Zookeeper coordination services which Hbase launches automatically. HBase, like HDFS, includes a MasterNode to govern the cluster and slaves to store table sections and perform data operations (Oussous et al., 2018).

6. ROLE OF AI AND MACHINE LEARNING IN DBMS FOR BIG DATA

6.1. Understanding AI for Database Management

AI in database administration is the application of machine learning, natural language processing, and predictive analytics to enhance and automate database processes. This integration aims to streamline operations, from regular maintenance activities to complicated data analysis, by utilizing AI's ability to learn from data, find trends, and forecast outcomes (Oussous et al., 2018).

6.2. Benefits of AI for Database Management

Automated Database Tuning: Artificial intelligence technologies are transforming database tuning by automating the optimization of database parameters to assure top performance while reducing the need for manual modifications (Menezes, 2024).

Enhanced Data Quality: AI-powered technologies can help databases improve data quality by automatically finding and fixing discrepancies, assuring data reliability and integrity (Menezes, 2024).

Predictive Analytics: AI's predictive capabilities enable firms to estimate future patterns and behaviors based on historical data analysis, allowing for more informed strategic planning and decisions.

The technology named Natural Language Processing (NLP) helps users communicate with databases through normal spoken language while reducing technical barriers for nontechnical persons (Menezes, 2024).

Despite the potential benefits, integrating AI into database administration poses several obstacles. Concerns about data privacy, the need for high-quality training data, and the challenges connected with adopting AI solutions are significant. When picking AI solutions for database administration, interoperability, scalability, and security are critical to ensuring a successful integration (Menezes, 2024). Examples of Machine Learning Applications for Big Data A combination of supervised and unsupervised machine learning algorithms for efficiently processing large amounts of crime data was presented. The combination consists of three steps: dimensionality reduction, clustering, and classification. R statistical software was employed since it is an effective tool for dealing with large amounts of data. The specific work is described as follows: Assess the association between crime and social characteristics. This strategy decreases the dimensionality of criminal data. Divide crime data into categories using unsupervised machine learning, then use k-means clustering to categorize it as risky, average, or safe. Utilize supervised machine learning to forecast if a specific region is risky or safe; to perform predictions, utilize the decision tree classification algorithm. Big Data was used to analyze and mine social network data for societal issues. Social data mining is the process of evaluating, representing, and extracting useful patterns from social network data (Unuriode et al., 2023).

7. APPLICATIONS OF BIG DATA IN DBMS

Big data applications have grown ubiquitous, establishing themselves as a core technology with a widespread function across a wide range of fields, much like computers. This section organizes various applications carefully, examining their impact, problems, and opportunities (Badshah et al., 2024). Various big data technologies operate within sectors such as smart cities, hospitals, education, transport, social media, business operations, promotions, event organizations and e-commerce. Big data obtains its primary data sources from these sectors which produce vast volumes of information. Privacy alongside security become essential issues in systems involving big data applications (Rahul et al., 2023).

7.1. Big data in healthcare

A substantial amount of information exists in the health-care industry. Healthcare stands among the most vital industries because medical services need to be accessible for all people at any social standing and throughout all caste and religious groups. Patient-related data generated by healthcare sensors acts as a foundation for disease prediction services critical functions related to both privacy protection of hospitalized patients and healthcare security measures and patient care administration and treatment delivery (Rahul et al., 2023). Analysing biomedical omics and clinical data presents both problems and potential for healthcare development. The healthcare business creates a large amount of data, such as hospital records, medical exams, and research, which must be managed properly in order to provide relevant insights. BDA in healthcare can enable individualized medicine, clinical risk management, and forecasting, as well as the standardization of medical language and patient registration (Badshah et al., 2024). This section demonstrates the vital need to integrate cloud computing with IoT by establishing multiple explanations based on different devices. The statistical assessments running under big data technologies within healthcare information systems (HISs) analyze diabetes through accuracy enhancement and F measure optimization. The study investigates methods for sustainable environment healthcare supply chains through the examination of their green supply chain strategies. Two evaluation systems named CAIE and TAMS perform the extraction evaluation and pattern mining assessment. Pateint-led healthcare systems implement extensive data influence through the integration of insures with service provides and pharmaceutical businesses together with medical practitioners and patients (Rahul et al., 2023). Big data is used in health analytics to help design effective medical policies, improve healthcare services, and improve disease prognosis, medicine recommendations, and treatment outcomes. The Gastroenterology Department at Xiangya Hospital in China has a powerful BDA platform that allows for full digestive medicine analysis. The technology incorporates electronic medical records and colonoscopy data, providing insights for optimal colorectal cancer screening ages and enhancing healthcare management. Using prescription big data can help with dosage prediction in pediatric medication. Traditional clinical decision support systems are frequently deficient in reliable pediatric data (Badshah et al., 2024). One of the most common processing tools in Big Data, including those used in health care, is Hadoop, which was invented by Apache and employs the notion of dispersion to handle massive amounts of data. Big Data sources in health care are divided into four groups. There is a scarcity of tools for analysing data from these diverse sources. A German calciphylaxis registry presented a framework and developed a platform for integrating medical records, imaging data, and signal data to improve understanding of uncommon disorders. A kernel-based framework is used to analyse heterogeneous data in the medical arena, addressing the missing data issue given by patients with sparse or absent data modalities. The kernel approach can be used to do regression and classification on heterogeneous medical data (Hong et al., 2018).

In the future, healthcare organizations will increasingly rely on big data for success. The utilization of HBD will improve marketing methods, particularly as wearable technology and the Internet of Things gain traction. The integration of continuous patient monitoring data from multiple sources will yield significant insights, allowing healthcare marketers to better identify and engage patients (Badshah et al., 2024).

7.2. Big data in Education sector

The education sector depends heavily on information and communication technology (ICT) which receives government backing at all times since its inception. ICT serves to increase the effectiveness and efficiency that education obtains from teaching and learning methods. The governmental support for assessment implementation and enhancement allows stakeholders and residents to receive mutual benefits alongside improved productivity across multiple levels. Innovative educational frameworks enable stakeholders to engage in active learning so society develops awareness of the value of progress in each aspect while undergoing operational learning stages. Big data brings together information and communication technology instruments to disseminate data and knowledge to society and to grant access to resources for implementation projects (Rahul et al., 2023). Personalized learning is currently the most researched topic in educational technology. individualized learning recommends individualized content or subjects to learners, allowing them to learn in their own environment. Education is intricately related to big data as both a producer and consumer. Millions of people, including students, instructors, and administrators, are active participants in this dynamic area. The demand for virtual classrooms has increased during the COVID-19 epidemic, stressing the importance of big data in satisfying these changing educational requirements. Personalized learning and home-based schooling are becoming increasingly popular, relying exclusively on the insights and capabilities supplied by big data. While there are numerous benefits, using big data in education presents a number of problems. The most significant of them is the potential of misuse, as thousands of learners' data, including institutional geography and student locations, could be mismanaged. Furthermore, questions concerning data bias and algorithmic bias raise possible issues that must be carefully considered in order to achieve fair and equal outcomes (Badshah et al., 2024).

7.3. Big data in transport and logistics

The use of big data in logistics and transportation has received a lot of attention. Researchers have investigated BDA in SCM and discovered its ability to correct flaws, improve efficiency, and reduce costs. Logistics organizations used big data and supply chain integration (SCI) to optimize supply chain performance, especially during the COVID-19 epidemic. Furthermore, the interaction of big data analytics technological capability (BDATC) with SCI has been shown to improve supply chain performance by promoting proactive and reactive skills, as well as resource reconfiguration. Blockchain technology is also making inroads into logistics and supply chain systems, providing technological support and lowering hazards (Badshah et al., 2024). The increase in global airplane passenger data requires processing for the delivery of optimal infrastructure alongside better customer service and component predictions such as incorrect road conditions alongside traffic patterns and passengers availability and more. The data processing approach involves data amalgamation of numerous points obtained from various sources following a refinement step before combining them for processing. An advanced cloud blockchain and Internet of Everything (IoE)-enabled quality control technology aims to improve quality management and boost consumer confidence in perishable supply chain logistics, as detailed in. This platform offers rapid sensor data collecting while maintaining authentication and transparency in cold supply chain operations (Badshah et al., 2024). The integration of big data in logistics and transportation offers great promise in the future. According to a recent study, last-mile delivery, which accounts for a significant amount of total shipping costs, has problems like as carrier coordination, manual processes, driver retention, fuel costs, WISMO calls, and return costs. These difficulties create chances for optimization through the proper use of big data solutions (Rahul et al., 2023).

8. FUTURE TRENDS IN BIG DATA AND DBMS

Quantum computing has the potential to revolutionise database administration by solving complicated optimization issues and processing massive datasets more quickly. Cryptography and data analysis are two areas that quantum computing has an impact on. As data quantities expand, good data governance and stewardship become increasingly vital. Future database systems will have to include advanced frameworks for controlling data quality, lineage, and compliance. Zhang et al.'s (2022) research highlights the importance of comprehensive data governance strategies to maintain data integrity and accountability (Tama & Mark, 2024).

Quantum computing has the ability to dramatically accelerate some sorts of issues when compared to traditional computers. As technology progresses, quantum approaches are being employed in practical applications like databases. In some cases, quantum databases can execute queries much quicker than standard databases due to quantum parallelism and entanglement. They store data using quantum bits, also known as qubits, which can represent a 1, 0 or a combination of the two. This one-of-a-kind feature allows you to query data at the same time. Quantum databases have the potential to revolutionize search, particularly for optimization problems such as finding the shortest path in a graph. However, quantum computers are prone to errors and have restrictions in the number of qubits they can manage. Quantum databases may initially function as accelerators, completing certain queries much quicker than classical databases, but they may not totally replace existing databases. Hybrid techniques that mix classical and quantum processing, particularly for tasks such as graph machine learning, are being investigated. As quantum technology advances, it has the potential to revolutionize chemical simulations, financial modeling, and artificial intelligence. However, due to stability and scaling issues, quantum databases may not fully duplicate all functionalities of regular databases, such as transactions and joins. To summarize, while quantum techniques have the potential to improve database performance, traditional procedures will continue to be critical for overall data management. As technology advances, unleashing the promise of quantum computing will bring fascinating new possibilities to the data-driven world (Kotha, 2024).

8.1. Anticipated Challenges

Challenges to Expect

i. Data security presents dual implications from technological development because emerging technology advances extremely fast. Organizations should expect problems which emerge from developing technologies such as quantum computing artificial intelligence and IoT integration alongside their impact on data security protocols.
ii. Regulatory Changes: Stay up to date on changing data protection requirements and anticipate the issues that new legislation will bring. Discuss how regulatory changes could affect data security policies and organizational compliance efforts.
iii. Cyber Threat Sophistication: Recognize the changing nature of cyber threats and prepare for rising attack vector sophistication. Discuss the problems faced by APTs, ransomware attacks, and insider threats, as well as ways for reducing these risks (Adenubi et al., 2024).

9. OVERVIEW OF SECURE DATABASE DESIGN AND CODING PRACTICES

Secure database architecture and secure coding standards are essential components in creating secure, resilient systems that can withstand attacks and protect sensitive information. Implementing architectural principles and best practices for secure database architecture helps to ensure data privacy, integrity, and availability. The picture includes data classification, rights, encryption, and auditing. Following best practices for secure database design can help organizations protect their data from theft, loss, and manipulation. The purpose of secure coding methods is to produce reliable applications. Common security problems, such as SQL injection, cross-site scripting, and buffer overflows, can be avoided or managed by following coding standards and employing secure programming techniques. Adopting secure coding techniques can help organizations limit the likelihood of generating vulnerabilities in the application layer that could be used to compromise the database. Implementing access restrictions is a critical component of any safe database architecture. Setting and enforcing user roles and permissions can help control access to sensitive information. To guarantee that users have only the information and functionality they require to perform their assignments, access restrictions should be fine-grained and based on the principle of least privilege. The database and any associated programs should also be safeguarded against unauthorized access by employing secure coding principles that enforce proper authentication and permission processes. Encryption is critical for creating secure database systems. Data must be encrypted both at rest and in transit to prevent theft or interception. Secure database architecture uses encryption and key management strategies to keep sensitive information out of the wrong hands. Safe coding standards should cover the encryption of sensitive data within application code, in addition to ensuring that encryption techniques are performed correctly and securely. Strong auditing and logging systems are also essential for secure database architecture and coding methodologies. In order to detect potential security incidents and ensure accountability, relevant security events, system activities, and user interactions must be recorded and analysed. It is impossible to build reliable and trustworthy systems without using secure database architecture and coding methods. Company data protection against security threats requires organizations to use strong entrance controls together with encryption methods and audit procedures and logging systems. Applications which reach into the database need to implement security-design principles before deployment so secure development practices minimize this threat. A secure database needs ongoing standard adherence and security posture evaluation and training (Omotunde & Ahmed, 2023).

10. CONCLUSION

Database management systems experience significant changes because of big data evolution. The distinguishing features of big data remain volume, variety and velocity in addition to all the data that originates from social media platforms and sensors alongside gadgets. Relational databases have the unique capability to store structured data forms in their databases. The magnitude complexity, and real-time nature of big data necessitate the development of new technologies and methods. Hadoop, MapReduce, and NoSQL databases have emerged to solve the limitations of traditional database management systems when dealing with large amounts of data. These big data platforms offer distributed, scalable, and fault-tolerant storage and processing capabilities. The incorporation of upcoming technologies such as quantum computing, artificial intelligence, and machine learning into database management systems is projected to further transform the field. Quantum databases have the potential to significantly improve query performance. AI-powered automation and predictive skills can help with database tuning, data quality, and decision support. Finally, the rise of big data has radically changed the DBMS landscape. The future of database administration will rely on big data technologies, quantum computing, and AI-driven capabilities to fully fulfil the potential of the data-driven world, but traditional databases will still play a significant role.

REFERENCES

Acciarini, C., Cappa, F., Boccardelli, P., & Oriani, R. (2023). How Can Organizations Leverage Big Data to Innovate Their Business Models? A Systematic Literature Review. *Technovation*, *123*, 102713. DOI: 10.1016/j.technovation.2023.102713

Adenubi, A. O., Oduroye, A., & Akanni, A. (2024, February). Data Security in Big Data: Challenges, Strategies, and Future Trends. *International Journal of Research in Education Humanities and Commerce*, *05*(02). Advance online publication. DOI: 10.37602/IJREHC.2024.5201

Adiba, M., Castrejon-Castillo, J.-C., Espinosa Oviedo, J. A., Vargas-Solar, G., & Zechinelli-Martini, J.-L. (2016). Big data management challenges, approaches, tools and their limitations. In S. Yu, X. Lin, J. Misic, & X. S. Shen (Eds.), Networking for big data (pp. 1–25). Chapman and Hall/CRC. https://hal.science/hal-01270335

Badshah, A., Daud, A., Alharbey, R., Banjar, A., Bukhari, A., & Alshemaimri, B. (2024). Big Data Applications: Overview, Challenges, and Future. *Artificial Intelligence Review*, *57*(290). Advance online publication. DOI: 10.1007/s10462-024-102713

Hong, L., Luo, M., Wang, R., Lu, P., Lu, W., & Lu, L. (2018, December 1). Big Data in Health Care: Applications and Challenges. *Data and Information Management*, *2*(3), 175–197. DOI: 10.2478/dim-2018-0014

Kotha, K. (2024). "Demystifying Quantum Databases." *Medium*. https://medium.com

Menezes, N. (2024, January). The role of artificial intelligence in modern database management. SQLOPS Blogs. https://www.sqlops.com/tag/artificial-intelligence

Omotunde, H., & Ahmed, M. (2023). A comprehensive review of security measures in database systems: Assessing authentication, access control, and beyond. *Mesopotamian Journal of CyberSecurity*, *2023*, 115–133.

Oussous, A., Benjelloun, F.-Z., Lahcen, A. A., & Belfkih, S. (2018, October). Big Data Technologies: A Survey. *Journal of King Saud University. Computer and Information Sciences*, *30*(4), 431–448. DOI: 10.1016/j.jksuci.2017.06.001

Rahul, K., Banyal, R. K., & Arora, N. (2023). A Systematic Review on Big Data Applications and Scope for Industrial Processing and Healthcare Sectors. *Journal of Big Data*, *10*, 133. DOI: 10.1186/s40537-023-00808-2

Riahi, Y. (2018, November). Big Data and Big Data Analytics: Concepts, Types, and Technologies. *International Journal of Research and Engineering*, *5*(9), 524–528. DOI: 10.21276/ijre.2018.5.9.5

Tama, E. H., & Mark, D. (2024, September). The Future of Database Management in the Era of Big Data and Cloud Computing. *International Journal of Information Technology and Computer Engineering*, *4*(45), 48–60. DOI: 10.55529/ijitc.45.48.60

Thelin, R. (2020). "What Is Big Data? Characteristics, Types, and Technologies." Last modified December 15, 2020.

Unuriode, A., Yusuf, B., Durojaiye, O., & Okunade, L. (2023, October). The Integration of Artificial Intelligence into Database Systems (AI-DB Integration Review). *International Journal on Cybernetics & Informatics*, *12*(6). Advance online publication. DOI: 10.5121/ijci.2023.1206012

Yang, C., Huang, Q., Li, Z., Liu, K., & Hu, F. (2016). Big Data and Cloud Computing: Innovation Opportunities and Challenges. *International Journal of Digital Earth*, *10*(1), 13–53. DOI: 10.1080/17538947.2016.1239771 [illegible] Big Data and Big Data Analytics [illegible] Technology. *International Journal of Research* [illegible] [illegible] 2016.5 [illegible]

[illegible]

[illegible]

[illegible]

[illegible]

Chapter 2
Extracting Structure From Complexity:
Algorithms for Reliable Data-Driven Decision-Making

G. Aravind
https://orcid.org/0009-0005-5980-4008
KCG College of Technology, India

ABSTRACT

Modern decision-making depends on extracting reliable patterns from noisy, heterogeneous data—tables, time series, text, and graphs. After collecting and unifying these sources, data must be cleaned, transformed, and converted into model-ready features. Modeling techniques span time-series methods, gradient boosting for tabular data, neural models for text, and graph analytics, with probabilistic tools adding uncertainty estimates. Rigorous evaluation ensures models align with business goals. Decisions are then operationalized through transparent rules, and continuous monitoring detects drift and guides retraining. In banking, integrating repayment histories, income data, network links, and text enables credit scoring, cash-flow forecasting, fraud detection, and risk-aligned pricing. This framework provides a general blueprint for turning messy data into evidence-based intelligence.

1. INTRODUCTION: FOUNDATIONS OF DATA-DRIVEN DECISION-MAKING

In the contemporary digital era, organizations across industries increasingly rely on data as a strategic asset. The ability to extract meaningful structure from vast quantities of heterogeneous, messy, and rapidly evolving information has become

DOI: 10.4018/979-8-3373-6746-0.ch002

central to effective decision-making. Whether in banking, healthcare, logistics, manufacturing, energy, or public administration, institutions are operating within environments characterized by complexity, uncertainty, and continuous change. Traditional intuition-driven approaches are no longer sufficient. Instead, organizations must transform raw data into structured knowledge that can guide actions, optimize operations, and support long-term planning. This transformation is neither trivial nor automatic. It requires rigorous methods, systematic processes, and advanced computational tools capable of uncovering deep patterns hidden within high-noise environments.

Modern data sources present an immense diversity in structure, scale, and semantic richness. Structured financial tables, for example, contain clearly defined numerical variables such as balances, interest rates, transaction amounts, or customer profiles. High-frequency sensor streams produce time-dependent sequences representing temperature variations, machine vibrations, weather measurements, energy usage, or stock price movements. Meanwhile, unstructured textual artifacts—including emails, customer reviews, complaint logs, call transcripts, and social media posts—capture nuanced expression, sentiment, and intent but lack predefined organization. Lastly, complex relational networks such as customer–merchant interactions, supply-chain dependencies, communication graphs, and fraud rings introduce topological relationships that require specialized analytical techniques.

Each of these data types alone provides only partial insight into an operational environment. However, real-world systems are rarely separated into clean, isolated modalities. Instead, they function as interconnected ecosystems where numerical trends, temporal behaviors, semantic signals, and relational patterns interact in subtle ways. For example, a financial institution evaluating credit risk may observe numerical attributes such as income and repayment history (structured data), behavioral patterns such as spending fluctuations (time series), communication sentiment expressed in support emails (text), and social or transactional networks linking individuals or businesses (graphs). The ability to integrate and interpret these multimodal signals is essential for accurate, holistic decision-making.

Despite its profound value, raw data is often far from ready for analysis. It tends to be messy, incomplete, and inconsistent. Noise infiltrates datasets from multiple sources: sensor malfunctions, human error, system latency, incomplete reporting, ambiguous expressions in text, or misaligned time stamps. Missing values can distort patterns if left unaddressed. Outliers may reflect either errors or rare but critical events. Differences in scale, format, or meaning must be harmonized before models can interpret them. This inherent messiness makes data preprocessing one of the most time-consuming yet essential components of the analytical pipeline.

Because data alone cannot speak for itself, organizations must rely on mathematical modeling, statistical inference, and algorithmic computation to extract structure from complexity. Mathematical models help formalize assumptions about how systems behave. Statistical inference allows analysts to quantify uncertainty, estimate relationships, and test hypotheses about underlying processes. Machine learning algorithms detect patterns, classify behaviors, forecast trends, and identify anomalies. Together, these tools convert noisy inputs into meaningful structure and actionable insights.

However, building robust decision-support systems requires more than simply applying sophisticated models. The entire lifecycle—from data collection to integration, preprocessing, modeling, evaluation, deployment, monitoring, and governance—must be carefully orchestrated. Each stage plays a critical role in ensuring that decisions are grounded in reliable, relevant, and interpretable information. Neglecting any stage can compromise the integrity of the whole system.

This chapter organizes the research landscape into a set of interconnected themes that collectively form a comprehensive blueprint for extracting structure from complex, heterogeneous data sources. These themes include:

1. **Multimodal Data Integration**

Techniques for unifying structured, unstructured, temporal, and relational data into coherent analytical representations.

2. **Data Cleaning, Transformation, and Preprocessing**

Methods for handling missing values, noise, inconsistencies, scale differences, and domain-specific transformations.

Figure 1. Data Cleaning-Transformation and Preprocessing

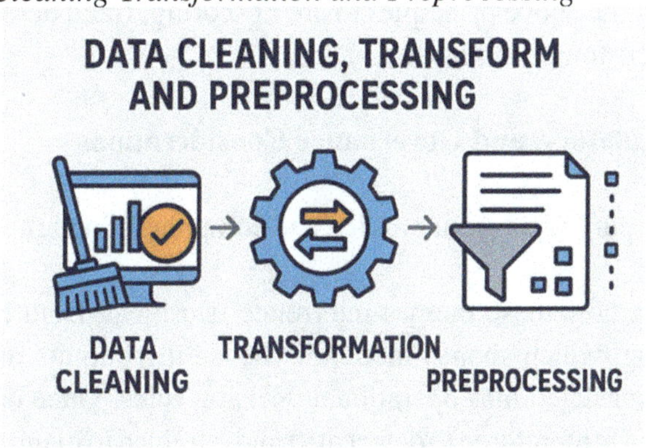

3. **Machine Learning for Complex Data Types**

Advanced modeling approaches tailored to time series, tabular data, text, and graphs, each requiring specialized algorithms and representations.

4. **Uncertainty Quantification and Probabilistic Modeling**

Frameworks for measuring and communicating uncertainty, enabling safer decision-making in high-risk environments.

5. **Model Evaluation, Robustness, and Fairness**

Techniques for assessing model performance under realistic conditions, ensuring reliability, resisting manipulation, and preventing bias.

6. **Decision Systems and Operationalization**

Approaches for converting model predictions into rules, policies, and automated workflows that govern real-world actions.

7. **Monitoring, Drift Detection, and Adaptive Learning**

Tools for ensuring long-term reliability by tracking changes in data, updating models, and testing alternatives.

8. **Domain-Specific Research in Banking**

Application of the above principles to credit scoring, fraud detection, cash-flow forecasting, and customer sentiment analysis.

9. **Ethical, Regulatory, and Governance Considerations**

Ensuring transparency, accountability, and compliance in high-stakes, regulated domains.

Understanding how these themes interrelate is critical. Data integration feeds into preprocessing, which shapes modeling. Modeling outputs must be carefully evaluated, then translated into operational decision rules. Once deployed, models must be continuously monitored to detect drift and retrained to maintain performance. Throughout this lifecycle, ethical and regulatory considerations impose constraints designed to ensure fairness, transparency, and accountability.

The need for a holistic framework becomes clearer when one considers real-world scenarios. In financial services, for instance, regulatory obligations demand that institutions maintain explainable credit decisions, robust fraud prevention systems, and accurate assessments of systemic risk. Each of these applications relies on the interplay between multimodal data processing, sophisticated modeling, and rigorous oversight. A credit model that predicts default risk may be technically accurate but still unacceptable if it cannot be explained to regulators or if it inadvertently embeds bias. A fraud detection system may be highly sensitive but useless if it generates excessive false alarms or fails to adapt to new fraudulent behaviors. Similarly, a forecasting model may provide strong historical performance but quickly degrade if underlying market conditions shift.

Beyond financial institutions, other domains exhibit similar complexities. In healthcare, integrating clinical measurements, diagnostic images, patient histories, genomic data, and physician notes requires a multimodal approach. Decisions involving patient risk, treatment planning, or disease progression must account for uncertainty and ethical considerations. In logistics, demand forecasting relies on time series data, weather patterns, transportation networks, and textual order instructions. In energy systems, sensor networks, consumption logs, grid topology, and maintenance reports must be analyzed together to ensure reliability and efficiency.

Given this complexity, the extraction of meaningful structure from heterogeneous data becomes both a scientific challenge and a strategic necessity. It demands interdisciplinary research spanning computer science, statistics, optimization, operations research, domain expertise, and regulatory understanding. No single algorithm or technique can address all challenges. Instead, solutions must be layered, iterative, and integrated into a coherent workflow.

This chapter therefore serves as both a conceptual overview and a roadmap for future exploration. It identifies key research areas that converge to support robust, reliable, and ethical data-driven decision-making. By systematically examining topics such as multimodal integration, preprocessing, advanced machine learning, uncertainty quantification, evaluation frameworks, operational deployment, adaptive monitoring, banking applications, and governance, the chapter establishes a foundational understanding of what is required to build effective data intelligence systems.

Ultimately, the goal of extracting structure from complexity is not merely analytical—it is fundamentally operational. Organizations must act on the patterns they uncover. These actions influence profitability, customer satisfaction, regulatory compliance, safety, and strategic direction. Therefore, the quality of decisions hinges on the quality of the analytical systems that support them. As data environments continue to expand in scale and complexity, the importance of reliable, interpretable, and adaptive decision-making systems will only intensify. By embracing the research themes outlined in this chapter, organizations and researchers can develop solutions capable of navigating modern challenges with precision, accountability, and foresight.

2. TECHNICAL FOUNDATIONS

2.1 Multimodal Data Integration

Modern analytical ecosystems rarely rely on a single type of data. Instead, organizations collect information from financial ledgers, operational logs, transactional events, customer interactions, and sensor-based monitoring systems. Each of these data sources carries unique structural and statistical properties. The ability to merge them into a unified analytical substrate lies at the heart of advanced decision-making systems.

Figure 2. Multimodal Data Integration

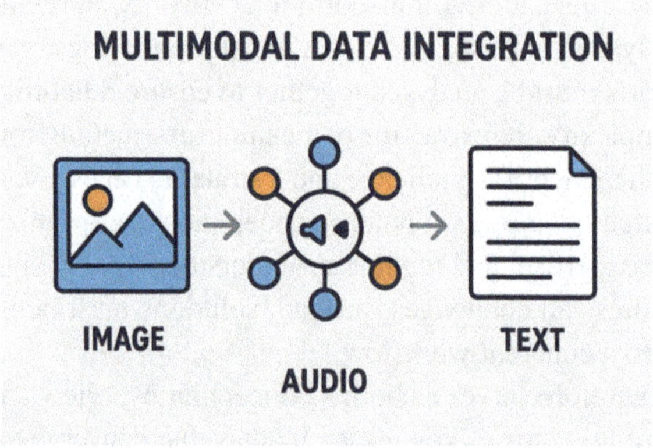

2.1.1 Cross-modal representation learning

addresses the fundamental challenge of representing heterogeneous data in a way that preserves useful structure while enabling interactions across formats. For example, tabular financial attributes can be embedded into vector representations, time series can be converted into latent temporal signatures, and textual customer notes can be transformed using semantic encoders such as BERT or LLaMA-derived models. Graph-based data, such as transaction networks or organizational hierarchies, can similarly be mapped to vector spaces using graph neural network encoders. The goal of cross-modal representation learning(Zeng et al., 2023) is to ensure that these embeddings occupy compatible or jointly learnable spaces, enabling downstream algorithms to reason across modalities.

2.1.2 Multimodal fusion architectures

form the blueprint for how these distinct representations are eventually combined. Techniques like *early fusion* involve merging raw features before model training, which can be advantageous when interactions between modalities are strong. *Late fusion* combines predictions from separate models, often yielding more interpretable pipelines and allowing specialized models to handle each modality independently. *Joint embedding space* approaches, often built on transformer or contrastive learning frameworks, create unified vector representations where semantically aligned data points such as a customer's transaction history, service tickets, and account graph—are brought closer together. These architectures not only allow richer predictions but also support tasks like anomaly detection, entity resolution, or cross-domain retrieval.

A particularly important subarea involves schema-free or schema-flexible integration. Modern enterprises are confronted with varied and evolving data structures whose fields and formats change over time. Rigid schema-dependent pipelines often break when new attributes are introduced or when text logs shift format. Schema-free integration leverages pattern recognition, natural language understanding, autoencoders, and statistical profiling to ingest data regardless of structural consistency. This makes pipelines robust to real-world variability.

Finally, the linkage of disparate data streams—for instance, synchronizing financial transaction tables with sensor logs, call center transcripts, or credit-bureau reports—requires methods for entity matching, temporal alignment, and feature-level synchronization. These mechanisms provide the backbone for building comprehensive profiles of customers, assets, or operational processes. The result is a unified analytical input space where the full richness of multimodal data can be exploited.

2.2 Data Cleaning, Transformation & Preprocessing

Raw data, no matter how valuable, is almost always messy. It contains missing values, inconsistencies, unusual outliers, and noise. Before any modeling can occur, the data must be subjected to carefully designed cleaning and preprocessing pipelines that maximize signal quality and minimize distortion.

2.2.1Automated data quality assessment

has become increasingly important as scale outpaces manual inspection. These methods evaluate completeness, accuracy, consistency, and validity across large datasets. Automated systems may detect anomalies in distributions, flag improbable values (e.g., negative transaction amounts), or identify mismatches in units or data types. Machine learning–driven approaches can even learn typical patterns from historical data and highlight deviations.

2.2.2 Imputation strategies

are essential for addressing missingness. Simple techniques such as mean or median imputation often perform poorly in high-stakes domains. More sophisticated methods multiple imputation, k-nearest neighbors(Halder et al., 2024), or model-based imputation(Enders et al., 2020) using gradient boosting—are required to preserve statistical integrity. Domain-specific constraints, such as known ranges for financial ratios or temperature sensors, must be integrated into the imputation logic to avoid producing unrealistic values.

2.2.3 Outlier detection

is an equally critical component. Outliers may signal fraud, instrumentation errors, seasonal bursts, or simply rare but valid behaviors. Using domain constraints helps distinguish harmful noise from insightful anomalies. For example, extremely high transaction volumes may indicate fraud in retail banking but may be normal for certain institutional clients. Techniques include isolation forests, robust statistical measures (e.g., median absolute deviation), and density-based clustering methods such as DBSCAN(Deng, 2020).

Standardization and normalization techniques (Huang et al., 2023) ensure that features measured on different scales can be meaningfully compared by models. Log normalization may be appropriate for skewed financial amounts, z-scoring for standardized metrics, and min–max scaling for bounded variables. Mixed-scale datasets, such as those with both counts and ratios, require thoughtful preprocessing so that the learning algorithm interprets each feature appropriately.

For textual data, NLU(Du et al., 2021)-focused preprocessing pipelines convert raw text into structured semantic representations. Tokenization, stemming, lemmatization, and the construction of embeddings—from classical TF-IDF matrices to transformer-based contextual embeddings—allow models to capture linguistic meaning. These representations enable downstream tasks like sentiment scoring, intent classification, and anomaly detection in communication logs.

Finally, time series preprocessing(Cortés-Ibáñez et al., 2020) involves decomposing temporal patterns into trend, seasonality, and residual components. Denoising filters, differencing, rolling-window feature creation, and spectral transformations (e.g., wavelet or Fourier transforms) make temporal structure more accessible to machine learning models. Proper temporal alignment ensures that features derived from different sources are synchronized, enabling coherent joint modeling.

Together, these preprocessing strategies transform messy raw data into standardized, meaningful, and reliable inputs that can support robust predictive modeling.

2.3 Machine Learning for Complex Data Types

Different data modalities require model classes tailored to their inherent structure. As a result, the field of machine learning has evolved into a diverse ecosystem, each branch specializing in the statistical and structural nuances of particular data types.

Figure 3. Machine Learning for Complex Data Types

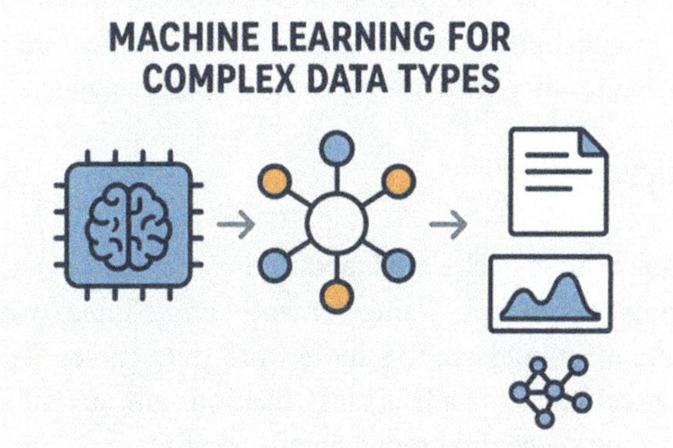

2.3.1 Time Series Modeling

Time series are defined by temporal dependencies—patterns that evolve over time. Capturing these dynamics requires specialized models. Modern approaches include:

- **Transformers**, which excel at long-range dependencies thanks to attention mechanisms.
- **State-space models**, which generalize classical methods like ARIMA with probabilistic latent states and flexible transition dynamics.
- **Hybrid statistical–ML frameworks**, which merge interpretable components (trend, seasonality) with nonlinear predictive layers.

Beyond forecasting point values, time series analysis frequently requires volatility modeling (e.g., GARCH variants, stochastic differential equations) to quantify uncertainty in dynamic environments such as financial markets, energy consumption, or risk exposures. Understanding not only the expected value but also the variability of future outcomes is essential for informed decision-making.

2.3.2 Tabular Data Modeling

Tabular datasets remain the backbone of enterprise analytics. Despite the rise of deep learning, tree-based boosting models(James et al., 2023) continue to dominate due to their ability to handle missing values, nonlinear interactions, and heterogeneous feature types. Variants of XGBoost, LightGBM, and CatBoost offer explainability through feature importance scores, SHAP values, and monotonicity constraints.

In sparse or noisy environments—common in credit scoring or healthcare analytics—regularized regression methods(Obi & Jecinta, 2023) (Lasso, Elastic Net) provide stable coefficient estimates and embedded feature selection. These models blend interpretability with strong baseline performance.

2.3.3 Text Analytics

Unstructured text contains rich signals about customer sentiment, legal compliance issues, or operational anomalies. Using semantic embeddings(Chen et al., 2021), systems can capture meaning beyond simple word frequencies. Transformer-based language models excel in tasks such as classification, summarization, and anomaly detection by producing contextual representations of text.

Figure 4. Text Analytics

2.3.4 Topic Modeling

Using methods like LDA or neural topic models, allows organizations to discover latent themes in large corpora of customer complaints or support conversations. These insights support product improvement, regulatory reporting, and operational optimization.

2.3.5 Graph Modeling

Graphs encode relationships between entities—accounts, transactions, devices, employees, or suppliers. Graph neural networks (GNNs) (Corso et al., 2024) can learn patterns of interaction that are invisible to traditional models. They are particularly effective in:

- Fraud detection, where criminal networks form distinctive topological structures.
- Recommendation systems based on network similarity.
- Credit risk propagation through corporate ownership networks.

Graph theory metrics—centrality, clustering coefficients, community detection—provide interpretable summaries of relational structure. These tools help identify influential nodes, detect anomalous subgraphs, and uncover hidden communities in complex systems.

Taken together, these machine learning methods enable organizations to extract structured insights from diverse, high-dimensional, and complex data landscapes.

2.4 Uncertainty Quantification & Probabilistic Modeling

In high-stakes decision-making—credit underwriting, fraud detection, medical treatment planning—understanding uncertainty is as important as predicting outcomes. Probabilistic modeling provides a formal framework for capturing uncertainty and integrating it into decisions.

2.4.1 Bayesian inference

is central to this approach. By treating model parameters as distributions rather than fixed values, Bayesian models(Van de Schoot et al., 2021) provide credible intervals, posterior distributions, and probabilistic interpretations of trends. This is crucial when data is scarce, noisy, or expensive to collect.

2.4.2 Probability calibration

ensures that predicted probabilities match real-world frequencies. For example, if a model assigns a 0.2 default risk to a set of customers, approximately 20% of them should default. Calibration techniques Platt scaling, isotonic regression, temperature scaling—are vital for regulated domains where model outputs must be interpretable and trustworthy.

2.4.3 Decision-making under uncertainty

involves frameworks like expected utility theory, risk-sensitive optimization, and robust optimization. These frameworks explicitly incorporate uncertainty into decision rules, enabling organizations to choose actions that minimize downside risks while balancing expected rewards.

Finally, uncertainty estimates such as confidence intervals and credible intervals provide transparency around predictions. Stakeholders can evaluate not only what the model predicts but also how confident it is. This is critical in environments where errors have substantial financial, ethical, or legal consequences.

2.5 Model Evaluation, Robustness & Fairness

A model is only as reliable as its performance under real-world conditions. Robust evaluation ensures that models are accurate, stable, fair, and secure against malicious or unintended misuse.

2.5.1 Performance metrics

must be tailored to the business context. Simple accuracy is rarely sufficient. Cost-sensitive metrics such as precision under budget constraints, recall thresholds for fraud detection, or profit-based KPIs—align model behavior with organizational objectives. Proper evaluation also involves examining performance across subgroups, time periods, and scenarios to detect hidden weaknesses.

2.5.2 Stress-testing and robustness analysis

evaluate how a model behaves under distributional shift. Real-world data is nonstationary: customer behavior changes, economic conditions fluctuate, and fraud tactics evolve. Stress tests simulate adverse conditions, such as sudden spikes in transaction volumes or changes in data quality, allowing teams to identify vulnerabilities before deployment.

2.5.3 Fairness analysis

is especially important in domains like lending, hiring, and insurance. Techniques such as disparate impact measurement, counterfactual fairness, and individual fairness metrics help ensure that models do not inadvertently discriminate against protected groups. These methods support compliance with regulatory frameworks like the Equal Credit Opportunity Act or GDPR.

Finally, robustness against adversarial manipulation(Tao et al., 2022) has become a central concern. Fraudsters and strategic actors may attempt to game the system by adjusting behaviors or manipulating inputs. Defensive techniques—including adversarial training, anomaly detection, and constraint-based modeling—help ensure that decision systems remain resilient under adversarial pressure.

3. FROM MODELS TO PRACTICE: DECISION SYSTEMS & APPLICATIONS

Predictive models, no matter how accurate or sophisticated, create business value only when they influence decisions. Building a high-performance model is just the first step; translating predictions into operational actions requires engineering, governance, interfaces, and continuous oversight. Section 3 covers two major pillars of this translation: decision systems and monitoring systems. Together they create the infrastructure for using machine learning responsibly, effectively, and sustainably in real-world environments.

3.1 Decision Systems & Operationalization

Organizations increasingly rely on automated or semi-automated decision processes. These might determine whether a customer is approved for credit, whether a transaction is flagged for fraud review, or how much liquidity a treasury team allocates on a given day. The challenge is not merely to generate predictions but to embed them in consistent, transparent, and auditable workflows.

3.1.1 Algorithmic Decision Rules

Decision rules formalize how predictions translate into actions. In risk scoring, for example:

- A predicted probability of default might feed into a set of thresholds determining whether to approve, decline, or manually review an application.

- Fraud detection systems may use dynamic thresholds that depend on transaction context, channel, or customer profile.
- Loan pricing algorithms can convert risk estimates into interest rates using pricing curves aligned to regulatory and profitability constraints.

These decision rules require rigorous calibration. They must account for nonlinearities between prediction and business impact, such as how missing a high-risk customer might generate losses far more severe than incorrectly declining a low-risk one. Research in this area focuses on developing frameworks that balance statistical accuracy with domain-specific utility functions.

3.1.2 Transparent and Auditable Decision Engines

In regulated industries such as banking, healthcare, or insurance, systems must produce explanations and audit trails. A decision engine is the infrastructure layer that:

- Implements the logic linking predictions to actions.
- Documents how each decision was made.
- Records model version, input data, and justification.
- Produces human-readable explanations for compliance and review.

Transparency also enhances trust among stakeholders—regulators, customers, and internal teams. As a result, research explores interpretable ML techniques, model documentation frameworks, and policy-enforced decision flows where every outcome can be reconstructed after the fact.

3.1.3 Policy Mechanisms for Approvals, Holds, and Collections

Automated policy mechanisms streamline operations by embedding business rules directly into the decision pipeline. Examples include:

- **Auto-approval policies**, which green-light low-risk applications instantly to improve customer experience.
- **Fraud holds**, triggered when anomalies or risk scores exceed threshold levels.
- **Automated collections workflows**, which adjust treatment strategies based on behavioral predictions (e.g., likelihood of repayment).

These policies must balance efficiency with caution. Over-automation can introduce systemic risks if models drift or perform unexpectedly. Thus, policy engines often incorporate:

- Confidence score thresholds.
- Fallback logic when data is missing or inconsistent.
- Safeguards preventing extreme decisions without human review.

The design challenge lies in creating mechanisms that are responsive, scalable, and aligned with institutional risk appetite.

3.1.4 Human-in-the-Loop Oversight

While automation can streamline decision-making, high-stakes environments require a human-in-the-loop (HITL) (Kumar et al., 2024) approach. This involves:

- Manual review queues for borderline or high-risk cases.
- Supervisory portals where analysts can override automated decisions.
- Audit processes that examine whether the system behaves as expected.
- Feedback loops that allow human insights to retrain or correct models.

Human oversight is crucial for catching nuanced, context-sensitive issues that models might miss—such as emerging fraud patterns, unusual financial behaviors, or new regulatory interpretations. HITL systems combine computational efficiency with human judgment, creating hybrid decision-making frameworks that are both robust and adaptive.

3.1.5 Bridging Predictive Analytics with Real-World Operations

Operationalization closes the gap between model outputs and real organizational processes. It requires infrastructure, cross-functional coordination, and thoughtful system design. Key components include:

- **API-based prediction services** that allow models to be accessed by business applications.
- **Orchestration layers** that manage workflows, decision rules, and routing logic.
- **Governance frameworks** to define roles, responsibilities, and escalation paths.
- **Version control systems** for managing updates, rollbacks, and model lineage.

The ultimate goal is not simply to score data but to embed predictions into coherent, stable decision-making pipelines that function reliably under operational constraints. This transformation—from model to system—is where the real value of machine learning is realized.

3.2 Monitoring, Drift Detection & Adaptive Learning

Even the best-designed models degrade over time. Customer behavior changes, market conditions shift, competitors influence new dynamics, and underlying data patterns evolve. Without continuous monitoring, a well-calibrated model can silently become unreliable or biased.

This section describes the tools and techniques that preserve a model's long-term reliability and enable systems to adapt to changing conditions.

3.2.1 Detecting Data Drift and Concept Drift

Two types of drift can degrade model performance:

1. **Data Drift**

Occurs when the distribution of input data changes—for example, shifts in transaction amounts, changes in customer demographics, or evolving text language patterns.

Techniques include:

- Statistical tests (Kolmogorov–Smirnov, Chi-square)
- Embedding drift detection using distance metrics
- Multivariate drift detectors (e.g., Mahalanobis distance, PCA-based monitoring)

2. **Concept Drift**

Occurs when the relationship between inputs and outputs changes. For example:

- Fraudsters adopt new tactics.
- Economic downturn alters repayment behavior.
- Regulations change acceptable risk levels.

Concept drift is harder to detect. Solutions involve:

- o Monitoring performance metrics over time.
- o Real-time feedback loops.
- o Model retraining triggers based on statistical alerts.

A mature monitoring framework evaluates both drift types and provides early warning signals before performance degradation becomes operationally costly.

3.2.2 Automated Retraining Pipelines

To maintain accuracy, models must periodically update their parameters. Automated retraining pipelines:

- Periodically ingest new batches of labeled or semi-labeled data.
- Retrain candidate models in controlled environments.
- Validate them against benchmarks.
- Deploy updated versions if they outperform existing ones.

In domains with rapid change—fraud detection, supply-chain forecasting, or network intrusion detection—online learning methods may be used. These allow models to update continuously or incrementally as new data arrives.

However, automated retraining must be carefully controlled. Blindly retraining may introduce instability or inadvertently encode short-lived noise patterns. Therefore, retraining frameworks incorporate:

- Safeguards such as validation thresholds.
- Simulated environments for stress-testing retrained models.
- Human approval steps for regulatory-critical models.

3.2.3 A/B Testing for Model Iteration

A/B testing provides a structured mechanism for comparing new models or decision rules against existing ones. In practice:

- A portion of traffic is routed to a new model version.
- Performance is monitored using pre-defined metrics (profit lift, precision, F1, cost savings).
- Statistical significance tests determine whether the new version offers reliable improvements.

A/B testing mitigates risk by ensuring that updates are data-validated rather than assumption-driven. More advanced techniques—multi-armed bandits, contextual bandits—allocate traffic optimally based on performance, accelerating learning while limiting exposure to underperforming variants.

3.2.4 Dashboarding and Monitoring Systems

Well-designed monitoring dashboards serve as the nerve center for operational ML. They help teams track:

- Prediction distributions
- Performance metrics
- Drift indicators
- Input data quality
- Operational statistics (latency, throughput, error rates)
- Alerts for anomalous system behavior

In financial decision systems, dashboards often include:

- Regulatory compliance indicators
- Stress-test results under synthetic scenarios
- Fairness metrics across demographic groups
- Threshold performance under varying economic conditions

Dashboards must be actionable. Analysts should be able to drill down into anomalies, trigger retraining processes, or escalate issues to risk and compliance teams.

3.2.5 Adapting Models to Market Conditions

Monitoring alone is not enough; adaptive learning is essential for long-term resilience. Models must survive and thrive under:

- Economic downturns
- Market booms
- Regulatory shifts
- Seasonal patterns
- Emergent fraud behaviors

Adaptive systems incorporate:

- Dynamic thresholds that adjust to volatility
- Reinforcement learning components for real-time optimization
- Scenario analysis for planning under uncertainty
- Metadata-aware training that accounts for context shifts

Over time, these adaptive mechanisms create systems capable of learning continuously—not just from labeled outcomes but from patterns, deviations, and real-time feedback.

3.2.6 Bringing Everything Together

The evolution from predictive modeling to practical deployment is complex. It demands expertise not only in machine learning but also in software engineering, risk management, policy design, and human–system interaction. Section 3 highlights this transition:

- **Decision systems** translate predictions into actions, ensuring transparency, consistency, and oversight.
- **Monitoring systems** protect models from decay, adapt them to new realities, and ensure they remain fair, robust, and compliant.

Together, these concepts form the operational backbone of modern AI-driven organizations.

3.3 Domain-Specific Research in Banking

The banking sector represents one of the most mature yet rapidly evolving domains for applied data-driven decision-making. Owing to the immense volume, velocity, and variety of financial data, as well as the high stakes associated with financial decisions, banks have become central laboratories for advancing multi-modal analytics, predictive modeling, and algorithmic governance. The intersection of heterogeneous data sources, complex regulatory expectations, and real-time operational demands makes banking an exemplary context for research across the themes discussed throughout this chapter.

3.3.1 Credit Risk Modeling Using Heterogeneous Multimodal Inputs

Credit risk is one of the foundational problems in financial analytics. Traditionally, risk assessment relied primarily on structured tabular data such as credit histories, repayment schedules, income verification reports, and borrower demographics. Today, however, risk modeling has evolved into a multimodal challenge.

Banks now incorporate a broader spectrum of inputs:

- **Time series data**, such as daily account balances, credit utilization trends, and repayment behaviors over time.
- **Unstructured text**, including customer inquiries, dispute letters, and customer-support transcripts, which can signal distress, intention, or financial misunderstanding.
- **Graph-structured data**, representing borrower–merchant networks, contact relationships, or transaction ecosystems.
- **Behavioral features**, such as login frequency, device patterns, or payment channel preferences.

Multimodal models leverage these varied signals to capture aspects of borrower behavior that traditional scores would miss. For example, early deviations in spending sequences may indicate emerging financial stress, while network-based signals—such as association with previously fraudulent merchant clusters—can provide early warnings of elevated risk. Textual sentiment extracted from customer communications may reveal dissatisfaction, dispute likelihood, or impending attrition, each of which may indirectly affect creditworthiness.

Moreover, the integration of uncertainty quantification (e.g., Bayesian credit scoring, probabilistic risk envelopes) allows institutions to gauge not only the predicted risk but also the confidence associated with that risk estimate. This is particularly valuable for borderline borrowers or novel customer segments where historical data may be limited.

3.3.2 Real-Time Fraud Detection Through Network Features and Text Signals

Fraud detection is arguably the most challenging and dynamic analytics function in banking. Fraud patterns evolve rapidly, adversaries adapt in near real time, and false decisions can have significant financial and reputational impact. The domain's complexity stems from the interplay of transactional, relational, and behavioral cues.

3.3.3 Network-Based Fraud Features

Fraud rarely occurs in isolation. Modern fraudsters operate in coordinated networks, exploiting weak links in financial ecosystems. Graph-based analytics—such as community detection, centrality analysis, and subgraph matching—enable banks to identify suspicious clusters or anomalous relational patterns.

Examples include:

- Rapidly emerging merchant networks with unusually dense interconnections.
- Rings of accounts sharing devices, addresses, or transactional pipelines.
- Outlier paths in transaction graphs indicative of layering or money laundering.

Graph neural networks (GNNs) extend traditional techniques by learning embeddings that propagate risk signals across connections, enabling models to detect fraud even among accounts with no prior suspicious activity.

3.3.4 Text-Based Fraud Features

Unstructured text adds a second layer of intelligence. Emails disputing unusual charges, chat messages with support teams, or even social media references (when available under compliance) can reveal important clues:

- Keyword and tone changes indicating account takeover.
- Customer-reported anomalies that flag early-stage fraud.
- Linguistic patterns associated with fraudulent claims or synthetic identity usage.

Combining network-based and text-based cues—together with classical transaction rules—creates multi-layered fraud detection systems that respond faster and more accurately than traditional rule-based approaches.

3.3.5 Cash-Flow Forecasting for Liquidity Planning

Liquidity forecasting is a strategic function in banking, helping institutions ensure adequate reserves, optimize capital allocation, and comply with regulatory liquidity coverage requirements. Predicting cash flows requires modeling temporal patterns at multiple scales:

- **Short-term flows** (daily, weekly), influencing ATM stocking, intraday credit exposure, and settlement risk.
- **Medium-term flows** (monthly), supporting deposit planning, loan disbursement cycles, and customer payment behaviors.
- **Long-term flows**, informing capital buffers, investment strategies, and risk-weighted asset optimization.

Time series forecasting techniques—from traditional ARIMA models(Elsaraiti & Merabet, 2021) to Transformer-based architectures—play a crucial role. However, multimodal enhancement is increasingly important. For example:

- Transaction-level time series paired with customer text complaints can signal impending attrition, affecting forecasted balances.
- Macroeconomic indicators(Iqmal, 2020), such as employment trends or inflation rates, influence long-term liquidity.
- Network features, such as corporate customer supply chain relationships, can anticipate shocks in payment flows.

Forecasts must include uncertainty intervals, enabling risk managers to prepare for worst-case liquidity scenarios. This is especially vital during market volatility, where small forecasting errors can cascade into significant exposure.

3.3.6 Sentiment and Complaint Analytics for Customer Service Optimization

Customer communication is an invaluable asset that reflects satisfaction, frustration, intent, risk, and emerging behavioral patterns. Banks receive millions of messages through email, call transcripts, chatbots, and surveys. Analyzing this text has become central to:

- Identifying customers at risk of attrition.
- Detecting unresolved disputes that could escalate into formal complaints.
- Improving service quality by uncovering systemic pain points.
- Assessing the emotional state of customers, which may correlate with credit stress or fraud vulnerability.

Techniques include:

- Sentiment analysis(Wankhade et al., 2022) using pretrained language models(Chen et al., 2023).
- Intent classification to route messages to appropriate channels.
- Topic modeling to uncover common themes without manual labeling.
- Early-warning signals derived from linguistic shifts (e.g., urgency, dissatisfaction).

When combined with transactional and behavioral data, sentiment analytics adds depth to risk models, enabling proactive service interventions, personalized communication strategies, and improved customer satisfaction.

3.3.7 Interplay of Multimodal Integration, Modeling, and Uncertainty Analysis

Banking is not simply a technical application of algorithms—it is a domain where multimodal data, advanced modeling, uncertainty quantification, and ethical decision-making converge. The complexity of financial systems, coupled with the regulatory and societal implications of algorithmic decisions, makes it essential that banks integrate:

- **Multimodal data fusion** to ensure decisions are based on comprehensive information.
- **Uncertainty quantification** to reflect model confidence and avoid overreliance on fragile predictions.
- **Explainability frameworks** to meet regulatory scrutiny.
- **Governance mechanisms** to ensure fair treatment of all customer groups.

Thus, banking stands as one of the most critical domains for developing responsible, effective, and future-ready data-driven decision systems.

3.4 Ethical, Regulatory & Governance Considerations

As predictive analytics increasingly influences financial decisions that affect individuals, businesses, and society, ethical and regulatory considerations have become central to model development and deployment. Banking, more than most industries, is tightly governed due to its profound social impact. Decisions such as loan approvals, fraud alerts, and credit scoring directly affect access to financial opportunity. Therefore, systems must be transparent, fair, accountable, and compliant with both legal frameworks and ethical norms.

3.4.1 Transparency and Explainability in Regulated Sectors

Regulators worldwide—including the Federal Reserve, the European Banking Authority, and financial conduct authorities—require that banks provide clear explanations for their automated decisions. Transparency is not optional; it is a prerequisite for legal compliance and public trust.

Explainability encompasses:

- **Model-level interpretability**, such as understanding which features influence predictions.
- **Instance-level explanations**, detailing why a specific customer received a particular decision.
- **System-level transparency**, documenting model behavior, assumptions, and limitations.

Methods such as SHAP values, counterfactual explanations, interpretable surrogate models, and monotonic constraints are commonly employed. However, regulatory expectations extend beyond technical explanations. Banks must ensure that explanations are *meaningful* to customers and understandable by auditors.

Explainability thus becomes a multidisciplinary challenge requiring:

- Data science (to produce interpretable models)
- Risk management (to assess compliance)
- Legal teams (to ensure explanations satisfy regulatory norms)
- User experience design (to present explanations accessibly)

3.4.2 Compliance-Aware Model Development

Compliance-aware model(Micheal, 2024) development integrates regulatory requirements into the entire model lifecycle. This includes:

- **Data governance**: ensuring that customer data is collected, stored, and processed lawfully.
- **Model governance**: implementing approval workflows, version tracking, validation reports, and signoffs.
- **Bias mitigation**: Bias mitigation(Hort et al., 2024) ensuring that protected attributes (race, gender, age, etc.) do not introduce unlawful discrimination.
- **Stress testing** (Kim et al., 2023): evaluating model resilience under adverse scenarios, including economic shocks or data shifts.
- **Documented model risk assessments**: detailing assumptions, limitations, and potential impacts.

Compliance is not a post-hoc activity. It must be embedded from the earliest stages of feature engineering to deployment and monitoring. The research frontier includes developing automated tools for compliance checks, scalable model documentation pipelines, and regulatory simulation systems that predict compliance risks before release.

3.4.3 Audit Trails for Algorithmic Accountability

Auditability is a cornerstone of responsible AI in finance. Every decision influenced by a model must be traceable. Audit trails include:

- Input data used at the time of decision.
- Model version and configuration.
- Feature transformations applied.
- Thresholds and policy rules triggered.
- Final recommendation or decision.
- Human override actions, if any.

These records enable internal audits, external regulatory reviews, and customer dispute resolution. They also support incident investigation: if a model misbehaves, audit trails help identify the root cause—whether it be data quality issues, drift, or misconfiguration.

Research in this space focuses on:

- Immutable logging mechanisms(Ijiga et al., 2025) (e.g., blockchain-based audit trails).
- Automated reproduction of past decisions.
- Lifecycle visualization tools.
- Cross-model comparison frameworks(Cohen et al., 2021).

As models grow more complex—especially deep learning systems—ensuring that every decision is reproducible becomes increasingly challenging and critical.

3.4.4 Ethical Frameworks for Automated and Semi-Automated Decision-Making

Beyond regulatory requirements, banks face ethical expectations from society. Decisions must be fair, respectful, and consistent with human values. Ethical frameworks guide how automated decisions should be designed, reviewed, and monitored.

Key principles include:

- **Fairness**: ensuring equal access to financial opportunities without discrimination.
- **Non-maleficence**(Bufacchi, 2020): avoiding harm to customers (e.g., misclassifying low-income customers in ways that increase financial burden).
- **Autonomy**: allowing customers to understand decisions and contest them.

- **Justice**: ensuring equitable distribution of risk and benefits.
- **Accountability**: ensuring that humans—not algorithms—bear responsibility for final decisions.

Operationalizing ethics requires:

- Fairness audits across demographic groups.
- Bias-detection tools that monitor shifts in model behavior.
- Human review for sensitive decisions.
- Governance committees overseeing algorithmic operations.
- Ethical impact assessments for new models or updates.

This area of research bridges AI ethics, legal theory, human-computer interaction, and machine learning.

3.4.5 Ensuring Trustworthiness, Compliance, and Social Responsibility

The future of financial analytics depends on trust. If customers perceive models as opaque, unfair, or discriminatory, institutions face legal, financial, and reputational risks. Ethical, regulatory, and governance frameworks thus form the backbone of safe AI adoption.

Trustworthy AI requires that models:

- Perform reliably under diverse conditions,
- Treat customers fairly,
- Produce transparent explanations,
- Support contestability and recourse,
- Operate within legal boundaries,
- Be continuously monitored for drift or bias,
- Include human oversight for sensitive decisions.

These objectives demand ongoing research into interpretable modeling techniques, robust monitoring systems, uncertainty-aware decision-making, and adaptive governance structures. The goal is to ensure that advanced analytics enhances—not undermines—financial stability, customer welfare, and societal trust.

4. FUTURE RESEARCH DIRECTIONS

As organizations continue to digitize their operations and generate increasingly rich data ecosystems, the research landscape in data-driven decision-making is evolving rapidly. The prior sections outlined the foundations of multimodal integration, modeling, uncertainty quantification, monitoring, and governance. Building upon these pillars, several forward-looking research trajectories are poised to shape the next generation of intelligent, reliable, and ethically grounded decision systems. These directions underscore the need for scalable architectures, autonomous pipeline management, enhanced robustness, and governance frameworks that match the complexity of real-world systems.

4.1 Advanced Multimodal Representation Learning

One of the most promising and challenging research directions lies in developing unified representations that can capture relationships across fundamentally different data types. Traditional systems often require extensive preprocessing—normalizing tabular fields, embedding text, aggregating time series, or constructing graphs—before any learning can occur. This creates significant bottlenecks in scalability, adaptability, and generalization.

4.1.1 Unified Embedding Spaces Without Manual Preprocessing

The ambition of advanced multimodal representation learning is to enable models to jointly process:

- Structured tables
- Sequential time series
- Unstructured text
- Relational graphs
- Even higher-order modalities like geospatial or image data

The research direction is analogous to what foundational models achieved in NLP, where raw text can be ingested without feature engineering. For multimodal systems, the challenge is magnified: each modality has distinct statistical characteristics, temporal dependencies, and semantic structures.

4.1.2 Key Areas of Innovation

1. **Cross-Modality Attention Mechanisms**

Extending transformer-style attention architectures so that signals from one modality can dynamically influence the representation of another. For instance, textual customer queries could influence the encoding of temporal payment patterns or transaction graphs.

2. **Joint Latent Spaces**

Developing shared latent manifolds where heterogeneous data points coexist meaningfully. This would support integrated reasoning—e.g., relating sensor anomalies to transaction irregularities or textual complaints.

3. **Self-Supervised Multimodal Objectives**

Creating pretraining tasks that leverage the natural alignment of real-world datasets: predicting missing modalities, reconstructing sequences, or inferring graph relationships solely from cross-modal signals.

4. **Handling Missing Modalities**

Real-world systems often encounter incomplete data. Next-generation representation learning must infer robust embeddings even when large portions of a modality are absent, corrupted, or delayed.

The long-term vision is a single architecture capable of seamlessly ingesting an organization's entire data universe and producing representations that generalize across tasks, domains, and temporal regimes.

4.2 Automated, Self-Healing Data Pipelines

A major source of risk—and a major cost center—in production machine learning systems is the maintenance of data pipelines. Errors in upstream systems, delayed feeds, missing records, and corrupt values often lead to downstream model failures, sometimes in silent and undetected ways.

The next generation of research focuses on pipelines that are not just automated but *self-healing*.

4.2.1 Self-Diagnosing Pipelines

Future pipelines should detect:

- Schema drift

- Unexpected value distributions
- Broken joins or referential inconsistencies
- Time series gaps
- Free-text anomalies

These issues should be flagged automatically via statistical, machine learning, and rule-based detectors.

4.2.2 Autonomous Remediation Mechanisms

Beyond detection, systems need the ability to *correct* issues without manual human intervention. This may include:

- Intelligent imputation based on historical patterns
- Reconstructing missing time segments
- Re-establishing broken keys or entity links
- Backfilling records using predictive models
- Automatically retrying failed ETL processes

Such systems reduce operational burdens dramatically and ensure model stability even when upstream sources are volatile.

4.2.3 Reinforcement Learning for Pipeline Optimization

A cutting-edge subfield involves reinforcement learning agents that:

- Learn which cleaning or transformation steps yield the most stable downstream predictions
- Adjust pipeline logic based on real-time performance feedback
- Optimize compute cost while maintaining model quality

These techniques convert data engineering into an adaptive, learning-driven process.

4.3 Foundation Models for Tabular and Multimodal Financial Data

While foundation models have revolutionized NLP(Zaremba & Demir, 2023) and are rapidly advancing in vision and code, their application to tabular and financial multimodal data(Yanglet et al., 2025) is still emerging. Financial datasets differ from text or images because they are:

- Highly structured
- Often sparse
- Governed by domain semantics
- Subject to regulatory scrutiny
- Sensitive to temporal and causal dependencies

4.3.1 Pretraining on Large Financial Corpora

There is growing interest in assembling large-scale pretraining datasets composed of:

- Banking transactions
- Credit histories
- Market data
- Customer interactions
- Risk signals
- Network relationships
- Operational text logs

The goal is to learn general-purpose representations that can transfer to specialized tasks such as fraud detection, credit scoring, operational risk forecasting, and liquidity planning.

4.3.2 Hybrid Architectures

A promising direction is developing architectures that integrate:

- **Transformer encoders** for text and sequence data
- **Implicit neural representations**(Molaei et al., 2023) for continuous signals
- **Graph neural networks** for relational structures
- **MLPs or attention mechanisms** specifically adapted to tabular features

These models would surpass the limitations of single-modality foundation models.

4.3.3 Regulatory-Aware Pretraining

Financial data cannot be treated like general-purpose internet corpora. Foundation models must incorporate:

- Interpretability layers
- Privacy-preserving training mechanisms
- Traceability of knowledge sources
- Calibration guarantees

Research in this area will shape the future of large-scale AI in regulated financial spaces.

4.4 Real-Time Probabilistic Decision Engines

Most current decision engines rely on deterministic score cutoffs or heuristic rules. However, real-world decisions—loan approvals, fraud alerts, pricing strategies—are inherently probabilistic.

4.4.1 Integration of Prediction and Optimization

Future decision engines will not simply output predictions; they will integrate:

- Uncertainty estimates
- Optimization constraints
- Risk–reward functions
- Real-time operational context

This creates a unified system that directly outputs *optimal actions* rather than raw predictions.

4.4.2 Key Research Directions

1. **Stochastic Optimization Under Uncertainty**
   Combining probabilistic forecasts with optimization models that can handle randomness, such as chance-constrained programming.
2. **Real-Time Decisioning**
   Meeting latency constraints for applications like fraud detection, algorithmic trading, or instant credit approval.
3. **Dynamic Thresholds**

Moving away from static rules toward thresholds that adjust based on market volatility, customer history, or institutional risk appetite.

4. **Human–AI Collaboration**
   In high-stakes contexts, the engine must provide uncertainty explanations so that human operators can intervene intelligently.

Figure 5. Human-AI Collaboration

This research direction signals a shift from predictive analytics to *decision analytics*.

4.5 Robustness Under Adversarial and Market Shift Pressures

Financial systems operate under constant pressure from evolving fraud patterns, strategic customers, regulatory changes, and macroeconomic conditions. Models that perform well today may degrade quickly.

4.5.1 Adversarial Robustness

Unlike adversarial attacks in image recognition, financial adversaries are strategic and adaptive. They may:

- Alter transaction behaviors
- Manipulate application details
- Obfuscate fraud networks (Zhu et al., 2024)
- Target model blind spots

Robustness research must incorporate adversarial simulation, game theory, and defensive learning methods.

4.5.2 Market Shift Resilience

Market conditions—interest rates, unemployment, spending habits—shift abruptly. Research must extend beyond typical drift detection to develop models that:

- Extrapolate reliably under structural break regimes
- Use causal inference to separate signal from noise
- Maintain calibration across macroeconomic cycles
- Adapt dynamically without catastrophic forgetting

This is critical for long-lived financial systems.

4.5.3 Stress Testing & Scenario Simulation

Borrowing from regulatory practices, machine learning systems must be able to:

- Run counterfactual scenarios
- Simulate rare-but-critical stress events
- Estimate tail risks

Future models will not just predict typical outcomes—they will predict *extreme* outcomes.

4.6 Transparent and Trustworthy AI Governance

As intelligent systems become deeply embedded in financial institutions, governance frameworks must evolve. The complexity of modern AI demands governance that is:

- Systematic
- Explainable
- Auditable
- Compliant with local and global regulations

4.6.1 Universal Frameworks for AI Trustworthiness

Emerging research seeks to harmonize:

- Fairness metrics
- Explainability techniques
- Compliance requirements
- Ethical guidelines

across diverse jurisdictions and business contexts.

4.6.2 Key Components

1. **Model Documentation Standards**

Templates and protocols for documenting model logic, data sources, and intended use.

2. **Algorithmic Accountability**

Systems that log decisions, providing full traceability for audits or investigations.

3. **Bias Mitigation Workflows**

Combining pre-processing, in-processing, and post-processing fairness strategies into unified governance pipelines.

4. **Explainability for Non-Experts**

Tools that translate complex model behavior into accessible explanations for regulators, customers, and executives.

In the next decade, governance research will determine which models are legally deployable in critical sectors.

4.7 Fully Autonomous Monitoring & Adaptive Learning Systems

Modern ML systems(Shethiya, 2023) are often brittle: once deployed, their performance decays over time. To maintain reliability, institutions must implement continuous monitoring and update cycles. Future research aims to fully automate these processes.

4.7.1 Autonomous Drift Detection

Next-generation systems will employ advanced statistical and machine learning techniques to:

- Detect feature drift
- Identify concept drift
- Recognize distributional anomalies
- Understand causal changes

without manual configuration.

4.7.2 Self-Updating Models

Autonomous systems will:

- Trigger retraining
- Select optimal model architectures
- Validate new models against old ones
- Deploy new versions based on performance and risk thresholds

This mirrors continuous integration/continuous delivery (CI/CD) pipelines (Zampetti et al., 2021) in software engineering but tailored for ML.

4.7.3 Policy Optimization Loops

In many decision systems, the policy—not just the model—must evolve. Autonomous systems could:

- Adjust approval policies
- Optimize fraud thresholds
- Modify pricing strategies
- Tune risk appetite dynamically

using reinforcement learning or bandit algorithms.

4.7.4 Minimizing Human Oversight

While human oversight remains crucial in high-stakes decisions, the amount of operational human labor can be dramatically reduced. Over time, operators shift from manual supervision to exception handling and strategic guidance.

REFERENCES

Bufacchi, V. (2020). Justice as non-maleficence. *Theoria*, *67*(162), 1–27.

Chen, G., Zheng, Y.-D., Wang, J., Xu, J., Huang, Y., Pan, J., Wang, Y., ... "Videollm: Modeling video sequence with large language models." *arXiv preprint arXiv:2305.13292* (2023).

Chen, J., Hu, H., Wu, H., Jiang, Y., & Wang, C. "Learning the best pooling strategy for visual semantic embedding." In *Proceedings of the IEEE/CVF conference on computer vision and pattern recognition*, pp. 15789-15798. 2021.

Cohen, S. M., Iyer, G. C., Brown, M., Macknick, J., Wise, M., Binsted, M., Voisin, N., Rice, J., & Hejazi, M. (2021). How structural differences influence cross-model consistency: An electric sector case study. *Renewable & Sustainable Energy Reviews*, *144*, 111009.

Corso, G., Stark, H., Jegelka, S., Jaakkola, T., & Barzilay, R. (2024). Graph neural networks. *Nature Reviews. Methods Primers*, *4*(1), 17.

Cortés-Ibáñez, J. A., González, S., Valle-Alonso, J. J., Luengo, J., García, S., & Herrera, F. (2020). Preprocessing methodology for time series: An industrial world application case study. *Information Sciences*, *514*, 385–401.

Deng, D. "DBSCAN clustering algorithm based on density." In *2020 7th international forum on electrical engineering and automation (IFEEA)*, pp. 949-953. IEEE, 2020.

Du, M., Manjunatha, V., Jain, R., Deshpande, R., Dernoncourt, F., Gu, J., Sun, T., & Hu, X. "Towards interpreting and mitigating shortcut learning behavior of NLU models." *arXiv preprint arXiv:2103.06922* (2021).

Elsaraiti, M., & Merabet, A. (2021). A comparative analysis of the arima and lstm predictive models and their effectiveness for predicting wind speed. *Energies*, *14*(20), 6782.

Enders, C. K., Du, H., & Keller, B. T. (2020). A model-based imputation procedure for multilevel regression models with random coefficients, interaction effects, and nonlinear terms. *Psychological Methods*, *25*(1), 88. PMID: 31259566

Halder, R. K., Uddin, M. N., Uddin, M. A., Aryal, S., & Khraisat, A. (2024). Enhancing K-nearest neighbor algorithm: A comprehensive review and performance analysis of modifications. *Journal of Big Data*, *11*(1), 113.

Hort, M., Chen, Z., Zhang, J. M., Harman, M., & Sarro, F. (2024). Bias mitigation for machine learning classifiers: A comprehensive survey. *ACM Journal on Responsible Computing*, *1*(2), 1–52.

Huang, L., Qin, J., Zhou, Y., Zhu, F., Liu, L., & Shao, L. (2023). Normalization techniques in training dnns: Methodology, analysis and application. *IEEE Transactions on Pattern Analysis and Machine Intelligence*, *45*(8), 10173–10196. PMID: 37027763

Ijiga, O. M., Balogun, S. A., Okika, N., Agbo, O. J., & Enyejo, L. A. (2025). An In-Depth Review of Blockchain-Integrated Logging Mechanisms for Ensuring Integrity and Auditability in Relational Database Transactions. *International Journal of Social Science and Human Research*, *13*(3), 245–259.

. Iqmal, Fariz. "Macroeconomic factors and influence on stock return that impact the corporate values." *International Journal of Finance & Banking Studies (2147-4486)* (2020).

James, G., Witten, D., Hastie, T., Tibshirani, R., & Taylor, J. (2023). Tree-based methods. In *An introduction to statistical learning: With applications in Python* (pp. 331–366). Springer International Publishing.

Kim, H., Sclar, M., Zhou, X., Bras, R., Kim, G., Choi, Y., & Sap, M. "FANToM: A benchmark for stress-testing machine theory of mind in interactions." In *Proceedings of the 2023 Conference on Empirical Methods in Natural Language Processing*, pp. 14397-14413. 2023.

Kumar, S., Datta, S., Singh, V., Datta, D., Singh, S. K., & Sharma, R. (2024). Applications, challenges, and future directions of human-in-the-loop learning. *IEEE Access : Practical Innovations, Open Solutions*, *12*, 75735–75760.

. Micheal, Lee. "Compliance-Aware Architectures for Deploying Large Language Models in Cloud Environments." (2024).

Molaei, A., Aminimehr, A., Tavakoli, A., Kazerouni, A., Azad, B., Azad, R., & Merhof, D. "Implicit neural representation in medical imaging: A comparative survey." In *Proceedings of the IEEE/CVF International Conference on Computer Vision*, pp. 2381-2391. 2023.

Obi, J. C., & Jecinta, I. C. (2023). A review of techniques for regularization. *International Journal of Research in Engineering and Science*, *11*(1), 360–367.

Shethiya, A. S. (2023). Learning to Learn: Advancements and Challenges in Modern Machine Learning Systems. *Annals of Applied Sciences*, *4*(1).

Tao, L., Feng, L., Wei, H., Yi, J., Huang, S.-J., & Chen, S. (2022). Can adversarial training be manipulated by non-robust features? *Advances in Neural Information Processing Systems*, *35*, 26504–26518.

Van de Schoot, R., Depaoli, S., King, R., Kramer, B., Märtens, K., Tadesse, M. G., & Vannucci, M.. (2021). Bayesian statistics and modelling. *Nature Reviews. Methods Primers*, *1*(1), 1.

Wankhade, M., Annavarapu, C. S. R., & Kulkarni, C. (2022). A survey on sentiment analysis methods, applications, and challenges. *Artificial Intelligence Review*, *55*(7), 5731–5780.

Yanglet, X.-Y. L., Cao, Y., & Deng, L. "Multimodal financial foundation models (mffms): Progress, prospects, and challenges." *arXiv preprint arXiv:2506.01973* (2025).

Zampetti, F., Geremia, S., Bavota, G., & Di Penta, M. "CI/CD pipelines evolution and restructuring: A qualitative and quantitative study." In *2021 IEEE International Conference on Software Maintenance and Evolution (ICSME)*, pp. 471-482. IEEE, 2021.

Zaremba, A., & Demir, E. (2023). ChatGPT: Unlocking the future of NLP in finance. *Modern Finance*, *1*(1), 93–98.

Zeng, Y., Mai, S., Yan, W., & Hu, H. (2023). Multimodal reaction: Information modulation for cross-modal representation learning. *IEEE Transactions on Multimedia*, *26*, 2178–2191.

Zhu, Z., Zhu, G., Zhang, Y., Shi, J., Huang, X., & Fang, Y. (2024). Eigenobfu: A novel network topology obfuscation defense method. *IEEE Transactions on Network Science and Engineering*.

Chapter 3
From Statistical Analysis to Deep Learning:
Modern Approaches in Quantitative Decision-Making

Ghada Nakhla
https://orcid.org/0000-0003-2235-5550
Edge Hill University, UK

ABSTRACT

This chapter explores the shift from traditional statistical analysis to advanced deep learning in quantitative decision-making. It argues for integrating classical paradigms—such as descriptive and inferential statistics conducted in tools like SPSS—with modern computational techniques to address complex, large-scale challenges. The discussion highlights the pivotal role of big data platforms, including Hadoop and Spark, in enabling real-time analytics and distinguishes core machine learning algorithms from deep neural architectures designed for unstructured data. Interdisciplinary case studies in healthcare, finance, and engineering demonstrate the practical synergy of these approaches. By embracing multi-paradigm strategies, the chapter offers insights into building transparent, scalable, and effective decision-support systems across diverse domains. It concludes by addressing challenges of ethics and governance and by pointing to future directions such as explainable AI and federated learning, emphasising that robust decision-making depends on a hybridised analytical framework.

DOI: 10.4018/979-8-3373-6746-0.ch003

INTRODUCTION

In the 21st century, the convergence of advanced numerical techniques with powerful computational resources has revolutionised quantitative decision-making. Organisations increasingly face vast, complex, and diverse datasets that outstrip the capabilities of traditional data analysis. This shift from intuition-driven to data-driven decision-making (DDDM) is reshaping how strategic choices are made across industries and disciplines (Rokach et al., 2023; IBM, 2024). DDDM involves systematically collecting, aggregating, and interpreting information from multiple sources—ranging from customer interactions and financial records to real-time sensor outputs. Unlike classical decision models that rely on subjective judgment, DDDM anchors decisions in empirical data, reducing uncertainty, minimising bias, and improving reliability (Cote, 2021).

Classical statistical analysis remains central to this paradigm, providing rigorous methodologies for hypothesis testing, inference, and summarising past trends (Hair, 2009). Tools like IBM Statistical Package for the Social Sciences (SPSS) and Analysis of Moment Structures (AMOS) facilitate structured data analysis and causal modelling, especially within well-defined, smaller datasets common in social sciences and management studies (Statistics Solutions, 2025). However, these classical methods encounter inherent limitations when confronted with the unprecedented scale, velocity, and heterogeneity of modern data (Chen et al., 2014).

To meet the demands of contemporary decision environments, big data analytics has emerged as a transformative paradigm. Powered by distributed computing frameworks such as Apache Hadoop and Spark, big data analytics excels in processing massive, diverse datasets—including unstructured formats like text, images, and sensor data—and supports real-time descriptive, diagnostic, predictive, and prescriptive analytics (IBM, 2024; Gandomi & Haider, 2015). This scalability enables organisations to uncover complex, hidden patterns inaccessible to classical statistical tools, expanding the scope and depth of insights available for decision-making.

Machine learning (ML) and artificial intelligence (AI) further amplify this shift by introducing adaptive, performance-driven models capable of learning from continuously evolving data streams (LeCun et al., 2015; Rokach et al., 2023). ML techniques not only enhance predictive accuracy but also facilitate the extraction of knowledge from unstructured and high-dimensional data spaces. In sectors such as finance, healthcare, and manufacturing, these AI-driven methods enable real-time risk detection, personalised treatment, and autonomously optimised operations (Esteva et al., 2017; Pure Search, 2025).

Deep learning, leveraging layered neural networks, represents an advanced subset of ML that excels in handling nonlinear and complex relationships inherent in big data. Its applications span image recognition, natural language processing, and autonomous system control—fields that were previously beyond the reach of traditional methods (Schmidhuber, 2015). Consequently, the integration of big data and AI techniques with foundational numerical analysis is not merely additive but synergistic, producing hybrid analytical frameworks that enhance decision robustness and precision.

As big data and AI systems become embedded within organisational workflows, challenges such as data quality management, interpretability trade-offs, and ethical governance surface, demanding multidisciplinary approaches and evolving expertise (IBM, 2024). Nonetheless, the trajectory is clear: quantitative decision-making must transcend traditional statistical confines, embracing the scalability, complexity management, and adaptability offered by big data analytics and AI. This evolution is vital to harness the full potential of contemporary data streams and to maintain competitiveness and responsiveness in a rapidly transforming data ecosystem. Figure 1 presents a roadmap for numerical analysis in decision-making, visually tracing the evolution from early statistical analysis paradigms through modern developments in big data analytics, ML, and deep learning. The figure helps clarify the layered and interconnected nature of decision-making tools and methodologies addressed throughout this chapter.

Figure 1. Showing a roadmap for numerical analysis in decision-making.

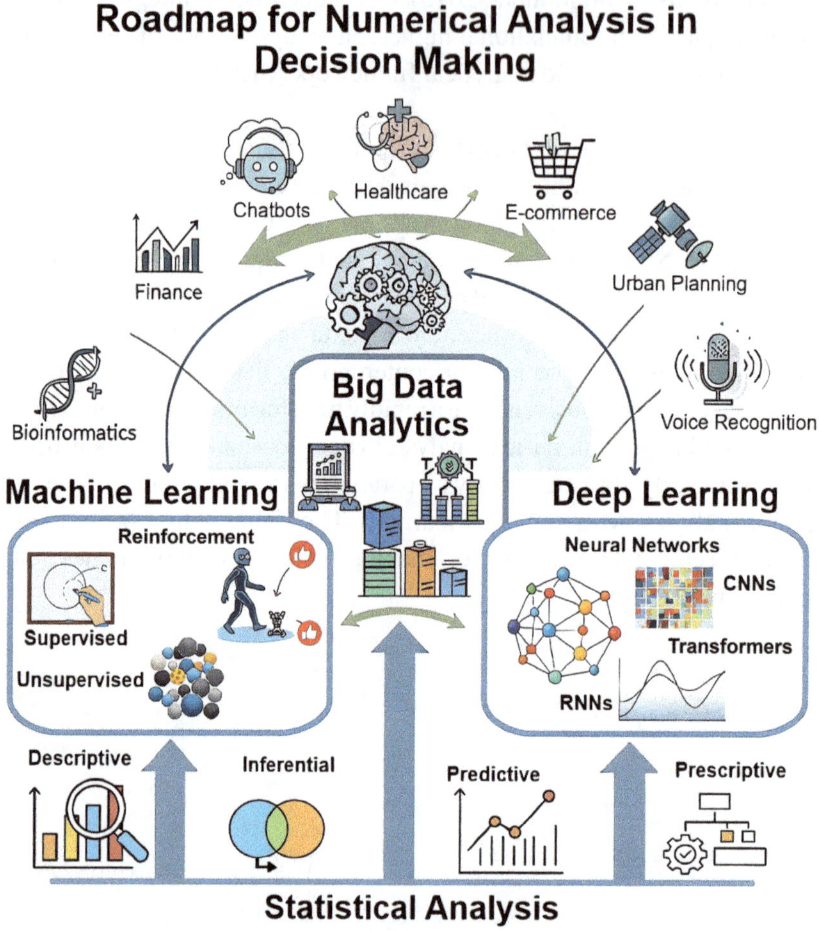

Aims of the Chapter

The aim of this chapter is to explore how data-driven approaches are transforming modern decision-making by integrating and bridging traditional statistical analysis, big data analytics, and AI. The chapter begins with an overview that highlights the importance of combining these paradigms to address increasingly complex challenges in various fields. It then introduces the foundational forms of statistical analysis—descriptive, inferential, predictive, and prescriptive—and demonstrates their pivotal role in supporting evidence-based decisions. The discussion transitions to practical applications, examining widely used analytical tools and software such as SPSS and AMOS, while evaluating their strengths, limitations, and integration with programming workflows in R and Python.

The chapter then expands on big data analytics, emphasising the defining characteristics of big data and the role of platforms and frameworks like Hadoop and Spark in enabling real-time, large-scale analysis. It distinguishes these techniques from conventional small-sample statistics and outlines the value of hybridising big data methodologies with traditional inferential approaches. Building on this foundation, the narrative presents the core paradigms of ML—supervised, unsupervised, and reinforcement learning—and discusses key algorithms including regression, decision trees, clustering, neural networks, and ensemble models. These are compared with classical statistical models to illustrate how ML enhances predictive accuracy and pattern recognition. The transition to deep learning is then explored, introducing neural architectures such as convolutional, recurrent, and transformer networks, with emphasis on their ability to process unstructured big data from sources including text, images, and sensors.

The integration of classical statistical techniques with AI and big data is discussed through comprehensive examples across disciplines, including clinical decision-making and the Internet of Medical Things in healthcare, predictive maintenance in engineering, policy modelling in economics, customer analytics in management science, and automated decision support in AI and data science. Case studies in predictive analytics for finance are also examined. Finally, the chapter identifies emerging challenges related to data quality, ethics, and governance. It closes by addressing future opportunities—including explainable AI, federated learning, and real-time analytics—reinforcing the necessity of bridging traditional approaches with advanced computational techniques to advance both research and practical outcomes in quantitative decision-making.

FOUNDATIONS OF STATISTICAL ANALYSIS FOR DECISION-MAKING

Statistical analysis is fundamental for transforming data into actionable insights that guide decision-making across various fields. It provides structured methods to collect, summarise, and interpret data, enabling decision-makers to evaluate evidence objectively and address uncertainty effectively. By applying statistical principles, organisations can better understand trends, assess risks, and optimise outcomes in complex environments. The core types of statistical analysis that underpin quantitative decision-making are outlined, setting the stage for exploring their specific roles and applications in modern practice.

Types of Statistical Analysis

Statistical analysis serves as a cornerstone of quantitative decision-making, providing a systematic set of methods for collecting, summarising, interpreting, and leveraging data to inform choices across various domains such as engineering, economics, healthcare, and management sciences (Hastie et al., 2009). Its primary strength lies in transforming raw data into actionable knowledge that supports evidence-based decisions. The foundational types of statistical analysis—namely descriptive, inferential, predictive, and prescriptive statistics— are explored to detail their relevance, methodologies, and applications in modern decision-making contexts. Understanding the foundations and distinctions of these techniques is essential for anyone engaged in quantitative decision-making, as each provides unique insights and tools that enhance the quality, precision, and impact of decisions. By empowering practitioners to tackle complexity and uncertainty, these methods promote effective decision-making and foster innovation and progress across diverse sectors.

Descriptive Statistics

Descriptive statistics is the first and most fundamental type of statistical analysis. It involves summarising and visualising data to represent the main features of a dataset clearly and succinctly (Dovetail, 2024). Essentially, descriptive statistics answer questions about "what happened" or "what is happening" by condensing large, complex datasets into manageable summaries without drawing conclusions beyond the data itself. Common measures include central tendency metrics such as the mean, median, and mode, which describe typical values within data, as well as measures of dispersion like range, variance, and standard deviation that characterise variability (Indeed Editorial Team, 2025).

Visualisation tools—such as histograms, bar charts, box plots, and scatterplots—are often employed alongside numerical summaries to provide an intuitive understanding of data patterns and distributions. For instance, in a marketing context, descriptive statistics might reveal seasonal sales trends or demographic distributions of customers, helping decision-makers grasp current conditions before deeper analysis. Although descriptive statistics does not extend to inference or prediction, this foundational step is indispensable as it lays the groundwork for more sophisticated analyses.

Inferential Statistics

Moving beyond description, inferential statistics allow practitioners to draw conclusions about a larger population based on a sample—thus addressing uncertainty, variability, and potential error in data (Dovetail, 2024). This type of analysis is crucial in decision-making when studying an entire population is impractical or impossible. Inferential methods use probability theory and sampling distributions to estimate population parameters and test hypotheses, providing measures of confidence for those estimates (Hastie et al., 2009).

Hypothesis testing, a key component of inferential statistics, evaluates competing claims about population characteristics by analysing sample data. For example, a company might test whether a new marketing strategy has significantly increased sales compared to a prior period, using tests such as t-tests, ANOVA (analysis of variance), or chi-square tests depending on data type and experimental design (Indeed Editorial Team, 2025; Coursera, 2024). Confidence intervals offer another estimation tool, defining a range within which the true population parameter likely lies with a given level of confidence, typically 95%. Inferential statistics underpin evidence-based research and policymaking by quantifying uncertainty and enabling generalisations which inform effective decisions. However, their validity depends on assumptions about sample representativeness, normality, and independence, which must be carefully evaluated in practical applications.

Predictive Statistics

The decision-making process often requires anticipating future events or behaviours, and predictive statistics address this need by leveraging historical data and statistical models to forecast outcomes (Dovetail, 2024). Predictive techniques range from relatively simple regression models to more advanced algorithms that can handle multiple predictors and time-dependent structures. Regression analysis, both simple linear and multiple regression, models the relationship between dependent (outcome) variables and one or more independent (predictor) variables. These models estimate how changes in predictors affect the expected value of the outcome, providing quantitative tools for forecasting (Indeed Editorial Team, 2025). For example, a retailer might use regression to understand how advertising spend, price changes, and competitor actions influence sales figures and then predict future sales under different marketing scenarios.

Time series models are another predictive tool, analysing data points collected sequentially over time to detect trends, seasonality, or cyclical patterns (Cote, 2021). Applications include stock market analysis, inventory demand forecasting, and energy consumption planning. Predictive analytics may incorporate probabilistic and ML methods, increasing accuracy and accommodating complex, nonlinear relationships, but predictions remain probabilistic and subject to uncertainties inherent in the data and modelling assumptions.

Prescriptive Statistics

Prescriptive statistics represents the most advanced statistical approach, going beyond forecasting to recommend optimal actions based on data-driven insights (Indeed Editorial Team, 2025). Employing techniques such as optimisation algorithms, simulation models, and complex decision rules, prescriptive analytics helps decision-makers choose the best course among alternatives given constraints and objectives. For instance, in supply chain management, prescriptive models optimise inventory levels, balancing costs associated with ordering, holding, and stockouts while considering predicted demand fluctuations (Dovetail, 2024). Simulation allows exploration of possible future scenarios and quantification of associated risks, enabling robust decision-making under uncertainty.

Techniques such as linear programming, integer optimisation, and Monte Carlo simulations are common prescriptive tools. By integrating statistical forecasts with optimisation and domain-specific knowledge, prescriptive analytics provide actionable recommendations, effectively answering "what should we do next?" This approach is increasingly supported by computational advances and AI-driven platforms that enable real-time, scalable decision support in complex environments.

Role in Supporting Evidence-Based Decisions

Statistical analysis plays a crucial role in supporting evidence-based decision-making by providing objective, data-driven insights that enable practitioners and policymakers to make informed choices. This approach prioritises quantitative and qualitative evidence over subjective judgment or intuition, thereby fostering decisions that are transparent, replicable, and justifiable (Angrist & Pischke, 2009). By leveraging statistical methods, decision makers can assess the strength of available evidence, estimate the impact of variables, and quantify uncertainty, which is vital for determining appropriate courses of action in complex environments (Brownstein et al., 2019).

One core function of statistical analysis in evidence-based decision-making is risk assessment and mitigation. By analysing historical data and identifying meaningful patterns or risk factors, organisations can forecast potential challenges or unintended consequences and implement pre-emptive strategies to address them (Technology Innovators, 2024). This is particularly important in fields such as public policy and healthcare, where decisions often affect large populations and resource allocations must be optimised to maximise benefits and minimise harms. Furthermore, statistical techniques support real-time monitoring and evaluation of implemented policies or business strategies. Continuous analysis of key performance indicators and feedback loops allow organisations to dynamically adjust their approaches in response to evolving circumstances or new data, thereby enhancing responsiveness and effectiveness (Technology Innovators, 2024). This adaptability is crucial in fast-moving domains such as supply chain management, where demand fluctuations and disruptions require timely interventions informed by timely data analytics.

Another essential contribution of statistical analysis is the facilitation of stakeholder engagement and effective communication. Data-derived insights provide a common language that helps bridge gaps between diverse actors—including managers, employees, customers, and external partners—making complex information accessible and actionable (Statistics Authority, 2024). This transparency fosters trust and buys-in from stakeholders, which is often critical for successful implementation and sustained outcomes. Statistical rigour also ensures decisions are unbiased and based on robust evidence rather than anecdotal experiences or cognitive biases. For example, controlled experiments and hypothesis testing can isolate the effects of interventions, allowing organisations to discern causal relationships and avoid erroneous conclusions drawn from spurious correlations (Select Statistics, 2024). By grounding decisions in sound statistical principles, organisations enhance their credibility and improve the likelihood of positive impacts.

Tools and Software for Statistical Analyses

Statistical software tools play a crucial role in modern quantitative research and decision-making, assisting practitioners from diverse disciplines to navigate data complexity, perform advanced analyses, and interpret findings effectively. This section focuses on two of the most widely used packages, IBM SPSS and AMOS, outlining their features, advantages, constraints, and compatibility with programming languages like R and Python. Additionally, it examines applied research examples within management science and psychological modelling that demonstrate the broad applicability of these tools in academic and organisational settings.

IBM SPSS

IBM SPSS remains a favoured option among researchers and data analysts due to its combination of powerful analytical capabilities and ease of use. Its user-friendly graphical interface facilitates complex statistical workflows through menu-based options and dialog boxes, accommodating users with limited programming expertise. SPSS is appreciated for supporting a variety of statistical procedures—including simple, multiple, and logistic regression—as well as factor analysis, which reveals underlying structures within data. This makes it particularly useful in exploratory research, scale development, and fields such as psychology, marketing, and education (Field, 2018; Hair, 2009).

In addition to regression and factor analysis, SPSS includes tools for cluster analysis, descriptive statistics, hypothesis tests, and non-parametric methods. Its data management capabilities cover tasks such as data transformation, aggregation, and treatment of missing values, enhancing its utility as a comprehensive analytical platform (Field, 2018). Extensions like SPSS Modeler incorporate ML algorithms and data mining functions for sophisticated predictive analytics. Integration with IBM's Watson AI platform further strengthens its capacity to handle big data environments (IBM, 2024).

AMOS

AMOS, an extension to SPSS, specifically supports SEM, a robust method for analysing complex associations between observed and latent variables (Statistics Solutions, 2025). It extends standard regression through integration of path analysis, factor analysis, and latent variable modelling within a unified framework (Nakhla, in press). AMOS features an intuitive drag-and-drop graphical interface, simplifying model specification and modification compared to traditional syntax-based SEM tools, making it accessible to professionals in psychology, social sciences, and management research (Hair, 2009). The software supports confirmatory factor analysis (CFA), mediation and moderation analyses, along with multi-group comparisons, enabling detailed exploration of theory-based hypotheses about latent constructs and their relationships (Statistics Solutions, 2025; Nakhla, in press). Outputs include key model fit indices such as CFI and RMSEA, standardised path coefficients, and modification indices that assist in improving model fit (Nakhla, in press).

Applied Analytics with SPSS and AMOS

The enduring relevance of SPSS and AMOS is reflected through their application in current research. SPSS is recognised for facilitating regression, factor analysis, and exploratory data analysis across operational studies, customer behaviour analytics, and credit risk evaluation (Hair, 2009; Sarker et al., 2024). Retailers, for example, utilise SPSS to identify drivers of customer loyalty and tailor marketing strategies accordingly, while the financial sector leverages regression and clustering to assess creditworthiness. AMOS addresses more intricate modelling needs involving latent variables central to psychological and organisational studies. For instance, the Theory of Planned Behaviour has been empirically validated using

SEM in AMOS, illustrating the connections between attitudes, social norms, and perceived behavioural control (Hair, 2009; Byrne, 2001).

Combining SPSS and AMOS yields a comprehensive research methodology that progresses from descriptive and exploratory analyses using SPSS to confirmatory SEM analysis in AMOS. This approach has elucidated factors influencing employee engagement, linking organisational culture and leadership to workforce retention and performance (Statistics Solutions, 2025). Such integrated use enhances validity and reliability in management, psychology, and educational research. Further developments improve these platforms' interoperability with R and Python, increasing scalability, visualisation capacity, and reproducibility when managing large datasets (IBM, 2024). In education and psychology, AMOS is frequently employed to confirm assessments related to academic motivation, anxiety, and learning attitudes, bolstering theoretical constructs with empirical data (Nakhla, in press; Byrne, 2001).

Strengths, Limitations, and Integration with R/Python Workflows

Both SPSS and AMOS offer distinct advantages that contribute to their widespread adoption. The intuitive graphical user interfaces reduce technical barriers, facilitating access to complex statistical techniques for users with diverse expertise levels. Extensive documentation and active user communities support effective use and problem-solving. Nonetheless, these packages experience limitations. SPSS efficiently handles classical statistical analyses but may encounter challenges with very large datasets and resource-intensive computational tasks typical of big data scenarios (Pure Search, 2025). AMOS, while adept at SEM, lacks the flexibility of specialised SEM software like M*plus* or the R package lavaan, especially when modelling complex structures or handling categorical variables (Hair, 2009).

Traditionally, SPSS and AMOS employ proprietary scripting and rely heavily on graphical interfaces, limiting automation relative to open-source languages. However, integration efforts with R and Python have increasingly bridged this gap by enabling sophisticated data manipulation, custom modelling, and enhanced visualisation. R packages such as 'foreign' and 'haven' facilitate importing and exporting SPSS datasets, while libraries like 'sem' and 'lavaan' enable SEM modelling directly in R (Dissertation India, 2024). Python similarly supports data exchange with SPSS through packages and application programming interface (API), enabling users to embed SPSS analytics within broader ML or data engineering pipelines. Such integration enhances reproducibility, scalability, and automation in analytical workflows, crucial for modern research and enterprise environments (IBM, 2024).

BIG DATA ANALYTICS

Big data analytics refers to the advanced process of examining large and diverse datasets, often generated at high speed and from a variety of sources, to uncover meaningful patterns, trends, and insights that support better decision-making (IBM, 2024). Unlike traditional data analysis, which typically focuses on relatively small, structured datasets, big data analytics handles massive volumes of structured, semi-structured, and unstructured information from social media, internet of things (IoT) devices, transactional databases, and more (Google Cloud, 2025). The growing digital landscape and technological advancements such as cloud computing, the IoT, and AI continually fuel the exponential increase in data generation. This shift has made conventional data storage and analysis methods inadequate for capturing the complexity and velocity of modern data flows (TechTarget, 2023). Consequently, big data analytics relies on specialised frameworks and scalable platforms capable of processing data in real time or near-real time to deliver actionable insights promptly.

Characteristics of Big Data

Big data is distinguished by five key characteristics commonly referred to as the "5 Vs": volume, velocity, variety, veracity, and value (IBM, 2024; Marr, 2021). The volume of big data refers to the massive scale of information generated from diverse sources such as social media, IoT devices, enterprise systems, and sensors. This data is measured in terabytes or petabytes, representing a scale far beyond traditional small-sample statistics. Velocity describes the rapid speed at which data is created and transmitted, often in real time or near real time. For instance, every minute, millions of social media posts, financial transactions, or sensor readings are generated and require instant processing (Teradata, 2022).

Variety captures the heterogeneity inherent in big data, which includes structured data (e.g., relational databases), semi-structured data (e.g., JSON, XML files), and unstructured data such as images, videos, and text documents (Google Cloud, 2025). This richness provides broader perspectives but also necessitates more sophisticated techniques to integrate and analyse diverse formats. Veracity concerns the quality, accuracy, and trustworthiness of data. Given big data's vast and varied sources, inconsistencies and noise are common, making rigorous data cleaning and validation essential to maintain analytical reliability (TechTarget, 2023). Finally, value underscores the ultimate goal of big data: to generate actionable insights that enable better decision-making, operational efficiency, and competitive advantage (Marr, 2021).

Distinct Approaches from Conventional Small-Sample Statistics

Traditional statistics focus on smaller, well curated datasets where assumptions about data distribution and sampling can be strictly adhered to. In contrast, big data analytics handle vast, fast-changing, and diverse datasets, requiring different conceptual and computational approaches (Marr, 2021). Classic hypothesis testing and inferential statistics rely on controlled experiments and representative sampling, while big data often emphasises exploratory analysis, pattern recognition, and predictive modelling using ML algorithms. The ability to continuously stream and analyse data in real time is another major departure — conventional methods struggle with these high-velocity, high-volume environments (Teradata, 2022).

Big data analytics frequently employ scalable, distributed computational frameworks to process massive datasets beyond the capacity of traditional single-machine software. These frameworks support parallel processing, real-time analytics, and advanced algorithms capable of managing diverse and noisy data (Google Cloud, 2025). Consequently, combining computational techniques from computer science with statistical theory is a hallmark of big data analytics.

Frameworks and Platforms

Two of the most transformative frameworks revolutionising big data analytics are Apache Hadoop and Apache Spark (IBM, 2024). Hadoop, introduced in the mid-2000s, provides an open-source distributed file system and a MapReduce programming model capable of handling petabyte-scale datasets across clusters of inexpensive commodity hardware. Its design enables storage and batch processing of massive volumes with fault tolerance and scalability. Apache Spark builds on Hadoop's limitations by offering in-memory processing, which significantly accelerates analytics, especially iterative algorithms integral to ML and real-time stream processing (IBM, 2024). Its support for diverse workloads—batch, streaming, interactive queries, and ML—makes it a versatile platform in big data ecosystems. Spark's APIs for languages such as Python, R, and Scala facilitate integration with statistical tools and frameworks, enabling more sophisticated hybrid analytic workflows.

Other complementary technologies include databases designed for big data, such as NoSQL (Not Only SQL) stores and data lakes, which can handle diverse data types without the need for the fixed schemas of traditional relational databases. This flexibility accommodates the variety and velocity of input data, further supporting the wide-ranging data ingestion and retrieval needs (Google Cloud, 2025).

Hybridisation with Inferential Statistics for Enriched Insights

Although big data analytics often rely on predictive and descriptive methods, integrating traditional inferential statistics offers significant advantages in strengthening the rigour and interpretability of findings. Hybrid approaches harness the power of big data's volume and velocity with the formal hypothesis testing and uncertainty quantification embedded in inferential statistics. For example, in marketing analytics, big data-driven ML models can generate predictions about consumer behaviour patterns, while inferential statistics help quantify confidence intervals and test the significance of observed effects, ensuring that results are both actionable and statistically valid (Marr, 2021).

This fusion addresses critiques of big data approaches as sometimes producing correlation-heavy but causally weak insights by embedding robust experimental or quasi-experimental designs where possible. Moreover, hybrid analytics facilitate transferability of findings beyond the sampled dataset, providing evidence that patterns hold across different populations or temporal spans, which is critical for strategic decision-making in areas such as finance, healthcare, and engineering (Marr, 2016).

MACHINE LEARNING FOR QUANTITATIVE DECISION-MAKING

Quantitative decision-making increasingly leverages advances in ML to enhance and expand analytical capabilities. Unlike data analysis techniques that primarily focus on fixed-model assumptions, contemporary approaches emphasise adaptive frameworks that learn from complex datasets with minimal human intervention. This evolution enables nuanced understanding and optimisation across a wide array of applications from finance and healthcare to marketing and operations. A foundational step in harnessing this potential is understanding the core ML paradigms, each representing a unique way of deriving insights and informing decisions through data-driven learning processes.

Machine Learning Paradigms

ML has emerged as a transformative approach for quantitative decision-making, offering new paradigms that extend far beyond traditional statistics and econometrics. The three foundational paradigms—supervised, unsupervised, and reinforcement learning (RL)—provide distinct frameworks for extracting knowledge from data and optimising decisions (IBM, 2025; Pecan AI, 2025).

Supervised learning

Supervised learning is a core paradigm of ML characterised by the use of labelled datasets, where each input sample is paired with a known target or output (GeeksforGeeks, 2024). The goal of supervised learning is to train a model to learn the mapping from inputs to outputs, thereby enabling it to make accurate predictions on new, unseen data. This learning process involves feeding the algorithm data with clearly defined input-output pairs, allowing the model to iteratively optimise its parameters to minimise prediction errors. Common supervised learning tasks include regression and classification. Regression predicts continuous numerical values, such as forecasting sales or housing prices, by modelling the relationship between input features and output targets. Classification involves categorising data points into predefined classes—for example, identifying whether an email is spam or legitimate or detecting fraudulent transactions (IBM, 2024). The algorithm learns from past examples where the outcomes are known and applies this knowledge to classify or predict future cases. Supervised learning supports applications in healthcare for disease diagnosis, finance for credit risk evaluation, marketing for customer targeting, and many more (Google Cloud, 2025).

Unsupervised learning

Unsupervised learning is a branch of ML that focuses on analysing data without pre-existing labels or categories. It operates on raw, unlabelled data and seek to uncover hidden patterns, structures, or relationships within it (GeeksforGeeks, 2024; IBM, 2021). The primary objectives of unsupervised learning include grouping similar data points—known as clustering—detecting anomalies and reducing the dimensionality of data to simplify complex datasets while preserving essential information. Clustering is particularly useful for market segmentation, customer profiling, educational research, and identifying latent groupings in various domains without assuming any prior category structure (Nakhla, in press).

Common algorithms for clustering include k-means, hierarchical clustering, and density-based methods, each suited to different data types and analysis goals (Oracle, 2024). Dimensionality reduction techniques such as principal component analysis and autoencoders help visualise high-dimensional data, reduce noise, and improve downstream analysis. Unsupervised learning is a powerful tool for exploratory data analysis because it reveals insights that might remain hidden in labelled analyses. For instance, retailers can discover new customer segments based on purchasing behaviour, while cybersecurity systems use anomaly detection to flag unusual activity.

Reinforcement Learning

RL is a branch of ML where an agent learns to make decisions through ongoing interaction and feedback with its environment, rather than relying on static datasets. At each step, the agent observes its state, chooses an action, and receives a reward or penalty as feedback, with the ultimate goal of maximising cumulative rewards over time (Milvus, 2025). This process balances exploration—testing new strategies to discover better outcomes—and exploitation—pursuing actions that have previously yielded high rewards.

Real-world examples include a robot navigating a maze, an AI playing chess, or an autonomous system managing resources, where feedback can be immediate or delayed and connections between actions and rewards are not always obvious. Because the sequence of actions shapes future states and opportunities, RL algorithms like Q-learning and temporal difference learning incrementally update policies to improve performance over time (IBM, 2024; Milvus, 2025). This feedback-driven, trial-and-error approach enables RL agents to adapt to complex, uncertain, or changing environments, excelling in domains that require continual adjustment and long-term strategy, such as advanced robotics, dynamic gaming, or resource management tasks (Milvus, 2025).

Key Algorithms

ML's power is actualised through a repertoire of algorithms, each serving specific analytical or predictive tasks (SAS, 2023; Sarker, 2021). This diverse set—including regression, decision trees, clustering, neural networks, and ensemble methods—provides flexible tools capable of both prediction and pattern discovery. Such adaptability complements classical methods, enabling more robust quantitative decision-making across various industries (SAS, 2023; Sarker, 2021).

Regression algorithms

Regression algorithms play a fundamental role in ML by modelling relationships between variables and enabling both predictive analytics and causal inference. Classical linear regression models a linear relationship between input features and a continuous output, making it straightforward and interpretable. Logistic regression extends this framework for classification problems, predicting the probability of categorical outcomes such as fraud detection or disease diagnosis (SAS, 2023).

Regularised regression techniques like ridge regression and LASSO (Least Absolute Shrinkage and Selection Operator) enhance traditional regression models by addressing multicollinearity and overfitting. Ridge regression penalises large coefficients to shrink them towards zero, improving model stability, especially with correlated variables. LASSO goes further by enforcing sparsity, effectively performing variable selection by shrinking some coefficients exactly to zero, which helps in high-dimensional datasets with many irrelevant features (Onlinemanipal, 2025). These algorithms balance interpretability and performance, making them essential tools for quantitative decision-making across sectors such as finance, healthcare, and marketing. Their adaptability to diverse data scenarios allows more precise and reliable predictions than classical models alone.

Decision trees

Decision trees are a popular and intuitive supervised ML algorithm used for both classification and regression tasks. The structure of a decision tree mimics a flowchart, consisting of a root node, internal decision nodes, and leaf nodes that represent final outcomes or predictions (Seldon, 2021; IBM, 2021). At each decision node, the data is partitioned based on the value of an input feature, splitting the dataset into subsets that are more homogeneous with respect to the target variable. This hierarchical, tree-like model makes decision trees highly interpretable, as the path from root to leaf clearly maps how decisions are made based on feature values.

For example, a decision tree predicting customer purchases might first split data based on income level, followed by age or previous purchase history, allowing stakeholders to easily understand the decision criteria (GeeksforGeeks, 2017). However, decision trees can be prone to overfitting, especially if allowed to grow deep and complex, leading to poor generalisation on new data. Techniques like pruning, which remove less important branches, and setting parameters such as maximum tree depth help mitigate this issue (Seldon, 2021).

Clustering algorithms

Clustering algorithms are a fundamental component of unsupervised ML, used to group data points with similar characteristics without needing pre-existing labels. Popular algorithms include k-means, which partitions data into k clusters by minimising the variance within each cluster; hierarchical clustering, which builds a tree-like structure of nested clusters based on data point similarities; and Gaussian mixture models (GMM), which assume data points are generated from a mixture of several Gaussian distributions and assign probabilistic cluster memberships (Sarker, 2021).

Beyond these, model-based methods like Expectation-Maximisation algorithms and grid-based methods such as STING and CLIQUE provide alternative ways to detect clusters in complex or high-dimensional datasets. Clustering is widely applied for customer segmentation in marketing, allowing businesses to tailor strategies based on identified groups. It's also critical for anomaly detection in cybersecurity and fraud prevention by highlighting unusual data points that deviate from normal clusters. Additionally, clustering aids exploratory data analysis by revealing intrinsic structures or patterns in data that inform subsequent modelling and decision-making.

Neural networks

Neural networks and deep learning architectures are designed to address complex, highly non-linear, and multidimensional problems that traditional algorithms struggle with. Inspired by the structure and function of the human brain, neural networks consist of layers of interconnected nodes, or artificial neurons, that process and transmit information through weighted connections (IBM, 2021). Data feeds into the input layer, passes through one or more hidden layers where increasingly abstract patterns are detected, and finally reaches the output layer to produce predictions or classifications.

Deep learning extends this concept by using many hidden layers, enabling the model to learn hierarchical feature representations automatically. This makes them especially powerful for tasks like image recognition, natural language processing, and speech recognition, where complexity and high dimensionality are inherent (Hardesty, 2017). By training on vast amounts of data, neural networks can generalise across domains, detecting intricate patterns in raw, unstructured inputs that other models might miss.

Ensemble algorithms

Ensemble algorithms are a powerful approach in ML that combines the predictions of multiple decision trees to enhance overall model performance and robustness (IBM, 2024; GeeksforGeeks, 2024). By aggregating the outputs of several individual trees, ensemble methods reduce the weaknesses of single models, such as high variance or overfitting. Random forests are one such ensemble technique that builds a large number of decision trees independently by training each tree on a different random subset of the data and features. The final prediction is made through majority voting (for classification) or averaging (for regression) across all trees, which reduces variance and leads to more stable and accurate models (Sarker, 2021).

In contrast, gradient boosting creates trees sequentially where each new tree attempts to correct the errors of its predecessors. This iterative approach often results in highly accurate models capable of capturing complex patterns, but it requires careful tuning to prevent overfitting (Sarker, 2021). Another ensemble method, bagging (bootstrap aggregating), involves training multiple models on bootstrapped samples of the training data and aggregating their outputs, improving stability and performance. These ensemble techniques have become integral to many predictive analytics tasks due to their balance of accuracy and robustness.

Decision-Making Optimisation through Predictive Accuracy and Pattern Detection

ML for decision optimisation centres on maximising predictive accuracy and detecting subtle patterns in data. Unlike manual rule-based systems, ML models continuously learn from new information, updating predictions and adapting to changing environments (ValueCoders, 2023). Increased accuracy translates directly to improved outcomes, whether classifying fraudulent transactions, forecasting demand, or optimising resource allocation. Pattern detection is equally critical: ML algorithms can autonomously discover regularities invisible to human analysts, flag anomalies, or identify market segments with unique needs.

For example, ML driven pattern recognition improves sales forecasting by discerning non-obvious seasonal or demographic influences (Codewave, 2024). In healthcare, predictive models can highlight abnormal patient trajectories and suggest early interventions. Crucially, ML driven decision processes can be deployed into real-time environments, continuously refining predictions as fresh data arrives. This allows organisations to respond proactively to risks, opportunities, or evolving market dynamics—often faster and with greater reliability than traditional analytics alone.

DEEP LEARNING

The transition from traditional ML to deep learning paradigms marks a significant evolution in the field of AI, mainly due to deep learning's ability to recognise complex patterns and scale effectively with large datasets. Traditional ML generally depends on human-engineered features and simpler algorithms designed to recognise relatively straightforward patterns. In contrast, deep learning models—particularly neural networks with multiple layers—learn hierarchical feature representations directly from raw data, enabling them to uncover intricate and abstract patterns without explicit feature engineering (Qadir, 2024; Elastic, 2024). This leap in capability facilitates their application to challenging real-world tasks involving unstructured data such as images, text, and sensor signals.

Neural Architectures in Deep Learning

Deep learning's remarkable capabilities in solving complex real-world problems largely come from the variety and sophistication of neural network architectures it employs. These architectures are designed to process different types of data and extract meaningful patterns automatically, which traditional ML often struggles to do without extensive manual feature selection.

Convolutional Neural Networks

Convolutional Neural Networks (CNN) are specially designed to process data with a grid-like structure—most notably images. CNNs scan these images using small, trainable filters that slide over the pixels in a process known as convolution. These filters act like "pattern detectors" that can recognise simple shapes such as edges, corners, or textures. As the data progresses deeper into the network through multiple convolutional layers, CNNs progressively combine these simple features to detect increasingly complex shapes—like eyes, ears, or cars—ultimately enabling the network to recognise entire objects. This hierarchical feature learning is akin to how the human visual system processes images, starting from basic lines and moving to complex forms. Besides convolutional layers, CNNs include a mechanism called pooling, which reduces the spatial size of the data, distilling the most important information while reducing computational load. Another key advantage of CNNs is the use of shared weights for their filters, meaning the same filter is applied across different parts of the image, allowing the network to scan for the same features regardless of their position. CNNs have revolutionised fields like medical imaging (e.g., automated tumour detection), facial recognition, and autonomous driving by providing precise, fast image analysis capabilities without handcrafted rules (IBM, 2021a).

Recurrent Neural Networks

While CNNs excel at spatial data, sequential data like text, speech, and time series require models that can remember information across time steps. Recurrent Neural Networks (RNNs) address this by including feedback loops in their architecture, allowing the output at one time step to influence processing at the next. This "memory" enables RNNs to capture temporal dependencies and context—a critical requirement for language and speech understanding. However, basic RNNs face challenges with long sequences because they tend to forget earlier inputs as new data arrives.

To overcome this, more sophisticated variants like Long Short-Term Memory (LSTM) networks were developed. LSTMs introduce specialised memory cells and gating mechanisms to control information flow, enabling the network to remember or forget data over longer sequences. This makes RNNs, particularly LSTMs, powerful for applications such as predicting the next word in a sentence (language modelling), transcribing speech to text, or forecasting stock prices based on historical trends (IBM, 2021b; IONOS, 2025).

Transformers

While RNNs improved sequential data processing, their inherently sequential architecture limits parallelisation and struggles with learning very long-range relationships effectively. Transformers, a revolutionary neural architecture introduced in 2017, overcome these issues through a self-attention mechanism, which allows the model to weigh the importance of different parts of the input sequence simultaneously. Imagine reading a sentence and instantly understanding which words relate to each other, even if they are far apart. Transformers mimic this by computing attention scores between all pairs of input positions, capturing dependencies without regard to their distance in the sequence. This ability has drastically accelerated training times and enhanced performance in natural language processing tasks like translation, summarisation, and question answering. Popular transformer-based models include Bidirectional Encoder Representations from Transformers (BERT), which focuses on understanding textual context (encoder-only), and Generative Pre-trained Transformer (GPT), which excels at text generation (decoder-only) (IBM, 2025).

Scalability in Handling Unstructured Big Data

One of the biggest strengths of deep learning is its ability to scale—that is, to handle and learn from extremely large and complex datasets, especially those that are unstructured. Unstructured data includes text, images, videos, audio, and sensor information, all of which don't fit neatly into traditional databases. Processing this kind of data is challenging because of its size and complexity. Deep learning models are built to manage this challenge by learning hierarchically. This means they process data in layers, extracting simple features first—like edges in an image or basic sounds in audio—and then combining those into more complex features, such as shapes or words, at deeper layers. This approach lets the models find meaningful patterns in massive, messy datasets without the need for humans to define specific rules or features beforehand (Zilliz, 2024).

The ability to train models on huge datasets is also supported by advances in specialised hardware, such as Graphics Processing Units (GPUs) and Tensor Processing Units (TPUs). These are powerful computing chips designed to run many calculations in parallel, drastically speeding up the training process. Distributed training frameworks allow this workload to be spread across multiple machines, making it possible to learn from billions of data points in relatively short time (Rane et al., 2024). To make sure models don't "overfit" or memorise the training data too closely, which would hurt their performance on new data, techniques like data augmentation (creating variations of data) and regularisation (adding constraints to the training) are used.

Additionally, to deploy large models efficiently for real-time applications, strategies like pruning (removing unnecessary parts of the model) and quantisation (reducing numerical precision) help reduce their size and computational cost. Because of these scalable features, deep learning is used widely across many industries. For example, consumer devices use it to recognise speech and images quickly and accurately, while businesses rely on it to analyse large volumes of text data or sensor inputs to gain valuable insights and make informed decisions.

Integration of Classical Statistical Methods with Machine Learning

Classical statistics and ML increasingly function as complementary pillars in modern data analysis. Statistical methods are foundational for supporting inference—explaining relationships, testing hypotheses, and providing interpretable results grounded in probabilistic frameworks (Hastie et al., 2009; Jaokar, 2025). By contrast, ML and AI emphasise prediction, excelling with large, high-dimensional, or unstructured data where the primary focus is on discovering patterns and generating accurate forecasts (Domingos, 2012). Table 1 highlights the principal differences between classical statistical models and ML models, including key aspects of assumption, regularisation, predictive accuracy, model complexity, overfitting, adaptability, and computational needs.

Table 1. Key differences between classical statistical models and ML models.

Aspect	Classical Statistical Models	ML Models
Assumptions	Assume data follows specific structures (e.g., linearity, independence, normality); number of features is fixed.	Make fewer assumptions; can handle non-linear, complex, and high-dimensional data; flexible with feature count.
Regularisation	Rare in traditional models (e.g., OLS regression); some extensions—like Ridge or LASSO regression—add penalties.	Regularisation is widely used (LASSO, Ridge, etc.) to improve generalisation and prevent overfitting.
Predictive Accuracy	Good for structured, simple, or "well-behaved" problems with known relationships; can serve as baseline.	Often superior for large, high-dimensional, or unstructured data (e.g., images, text); boosted by ensembles.
Model Complexity	Simple, interpretable equations; transparent parameters.	Models can be complex (deep neural nets, ensembles); many hyperparameters; may require extensive tuning

continued on following page

Table 1. Continued

Aspect	Classical Statistical Models	ML Models
Overfitting	Less prone if assumptions hold and model is specified correctly; simple models may underfit complex data.	Higher risk of overfitting complex models without careful regularisation or proper validation.
Parameter Recovery	Excellent for estimation, hypothesis testing, and identifying effect sizes or relationships.	Varies: some models prioritise variable selection or interpretability, others focus solely on prediction.
Flexibility & Adaptability	Limited in modelling complex or noisy relationships; best for simple, clean data.	Highly flexible; suitable for modelling complex, noisy, or evolving data structures
Computational Needs	Typically, fast and efficient to train and use; low resources required for small and moderate-sized data.	Can be resource-intensive, needing significant computational power and large datasets, especially for deep learning.
Use Case Preference	Preferred for inference, explainability, and causal analysis, especially with smaller datasets or clear relationships.	Preferred for maximising predictive accuracy on large or complex datasets where relationships are unknown or data is unstructured

The most effective analytic strategies recognise these approaches as collaborative rather than competitive. Statistical models are valued for interpretability and alignment with theory-driven inquiries, whereas ML provides scalability, flexibility, and improved predictive accuracy, particularly in dynamic or data-rich environments (Bzdok et al., 2018; Gentsch, 2019). Hybrid strategies—such as first using statistical methods for preliminary variable selection or feature engineering, followed by ML to refine prediction—are increasingly common across disciplines. This integration allows for both robust explanation and dynamic responsiveness to new or evolving data. Selection of analytic approach should be dictated by task requirements. For situations demanding inference, transparency, or parameter estimation, classical statistics remains preferred. In contrast, ML is favoured for tasks requiring predictive performance, adaptability, and the handling of complex or voluminous datasets (Bere et al., 2021). Embracing both paradigms empowers analysts to blend strengths and address diverse analytic goals effectively.

APPLICATIONS ACROSS DISCIPLINES

Statistics, ML, and big data methods are driving innovation in diverse fields by combining explanatory insight with powerful predictive capabilities. These integrated approaches support enhanced decision-making and practical solutions in areas such as healthcare, engineering, economics, management science, psychology, education, and finance. Key applications span clinical diagnostics, equipment maintenance, policy formulation, customer behaviour analysis, decision support tools, and real-world case studies demonstrating impact in both social sciences and financial sectors.

Healthcare: Clinical Decision-Making and the Internet of Medical Things

Deep learning and ML continue to revolutionise healthcare by enhancing diagnostic precision, personalising treatment, and optimising patient outcomes. Advanced medical imaging analysis using CNNs achieves expert-level accuracy in detecting abnormalities across modalities such as X-rays, CT scans, and MRIs (Paubox, 2025). These technologies reduce diagnostic errors, and accelerate review times (Khosravi et al., 2024). Additionally, AI algorithms analyse extensive electronic health records—incorporating both structured and unstructured data—to predict disease progression, optimise personalised treatment plans, and support continuous patient monitoring (Alhumaidi et al., 2025). AI-driven predictive analytics identify high-risk patients and forecast potential complications or readmissions, facilitating timely interventions and improved resource management (Gentsch, 2019). Furthermore, wearable devices with deep learning models enable real-time monitoring, offering actionable insights for patients and clinicians (Mahajan et al., 2025; Matthew et al., 2025).

A transformative dimension to this landscape is the Internet of Medical Things (IoMT), which encompasses interconnected medical devices and sensors that collect, transmit, and analyse health data seamlessly across healthcare settings (He et al., 2024). These devices range from wearable biosensors and implantable monitors to clinical equipment, all connected via WiFi, Bluetooth, and cellular technologies, ensuring continuous data flow (Ozcelik et al., 2025). The integration of IoMT enhances remote patient monitoring, facilitating proactive management of chronic diseases beyond traditional clinical environments while also improving hospital asset management and medication adherence (He et al., 2024).

IoMT devices gather vital signs such as heart rate, blood pressure, glucose levels, and respiratory metrics, feeding data into AI and deep learning algorithms that detect anomalies, predict disease exacerbations, and optimise treatment workflows (Matthew et al., 2025). This synergy supports sophisticated clinical decision support systems that combine expert knowledge with real-time data analytics, enhancing care quality, safety, and efficiency (Khosravi et al., 2024). While challenges including data privacy, system interoperability, and model interpretability persist, ongoing advancements in explainable AI and secure connectivity aim to strengthen healthcare providers' trust and facilitate broader adoption (Mahajan et al., 2025). As these converging technologies mature, their impact will increasingly enable precise, personalised, and proactive healthcare delivery, reshaping clinical decision-making and patient care paradigms worldwide.

Engineering: Predictive Maintenance Decision Frameworks

Predictive maintenance (PdM) represents one of the most impactful applications of ML and AI in engineering, particularly in manufacturing, energy, and transportation. Traditional strategies—reactive or time-based—often result in excessive downtime and unnecessary expense (Chupryna & Leleko, 2025). PdM instead uses sensor data, IoT devices, and AI algorithms to forecast failures before they occur, enabling proactive, data-driven decisions (Neural Concept, 2024). The PdM cycle involves continuous monitoring of operational parameters such as vibration, temperature, and pressure through embedded sensors. ML models—especially supervised deep learning architectures like LSTM networks and transformer-based models—analyse time-series data to detect anomalies and estimate remaining useful life (Chupryna & Leleko, 2025; Vasilchenko, 2024). These models generate alerts and recommend maintenance actions, allowing optimised scheduling that balances resource use while minimising interruptions.

The value of PdM lies in reducing unexpected failures, improving safety, and extending asset longevity, which directly supports efficiency and profitability. AI enabled PdM has been shown to decrease downtime by up to 20% and reduce costs, while increasingly integrated platforms now combine ingestion, analytics, and decision support to automate workflows and strengthen competitive advantage (Chupryna & Leleko, 2025).

Economics: Policy Modelling with SEM and ML

In economics, effective decision-making requires understanding cause-and-effect relationships while also predicting future outcomes in uncertain conditions. SEM provides a strong statistical framework for representing hidden variables and how they interact, offering clear insights that are important for policy design (Wooldridge, 2020). New approaches now combine SEM with ML, which can extract patterns that traditional methods might miss—for example, detecting complex relationships in economic data, market trends, or consumer behaviour (Alber et al., 2019). ML further strengthens analysis by handling large and varied datasets, from financial indicators to public sentiment, making forecasts more reliable. Tools like gradient boosting and deep neural networks support scenario testing, stress analysis, and evaluation of policy impacts (Nguyen et al., 2018). Together, these hybrid approaches combine theory and data-driven prediction, helping policymakers make more confident decisions about resource allocation, monetary policy, and targeted interventions.

Management Science: Customer Analytics

Customer analytics in management science increasingly integrates IBM SPSS, big data environments, and ML techniques. SPSS supports regression, factor analysis, and hypothesis testing, valued for transparency and interpretability in managerial decision-making (IBM, 2024). Extending these capabilities with high-volume data from social media, transactions, and online interactions scales insights beyond traditional limits. ML expands these functions by supporting segmentation, churn prediction, and personalisation. Random forests and support vector machines identify patterns in customer data to predict purchasing behaviour, derive propensity scores, and detect hidden preferences (Domingos, 2012). This process informs targeted campaigns, dynamic pricing, and customer value optimisation. Hybrid platforms allow iterative tuning and domain informed oversight, maximising accuracy while preserving managerial control, ultimately boosting sales growth and loyalty (Zheng & Casari, 2018).

AI and Data Science: Decision Support Systems

Automated decision support systems (DSS) represent a pinnacle of AI and data science: fusing expert systems, probabilistic models, and ML into tools that guide complex decision-making. These systems operate across domains from healthcare diagnostics to financial risk management. AI enabled DSS employ predictive models and optimisation to evaluate trade-offs and constraints, generating recommendations, alerts, or risk assessments. For example, clinical decision tools integrate patient data with guidelines to assist diagnostics and treatment planning (Khosravi et al., 2024).

In industry, DSS improve supply chain resilience by simulating trade-offs between cost, time, and risk, using forecasts that accelerate decision speed and accuracy (Gentsch, 2019). Crucially, design priorities emphasise explainability and trust: human-in-the-loop frameworks combine algorithmic decisiveness with flexibility, ensuring systems adapt to evolving data while reducing cognitive burden (Bzdok et al., 2018). DSS thus augment human judgment, representing a transformative step in applying AI for high-stakes decisions.

Case Studies: Predictive Analytics in Finance

In financial services, ML has transformed financial risk assessment by analysing vast and diverse datasets, including credit histories, market fluctuations, transactions, and social media sentiment. This enhances risk prediction accuracy beyond traditional methods (Hyperstack, 2025). Consequently, leading financial institutions have adopted ML algorithms to enhance fraud detection, credit assessment, market risk management, and regulatory compliance. These advancements underscore the critical role of predictive analytics in safeguarding financial stability, improving operational efficiency, and securing a competitive advantage.

Alibaba Cloud exemplifies ML's transformative role in fraud detection. Facing billions in annual fraud losses, Alibaba implemented a real-time, ML powered anomaly detection platform that analyses vast datasets including transaction patterns and user behaviour. The system instantly flags suspicious activities, reducing fraudulent transactions by 60% and enhancing security with AI-driven authentication, such as facial recognition. Seamlessly integrated across Alibaba's e-commerce and financial services, the platform uses advanced ML models and real-time stream computing for scalable, high-speed risk detection. By automatically blocking high-risk transactions and enabling manual reviews of flagged activities, Alibaba has significantly improved operational efficiency, reduced costs, and bolstered regulatory compliance (Hyperstack, 2025; Alibaba Cloud, 2024).

Similarly, Upstart's AI-powered lending system transforms credit assessment by evaluating creditworthiness based on a broad range of features, including education, employment history, and social profiles—beyond traditional credit scores. This comprehensive analysis improves fairness and expands loan access among underserved populations. The ML driven underwriting process automates decision-making, with over 67% of loans processed autonomously by the system, reducing default risk and operational costs while promoting financial inclusion. Upstart's AI models continuously learn and adapt, improving predictive accuracy and mitigating bias inherent in traditional credit scoring (Hyperstack, 2025).

Barclays uses advanced ML techniques, including gradient boosting and neural networks, to enhance market risk management. By analysing extensive datasets of historical price movements and news sentiment, its ML-powered predictive analytics improve risk forecast accuracy and responsiveness. These quantitative models detect subtle patterns often missed by traditional methods, boosting decision-making efficiency and financial stability. Barclays' AI-driven forecasting highlights the vital role of ML in modern financial services, balancing innovation with rigorous risk governance to maintain competitiveness and compliance (Hyperstack, 2025).

CHALLENGES

As AI, ML, and big data technologies increasingly influence decision-making processes across diverse sectors, a complex array of critical challenges has arisen. Addressing these challenges is vital not only for ensuring the effectiveness of AI-driven decision models but also for guaranteeing their ethical integrity and long-term sustainability in varied social contexts.

Data Quality and Algorithmic Bias

A foundational challenge is ensuring data quality, as the maxim "garbage in, garbage out" remains an enduring truth. Models trained on incomplete, inaccurate, biased, or inconsistent data inevitably generate unreliable, unfair, and potentially harmful outcomes (Dilmegani, 2025; Chen, 2025). Data quality encompasses accuracy, completeness, timeliness, consistency, and relevance, all of which critically shape model reliability. Many organisations grapple with data silos, fragmented governance, inconsistent data standards, and poor validation methods that exacerbate these issues (Alation, 2025). The surge in unstructured and real-time streaming data further complicates verification, adding layers of technical and ethical risk (Stack AI, 2025).

These data quality issues directly manifest as algorithmic bias. Biases embedded in training data—reflecting historical inequalities—can produce models that unjustly favour or disadvantage specific social groups. This risk is especially pronounced in sensitive domains like healthcare, hiring, and criminal justice, where biased decisions can inflict real harm, perpetuate systemic inequities, and trigger significant legal consequences (Dilmegani, 2025; Weiner et al., 2025). For example, healthcare algorithms using proxy variables like healthcare costs have been shown to under-allocate resources to marginalised populations, exacerbating disparities (Weiner et al., 2025).

Governance, Interpretability, and the Accuracy Trade-off

Mitigating these risks requires robust governance frameworks to ensure transparency, accountability, and fairness. A core dilemma within this is the trade-off between interpretability and predictive accuracy. High-performance models like deep neural networks often achieve exceptional accuracy but act as "black boxes," while simpler, interpretable models like linear regressions may lack sufficient power for complex data (Bzdok et al., 2018). Interpretability is crucial for trust and compliance, as stakeholders must comprehend decisions to validate them, particularly in regulated fields like healthcare and finance (Khosravi et al., 2024; Nguyen et al., 2018).

Therefore, effective governance must encompass strategies to resolve this tension. This includes regular auditing for bias and performance, and the use of explainable AI techniques like LIME and SHAP to provide clear explanations for complex model decisions, fostering trust without necessarily sacrificing accuracy (Barredo Arrieta et al., 2020). Furthermore, human-in-the-loop strategies allow for expert oversight and contestation of AI outputs, strengthening ethical safeguards (Kanerika, 2025). Governance must also uphold privacy and consent through techniques like federated learning and differential privacy to protect individual rights and ensure compliance

(Kanerika, 2025; Weiner et al., 2025). As generative AI becomes more pervasive, adaptive frameworks are also needed to address novel dilemmas around misinformation and intellectual property (SAS Blogs, 2024; Kanerika, 2025).

Integration Challenges of Legacy and Modern Systems

Legacy statistical software such as IBM SPSS or AMOS remains prevalent but poses significant technical and organisational challenges when integrated with modern big-data pipelines. These tools excel at traditional analytics but struggle with the high-velocity, high-volume data streams typical of big data applications, often lacking the scalability to interface with distributed systems like Hadoop (Zhang & Tang, 2025). This leads to data silos and workflow fragmentation. Bridging this gap requires hybrid pipelines that preprocess big data using frameworks like Apache Spark for feature engineering before analysis in legacy tools, or by piping statistical outputs into ML environments (Zheng & Casari, 2018). Effective integration demands sophisticated data orchestration, metadata management, and aligned governance policies. Overcoming this "legacy barrier" also necessitates addressing the skill divide between classical analysts and data scientists through cross-disciplinary collaboration and training.

By embedding ethics centrally in AI development and deployment—through continual bias mitigation, transparency enhancements, privacy protection, and multi-stakeholder inclusion—organisations can build AI decision models that are not only performant but socially just and trustworthy over the long term (UNESCO, 2024; Weiner et al., 2025).

FUTURE DIRECTIONS

The future of AI and data-driven decision-making will be shaped by exciting new developments that tackle current challenges. Explainable AI is expected to become a must-have feature, not just an extra option. As regulations grow stricter worldwide, especially in important fields like finance, healthcare, and law, AI systems will need to clearly explain their decisions to gain trust and meet legal requirements. Research is moving towards building AI models that are inherently easier to understand, combining symbolic reasoning with traditional AI, and creating interactive tools that help users see how AI works (Barredo Arrieta et al., 2020).

Another important advance is Federated Learning (FL), which lets AI models learn from data stored on many separate devices or locations without collecting all that data in one place. This protects privacy while still allowing AI to learn from a lot of information. FL is especially useful in areas like healthcare, finance, and IoT devices, where sharing raw data is sensitive. This approach will help make AI more accessible and privacy-friendly for many users (Hu et al., 2024).

Finally, real-time AI decision-making systems will become more common. These systems use streaming data, make quick AI predictions, and trigger automatic responses almost instantly. They will support things like dynamic pricing, fraud detection, self-driving cars, and smart factories. By combining event-driven computing, processing data at the network's edge, and continuous AI model updates, these systems allow organisations to react swiftly to changes and improve operational efficiency (Gentsch, 2019; Zilliz, 2024).

CONCLUSION

The landscape of quantitative decision-making has grown increasingly diverse by blending traditional statistical techniques with modern deep learning methods. Statistical analysis excels at explaining and testing relationships within structured data under clear assumptions. In contrast, deep learning and ML offer powerful solutions for uncovering patterns and making predictions from vast, complex, and often unstructured datasets. Combining these approaches allows practitioners to harness both the interpretability and insight of classical methods alongside the flexible, high-performance capabilities of contemporary computational algorithms, thereby expanding the scope and depth of data-driven decisions.

The integration of classical approaches with new-age techniques emerges as crucial for harnessing the full potential of data. Statistical principles inform feature engineering and preprocessing, ensuring that ML models are built on relevant and calibrated inputs, while hybrid models combining structures like SEM with deep learning further bridge inferential rigour and algorithmic flexibility. Organisational contexts benefit when long-standing tools such as SPSS and AMOS interoperate with scalable ML pipelines, facilitating holistic insight generation and fostering cross-disciplinary collaboration. Such integrative practices pave the way towards transparent, robust, and effective decision support systems.

Looking forward, opportunities abound at the convergence of numerical analysis, ML, and decision sciences. Continued advancements in explainable AI aim to illuminate the decision logic behind complex models, fostering trust and accountability. FL promises privacy-preserving collaboration across distributed datasets, explainable AI's reach without compromising sensitive information. The rise of real-time decision ecosystems embedded in IoT and edge computing environments will demand innovations in incremental learning and resilient automation. Furthermore, ongoing mathematical research will optimise algorithmic efficiency and scalability, reinforcing the theoretical foundations necessary for next-generation intelligent systems.

Ultimately, the future of quantitative decision-making lies in embracing a multi-paradigm, interdisciplinary approach that balances interpretability and predictive prowess, leverages advanced computation, and adheres to strong governance frameworks. By doing so, data science will continue to transform how organisations navigate uncertainty, derive insights, and drive impactful decisions across varied domains, from healthcare and finance to manufacturing and social sciences. This chapter underscores the value of blending statistical insight with cutting-edge AI, setting the stage for smarter, more transparent, and ethically sound data-driven decisions that shape the future of science and society alike.

Publisher's Note

Correction Notice: A correction to this manuscript has been published

Reason: Figure was incorrectly replaced

The publisher was notified after publication that the image was changed from proof but appears to be incomplete. We have updated the figure accordingly so the entire figure is now visible.

Correction Date: February 27th, 2026

REFERENCES

Alation. (2025). AI data quality explained: Tools, challenges, fixes. Alation Blog. https://atlan.com/know/ai-data-quality/

Alber, M., Buganza Tepole, A., Cannon, W. R., De, S., Dura-Bernal, S., Garikipati, K., Karniadakis, G. E., Lytton, W. W., Perdikaris, P., Petzold, L., & Kuhl, E. (2019). Integrating machine learning and multiscale modeling—Perspectives, challenges, and opportunities in the biological, biomedical, and behavioral sciences. *NPJ Digital Medicine*, *2*(1), 115. DOI: 10.1038/s41746-019-0193-y PMID: 31799423

Alhumaidi, N. H., Dermawan, D., Kamaruzaman, H. F., & Alotaiq, N. (2025). The use of machine learning for analyzing real-world data in disease prediction and management: Systematic review. *JMIR Medical Informatics*, *13*, e68898. DOI: 10.2196/68898 PMID: 40537090

Alibaba Cloud. (2024). *Safeguarding digital frontiers: A deep dive into Alibaba Cloud fraud detection*. Alibaba Cloud Blog https://www.alibabacloud.com/blog/safeguarding-digital-frontiers-a-deep-dive-into-alibaba-cloud-fraud-detection_602116

Angrist, J. D., & Pischke, J.-S. (2009). *Mostly harmless econometrics: An empiricist's companion*. Princeton University Press.

Barredo Arrieta, A., Díaz-Rodríguez, N., Del Ser, J., Bennetot, A., Tabik, S., Barbado, A., García, S., Gil-López, S., Molina, D., Benjamins, R., Chatila, R., & Herrera, F. (2020). Explainable artificial intelligence (XAI): Concepts, taxonomies, opportunities and challenges toward responsible AI. *Information Fusion*, *58*, 82–115. DOI: 10.1016/j.inffus.2019.12.012

Bere, A., Sithuba, G. H., Mabvuu, C., Mashabela, R., Sigauke, C., & Kyei, K. (2021). A COMPARISON OF LASSO AND GRADIENT BOOSTING. *South African Statistical Journal*, *55*(1), 29–44. DOI: 10.37920/sasj.2021.55.1.3

Blogs, S. A. S. (2024). 6 unique ethical challenges for generative AI in 2025. SAS Blogs. https://blogs.sas.com/content/sascom/?p=70982

Brownstein, N. C., Louis, T. A., O'Hagan, A., & Pendergast, J. (2019). The Role of Expert Judgment in Statistical Inference and Evidence-Based Decision-Making. *The American Statistician, 73*(sup1), 56–68. https://doi.org/DOI: 10.1080/00031305.2018.1529623

Byrne, B. M. (2001). *Structural equation modeling with AMOS: Basic concepts, applications, and programming* (3rd ed.). Routledge.

Bzdok, D., Altman, N., & Krzywinski, M. (2018). Statistics versus machine learning. *Nature Methods, 15*(4), 233–234. DOI: 10.1038/nmeth.4642 PMID: 30100822

Chen, J. (2025). Data quality quantized framework: Ensuring large-scale data integration in gig economy platforms. *Data quality quantized framework: Ensuring large-scale data integration in gig economy platforms*. https://www.researchgate.net/publication/388769132

Chen, M., Mao, S., & Liu, Y. (2014). Big data: A survey. *Mobile Networks and Applications, 19*(2), 171–209. DOI: 10.1007/s11036-013-0489-0

Chupryna, R., & Leleko, S. (2025). Predictive maintenance with machine learning: A complete guide. *SPD Technology*. https://spd.tech/machine-learning/predictive-maintenance

Codewave. (2024). Machine Learning Pattern Recognition: A Beginner's Guide. https://codewave.com/insights/pattern-recognition-machine-learning/

Cote, C. (2021). 4 types of data analytics to improve decision-making. Harvard Business School Online. https://online.hbs.edu/blog/post/types-of-data-analysis

Coursera. (2024). 7 statistical analysis methods beginners should know. https://www.coursera.org/articles/statistical-analysis-methods

Dilmegani, C. (2025). Top 10 applications of deep learning in manufacturing. AI-Multiple. https://research.aimultiple.com/deep-learning-in-manufacturing/

Dissertation India. (2024, February 20). What is the difference between SPSS and Amos? https://www.dissertationindia.com/article/what-is-the-difference-between-spss-and-amos-60

Domingos, P. (2012). A few useful things to know about machine learning. *Communications of the ACM, 55*(10), 78–87. DOI: 10.1145/2347736.2347755

Dovetail. (2024). 10 statistical analysis methods for research. https://dovetail.com/research/key-statistical-analysis-methods-explained/

Elastic. (2024). Deep learning vs. machine learning: Understanding the differences. Elastic Blog. https://www.elastic.co/blog/deep-learning-vs-machine-learning

Esteva, A., Kuprel, B., Novoa, R. A., Ko, J., Swetter, S. M., Blau, H. M., & Thrun, S. (2017). Dermatologist-level classification of skin cancer with deep neural networks. *Nature, 542*(7639), 115–118. DOI: 10.1038/nature21056 PMID: 28117445

Field, A. (2018). *Discovering statistics using IBM SPSS statistics*. Sage publications limited.

Gandomi, A., & Haider, M. (2015). Beyond the hype: Big data concepts, methods, and analytics. *International Journal of Information Management, 35*(2), 137–144. DOI: 10.1016/j.ijinfomgt.2014.10.007

GeeksforGeeks. (2017). Decision Tree in Machine Learning. https://www.geeksforgeeks.org/machine-learning/decision-tree-introduction-example/

GeeksforGeeks. (2024). Supervised vs Unsupervised vs Reinforcement Learning. https://www.geeksforgeeks.org/machine-learning/supervised-vs-reinforcement-vs-unsupervised/

Gentsch, P. (2019). *AI in marketing, sales and service: How marketers without a data science degree can use AI, big data and bots*. Springer.

Google Cloud. (2025). *What is big data?* Google Cloud. Retrieved September 2, 2025, from https://cloud.google.com/learn/what-is-big-data

Hair, J. F. (2009). *Multivariate data analysis*. Pearson.

Hardesty, L. (2017). Explained: Neural networks. *MIT News*. https://news.mit.edu/2017/explained-neural-networks-deep-learning-0414

Harvard Business School Online. (2019). The advantages of data-driven decision-making. https://online.hbs.edu/blog/post/data-driven-decision-making

Hastie, T., Tibshirani, R., & Friedman, J. (2009). *The elements of statistical learning: Data mining, inference, and prediction* (2nd ed.). Springer.

He, P., Cai, T., & Gu, J. (2024). A survey of Internet of Medical Things: Technology, applications, and challenges. *Sensors (Basel), 24*(2), 36. DOI: 10.1016/j.dcan.2024.11.013

Hu, K., Gong, S., Zhang, Q., Seng, C., Xia, M., & Jiang, S. (2024). An overview of implementing security and privacy in federated learning. *Artificial Intelligence Review, 57*(8), 204. DOI: 10.1007/s10462-024-10846-8

Hyperstack. (2025). Exploring risk assessment with machine learning in finance. https://www.hyperstack.cloud/blog/case-study/exploring-risk-assessment-with-machine-learning-in-finance

IBM. (2021). What is a Decision Tree? https://www.ibm.com/think/topics/decision-trees

IBM. (2021a). What are Convolutional Neural Networks? IBM Cloud Learn Hub. https://www.ibm.com/think/topics/convolutional-neural-networks

IBM. (2021b). What is a Recurrent Neural Network (RNN)? IBM Research Blog. https://www.ibm.com/think/topics/recurrent-neural-networks

IBM. (2024). What is Big Data? IBM. https://www.ibm.com/think/topics/big-data

IBM. (2024). What is data-driven decision-making? https://www.ibm.com/think/topics/data-driven-decision-making

IBM. (2025). What is Machine Learning (ML)? https://www.ibm.com/think/topics/machine-learning

IBM. (2025). What is a Transformer Model? IBM Cloud Learn Hub. https://www.ibm.com/think/topics/transformer-model

Indeed Editorial Team. (2025). 7 types of statistical analysis techniques (and process steps). Indeed. https://www.indeed.com/career-advice/career-development/types-of-statistical-analysis

IONOS. (2025). What are recurrent neural networks? IONOS Digital Guide. https://www.ionos.co.uk/digitalguide/websites/web-development/recurrent-neural-network/

Jaokar, A. (2025). Statistical inference vs machine learning inference vs deep learning inference. LinkedIn Pulse. https://www.linkedin.com/pulse/statistical-inference-vs-machine-learning-deep-ajit-jaokar-khfue

Kanerika. (2025). *How to address key AI ethical concerns in 2025*. Kanerika Blog. https://kanerika.com/blogs/ai-ethical-concerns/

Khosravi, M., Zare, Z., Mojtabaeian, S. M., & Izadi, R. (2024). Artificial intelligence and decision-making in healthcare: A thematic analysis of a systematic review of reviews. *Health Services Research and Managerial Epidemiology*, *11*. Advance online publication. DOI: 10.1177/23333928241234863 PMID: 38449840

LeCun, Y., Bengio, Y., & Hinton, G. (2015). Deep learning. *Nature*, *521*(7553), 436–444. DOI: 10.1038/nature14539 PMID: 26017442

Mahajan, A., Heydari, K., & Powell, D. (2025). Wearable AI to enhance patient safety and clinical decision-making. *npj. Digital Medicine*, *8*, 176. DOI: 10.1038/s41746-025-01554-w PMID: 40121336

Marr, B. (2016). *Big data in practice: how 45 successful companies used big data analytics to deliver extraordinary results*. John Wiley & Sons.

Marr, B. (2021). *What are the 4 Vs of Big Data?* BernardMarr.com. https://bernardmarr.com/what-are-the-4-vs-of-big-data/

Matthew, P., Mchale, S., Deng, X., Nakhla, G., Trovati, M., Nnamoko, N., Pereira, E., Zhang, H., & Raza, M. (2025). A Review of the State of the Art for the Internet of Medical Things. *Sci*, *7*(2), 36. DOI: 10.3390/sci7020036

Milvus. (2025). How does reinforcement learning differ from supervised and unsupervised learning? Milvus AI Quick Reference. https://milvus.io/ai-quick-reference/how-does-rl-differ-from-supervised-and-unsupervised-learning

Nakhla, G. (in press). Structural equation modelling. In Kaur, B., & Atkins, L. (Eds.), *Elgar encyclopaedia of educational research*. Edward Elgar Publishing., https://research.edgehill.ac.uk/en/publications/structural-equation-modelling

Nakhla, G. (in press). Cluster Analysis. In Kaur, B., & Atkins, L. (Eds.), *Elgar encyclopaedia of educational research*. Edward Elgar Publishing., https://research.edgehill.ac.uk/en/publications/cluster-analysis

Neural Concept. (2024). Applications of machine learning in mechanical engineering. https://www.neuralconcept.com/post/applications-of-machine-learning-in-mechanical-engineering

Nguyen, T., Zhou, L., Spiegler, V., Ieromonachou, P., & Lin, Y. (2018). Big data analytics in supply chain management: A state-of-the-art literature review. *Computers & Operations Research*, *98*, 254–264. DOI: 10.1016/j.cor.2017.07.004

Onlinemanipal. (2025, June 2). 10 Popular Regression Algorithms in Machine Learning. https://www.onlinemanipal.com/blogs/popular-regression-algorithms-in-machine-learning

Oracle. (2024, July 16). What is Unsupervised Learning? https://www.oracle.com/uk/artificial-intelligence/machine-learning/unsupervised-learning/

Ozcelik, M. M., Kok, I., & Ozdemir, S. (2025). A Survey on Internet of Medical Things (IoMT): Enabling Technologies, Security and Explainability Issues, Challenges, and Future Directions. *Expert Systems: International Journal of Knowledge Engineering and Neural Networks*, *42*(5), e70010. DOI: 10.1111/exsy.70010

Paubox. (2025). Deep learning in healthcare. Paubox Blog. https://www.paubox.com/blog/deep-learning-in-healthcare

Pecan, A. I. (2025). Three Types of Machine Learning You Should Know. https://www.pecan.ai/blog/3-types-of-machine-learning/

Pure Search. (2025). Quantitative analytics in the age of AI and big data. https://www.puresearch.com/blog/quantitative-analytics-in-the-age-of-ai-and-big-data

Qadir, M. (2024). Transition from traditional machine learning to neural networks. LinkedIn Pulse. https://www.linkedin.com/pulse/transition-from-traditional-machine-learning-neural-networks-qadir-ujowc

Rane, J., Mallick, S. K., Kaya, O., & Rane, N. L. (2024). Scalable and adaptive deep learning algorithms for large-scale machine learning systems. In *Future Research Opportunities for Artificial Intelligence in Industry 4.0 and 5.0* (pp. 39–92). Deep Science Publishing., DOI: 10.70593/978-81-981271-0-5_2

Rokach, L., Maimon, O., & Shmueli, E. (2023). *Machine learning for data science handbook*. Springer., DOI: 10.1007/978-3-031-24628-9

Ross, S. M. (2014). *Introduction to probability models* (11th ed.). Academic Press.

Sarker, B. K., Sarker, D. K., Shaha, S. R., Saha, D., & Sarker, S. (2024). Why apply SPSS, SmartPLS and AMOS: An essential quantitative data analysis tool for business and social science research investigations. [IJRISS]. *International Journal of Research and Innovation in Social Science*, *8*(9), 2688–2699. DOI: 10.47772/IJRISS.2024.8090224

Sarker, I. H. (2021). Machine learning: Algorithms, real-world applications and research directions. *SN Computer Science*, *2*, 160. DOI: 10.1007/s42979-021-00592-x PMID: 33778771

SAS. (2023). A guide to the types of machine learning algorithms. https://www.sas.com/en_gb/insights/articles/analytics/machine-learning-algorithms.html

Schmidhuber, J. (2015). Deep learning in neural networks: An overview. *Neural Networks*, *61*, 85–117. DOI: 10.1016/j.neunet.2014.09.003 PMID: 25462637

Seldon. (2021). *What is machine learning?* Seldon. Retrieved September 10, 2025, from https://www.seldon.io/what-is-machine-learning/

Select Statistics. (2024). Evidence-based decision making. https://select-statistics.co.uk/videos/evidence-based-decision-making/

Stack, A. I. (2025). The 7 biggest AI adoption challenges for 2025. Stack AI Blog. https://www.stack-ai.com/blog/the-biggest-ai-adoption-challenges

Statistics Authority. (2024). Analytical leadership: Achieving better outcomes for citizens. https://osr.statisticsauthority.gov.uk/publication/analytical-leadership-achieving-better-outcomes-for-citizens/pages/3/

Statistics Solutions. (2025). An introduction to the AMOS software. https://www.statisticssolutions.com/academic-research-consulting/dissertation-consulting-services/spss-statistics-help/amos/

Sutton, R. S., & Barto, A. G. (2018). *Reinforcement learning: An introduction* (2nd ed.). MIT Press., https://mitpress.mit.edu/9780262039246/reinforcement-learning/

Technology Innovators. (2024). Data analytics for evidence-based decision-making. https://www.technology-innovators.com/data-analytics-for-evidence-based-policy-making-and-decision-making/

TechTarget. (2023). What are the 5 V's of big data? TechTarget. https://www.techtarget.com/searchdatamanagement/definition/5-Vs-of-big-data

Teradata. (2022). What are the 5 V's of Big Data? Teradata. https://www.teradata.com/glossary/what-are-the-5-v-s-of-big-data

UNESCO. (2024). Ethics of artificial intelligence. UNESCO. https://www.unesco.org/en/artificial-intelligence/recommendation-ethics

ValueCoders. (2023). The Role of Machine Learning in Predictive Analytics and Decision-making. https://www.valuecoders.com/blog/analytics/role-of-machine-learning-in-predictive-analytics-and-decision-making/

Vasilchenko, A. (2024). Predictive maintenance in manufacturing using machine learning: Use cases, challenges and best practices. *MobiDev*. https://mobidev.biz/blog/predictive-maintenance-in-manufacturing-using-machine-learning

Weiner, E. B., Dankwa-Mullan, I., Nelson, W. A., & Hassanpour, S. (2025). Ethical challenges and evolving strategies in the integration of artificial intelligence into clinical practice. *PLOS Digital Health*, *4*(4), e0000810. DOI: 10.1371/journal.pdig.0000810 PMID: 40198594

Wooldridge, J. M. (2020). *Introductory econometrics: A modern approach* (7th ed.). Cengage Learning.

Zhang, Y., & Tang, Q. (2025). Integrating AI-generated content tools in higher education: A comparative analysis of interdisciplinary learning outcomes. *Scientific Reports*, *15*(1), 25802. DOI: 10.1038/s41598-025-10941-y PMID: 40670544

Zheng, A., & Casari, A. (2018). *Feature engineering for machine learning: Principles and techniques for data scientists*. O'Reilly Media.

Zilliz. (2024). How does deep learning scale to large datasets? Zilliz AI FAQ. https://zilliz.com/ai-faq/how-does-deep-learning-scale-to-large-datasets

KEY TERMS AND DEFINITIONS

Big Data Platforms: Distributed computing frameworks, such as Hadoop and Spark, designed to process, store, and analyse very large datasets efficiently in real time.

Deep Neural Networks (DNNs): Advanced ML architectures composed of multiple processing layers that excel at identifying complex patterns in unstructured data such as images, text, and signals.

Descriptive Statistics: Statistical techniques that summarise and present datasets through measures such as mean, median, standard deviation, and visual representations.

Ethics and Governance in AI: The frameworks, principles, and policies that address fairness, accountability, data privacy, and societal impacts of deploying AI and advanced analytics in decision-making.

Explainable AI: A subfield of AI focused on making model outputs transparent and interpretable, enhancing trust and accountability in decision-making.

Federated Learning: A decentralised ML approach that enables model training across distributed devices or servers while keeping raw data localised, enhancing privacy and security.

Inferential Statistics: Methods that use sample data to make generalisations or predictions about a larger population, often involving hypothesis testing and estimation.

ML Algorithms: Computational methods that enable systems to automatically learn patterns from data and improve performance without explicit programming.

Quantitative Decision-Making: The use of numerical data, statistical methods, and computational techniques to guide choices and evaluate alternatives in complex problem-solving.

Chapter 4
Explainable and Robust Multi-Criteria Decision Making:
An Integrated Pipeline of Predictive Analytics, Preference Learning, and Distribution

Silvio Andrae
https://orcid.org/0000-0002-8586-7812
Independent Researcher, Germany

ABSTRACT

Quantitative decision-making challenges in management, technology, and finance increasingly rely on data-driven forecasts and heterogeneous preferences, which are often only partially observable. This chapter presents an integrated decision-making framework that explicitly incorporates forecast and preference uncertainty to facilitate robust and explainable decisions. The proposed explainable and robust multi-criteria decision-making (xMCDM) approach integrates a predictive layer with uncertainty quantification, employs a Bayesian preference-learning method to model uncertain weights, and uses distribution-robust decision aggregation. Empirical case studies in supplier selection, renewable energy site planning, and credit portfolio allocation demonstrate that xMCDM enhances rank stability, reduces regret, and enables transparent evaluation of robustness trade-offs. Consequently, xMCDM advances the focus from score-based evaluation to robust decision-making under uncertainty.

DOI: 10.4018/979-8-3373-6746-0.ch004

INTRODUCTION: FROM SCORES TO ROBUST DECISIONS

Quantitative decision support systems are increasingly being used in application areas where decisions have high economic, technical, or social consequences, but are based on data-driven forecasts and heterogeneous preferences. Examples range from supplier selection in operations management to location decisions in energy technology to portfolio allocation in the financial sector. In all these contexts, decision-makers are confronted with multiple, sometimes conflicting criteria and must reconcile assessments derived from uncertain data sources. Classic multi-criteria decision-making (MCDM) methods were developed precisely for this purpose: they structure complex decision-making problems by evaluating and aggregating alternatives along defined criteria (Belton & Stewart, 2002; Keeney & Raiffa, 1993).

Despite their widespread use, however, many established MCDM approaches remain score-based at their core. They typically use point values for criterion characteristics and fixed preference weights, which are assumed to be given. Forecast uncertainty, model risks, or ambiguity in stakeholder preferences are either ignored or only considered in downstream sensitivity analyses. This simplification was often acceptable in classic applications but is proving increasingly problematic in data-intensive, dynamic decision-making environments.

Two developments are exacerbating this problem. First, the use of modern statistical learning methods and machine learning models has fundamentally changed the nature of decision inputs. Criteria values—such as default probabilities in lending, expected energy yields from renewable energy plants, or supplier delivery reliability—are now often the result of complex forecasting models and are no longer deterministic variables (Breiman, 2001; Friedman, 2001). Such models are inevitably subject to uncertainties due to finite samples, model uncertainty, and possible non-stationarity. Second, preferences in real-world decision-making situations are rarely stable and fully observable. They must be inferred from limited interactions, pairwise comparisons, or explicit weighting specifications and are therefore also subject to uncertainty (Keeney & Raiffa, 1993). If these uncertainties are ignored, seemingly precise rankings emerge, but they prove unstable when minor changes are made to the inputs.

Research has addressed these shortcomings through various extensions. Approaches to uncertainty-aware MCDM integrate sensitivity analyses, scenarios, or stochastic elements to assess the robustness of decisions (Belton & Stewart, 2002; Roy, 1996). However, these extensions often remain fragmentary: preference weights are varied ex post, distributions are assumed to be known, or robustness is approximated through discrete scenarios rather than being an integral part of the decision formulation. At the same time, powerful building blocks have become established in other disciplines. In predictive analytics, bootstrap methods and Bayesian models enable explicit quantification of forecast uncertainty (Efron & Tibshirani, 1993; Gelman et al., 2013). In preference modeling, probabilistic models such as Bradley–Terry or Luce models allow latent utility structures to be estimated from noisy pair comparisons (Bradley & Terry, 1952; Luce, 1959; Chu & Ghahramani, 2005). In optimization, distributionally robust optimization (DRO) has established itself as a strong tool for safeguarding decisions against ambiguity in data and model worlds (Ben-Tal et al., 2009; Shapiro et al., 2014).

What has been largely missing so far is a fully integrated set of these building blocks into a consistent MCDM approach. In practice, forecasting models, preference elicitation, and decision aggregation are often treated as separate steps. The consequences of this separation are significant: uncertainties propagate through the decision chain in an opaque manner, robustness is assessed only heuristically, and decision-makers rarely receive information about how reliable a particular ranking actually is. The result is decisions that appear optimal in formal terms, but whose stability in the face of plausible deviations in data or preferences remains unclear.

This chapter addresses this gap with the concept of explainable and robust multi-criteria decision making (xMCDM). The goal is to develop an integrated, end-to-end decision-making framework that systematically combines predictive model uncertainty, preference uncertainty, and distribution-robust decision aggregation. The central change in perspective is to no longer understand MCDM as pure score aggregation, but as a problem of robust decision-making under uncertainty. The focus is not only on the question of which alternative maximizes aggregate utility under nominal assumptions, but also on which alternatives perform consistently well and cause little regret even under plausible deviations from the underlying assumptions.

In the proposed xMCDM framework, criterion values are first estimated using statistical or machine learning methods, with uncertainty explicitly modeled using bootstrap distributions or Bayesian credible intervals (Breiman, 2001; Efron & Tibshirani, 1993; Gelman et al., 2013). The preferences of decision-makers or stakeholder groups are recorded probabilistically, either via a priori distributed weights or via Bayesian preference learning based on pair comparisons (Bradley & Terry, 1952; Chu & Ghahramani, 2005). These two sources of uncertainty are then jointly embedded in a distribution-robust decision formulation. Classic MCDM aggregators—including additive utility functions, TOPSIS- or VIKOR-type distance measures, and ELECTRE-based outranking approaches—are reformulated to be robust to ambiguity in the underlying distributions (Ben-Tal et al., 20091; Blanchet & Murthy, 2017). The results are rankings based not only on point forecasts but also on explicit robustness and regret measures.

A central design principle of xMCDM is the explainability of decisions. In highly relevant fields of application, it is not enough to make robust decisions; they must also be comprehensible and communicable. While explainable AI methods were initially developed to interpret individual prediction models (Ribeiro et al., 2016; Lundberg & Lee, 2017), this chapter shifts the focus to decision-related explainability. The focus is on the contributions of individual criteria, explicit trade-offs, and counterfactual analyses that show which changes in data or preferences would lead to different decisions (Molnar, 2022). Explainability thus functions not only as a transparency tool, but also as an integral part of robustness diagnostics.

Against this background, the chapter pursues three overarching objectives. First, a numerically stable, modular pipeline is developed that consistently integrates prognostic and preference-based uncertainties into MCDM problems. Second, explainable artifacts are presented that make the robustness and sensitivity of decisions transparent, thereby promoting decision acceptance. Third, the chapter provides empirical evidence from different domains showing the extent to which distribution-robust approaches outperform classical MCDM methods in terms of stability and out-of-sample performance.

These objectives are specified in the following research questions: (i) How do ambiguity sets in distribution-robust formulations influence rankings and regret in MCDM problems? (ii) Which combinations of model complexity, regularization, and robustness parameters lead to maximized out-of-sample robustness? (iii) Which explainability metrics are related to the acceptance of decisions by stakeholders?

The rest of the chapter is structured as follows. Section 2 places the approach within the existing literature on MCDM, distribution-robust optimization, preference learning, and explainable AI. Section 3 introduces the xMCDM framework and describes the predictive and preference-based levels. Section 4 develops distribution-robust decision-aggregation and numerical-solution approaches. Section 5 is devoted to explainability, sensitivity analyses, and robustness diagnostics. Section 6 presents empirical case studies from operations management, engineering, and finance. Section 7 concludes with best practice recommendations, limitations, and an outlook on future research directions.

RELATED WORK AND CLASSIFICATION

This chapter situates the proposed approach of explainable and robust multi-criteria decision making (xMCDM) within the context of four central strands of literature: (i) classical multi-criteria decision-making methods, (ii) preference learning and probabilistic utility specification, (iii) distribution-robust optimization, and (iv) explainable AI. The aim is to identify conceptual connections and systematic gaps that highlight the need for an integrated xMCDM framework.

Classic MCDM Approaches and Their Limitations Under Uncertainty

Multi-criteria decision analysis (MCDA or MCDM) has a long tradition and is used to structure the evaluation of alternatives under multiple, often conflicting objectives. Fundamental contributions come from Keeney and Raiffa (1993), who established the concept of utility as a consistent aggregation principle, and from Roy (1996), who developed the ELECTRE family as an outranking-based alternative to pure utility aggregation. Overview works such as Belton and Stewart (2002) systematize MCDM methods according to their aggregation logic, normative assumptions, and requirements for data and preferences.

Standard MCDM methods can be divided into three classes. Additive utility models, including weighted sums or the Analytic Hierarchy Process (AHP), aggregate criteria linearly using fixed weights. Distance-based methods such as TOPSIS or VIKOR evaluate alternatives based on their proximity to ideal and anti-ideal solutions. Outranking approaches such as ELECTRE or PROMETHEE dispense with complete compensation and model partial preference relationships between other options. Despite methodological differences, these approaches share a central commonality: they operate predominantly on deterministic inputs. Criterion values and preference weights are assumed to be known and stable or, at most, vary selectively.

Sensitivity analyses and extensions for dealing with uncertain weights or fuzzy criteria have been around for some time, for example, in the form of fuzzy MCDM or scenario analyses (Belton & Stewart, 2002). However, these approaches often remain additive to the actual decision model and do not fundamentally change its structure. In particular, uncertainties are not systematically modeled as probability distributions, but are approximated using heuristic ranges or discrete scenarios. This results in limited information about ranking stability and the expected regret in the event of an error.

Preference Learning, DRO, and Explainable AI as Complementary Building Blocks

Parallel to the classic MCDM literature, methodological strands have developed in related disciplines that address sub-problems, but these have rarely been brought together consistently.

In the fields of preference modeling and preference learning, utility and weights are not assumed to be given; they are inferred as latent variables from observed decisions or pair comparisons. Early models such as the Bradley–Terry model (Bradley & Terry, 1952) or the Luce model (Luce, 1959) form the basis for probabilistic choice models in which preferences are interpreted stochastically. More recent work extends these approaches to incorporate Bayesian perspectives, enabling explicit quantification of preference uncertainty and its combination with prior knowledge (Chu & Ghahramani, 2005). Despite these advances, preference learning has so far found only limited application in classical MCDM practice and is rarely linked to the uncertainty of the criterion evaluations themselves.

A formal approach to explaining MCDM results is presented by Erwig and Kumar (2022). The authors develop an explanation-oriented framework for linear and hierarchical MCDM methods, such as WSM, WPM, and AHP, based on a fine-grained value decomposition of the decision-making process. At the core of the approach are so-called minimal dominating sets (MDS), which serve as contrastive explanations showing which subsets of criteria are sufficient to justify the preference of one alternative over a reference alternative. Methodologically noteworthy is that the explanations are derived directly from the internal calculation structure of the MCDM methods and are therefore not based on external approximations or surrogate models. At the same time, the approach is limited to deterministic decision models with fixed criterion values and weights. Uncertainties in data, preferences, or model assumptions are not accounted for, nor are issues of robustness or rank stability. These limitations mark a central starting point for the extension developed in this chapter toward explainable and distribution-robust xMCDM methods.

Recent work has advanced explainable decision-making beyond model-level feature interpretation toward decision-oriented explanations. Zuheros et al. (2023) propose an explainable method for decision-making in the crowd that integrates sentiment analysis, attention mechanisms, and subgroup discovery to justify multi-criteria rankings derived from large-scale social media opinions. Their approach provides multilevel explanations that identify influential criteria, salient aspects, and representative textual evidence supporting both the highest- and lowest-rated alternatives. However, criterion ratings and weights are aggregated deterministically, and neither the uncertainty in the ratings nor the ambiguity of preferences is explicitly modeled. Consequently, robustness is implicitly assessed through aggregation rather than enforced by design. This distinction motivates the xMCDM framework, which combines decision-oriented explainability with probabilistic uncertainty modeling and distribution-robust optimization to ensure the ranking's stability under plausible perturbations.

Recent surveys show that increasing efforts are being made to integrate explainability into optimization-oriented decision-making systems, with a focus on the transparency of heuristic search, mathematical programming, and learning-based optimization pipelines. Systematic reviews emphasize approaches to explainability by design, such as interpretable constraints, certificates, and rule-based solvability, particularly in high-risk areas such as finance, healthcare, and energy systems (Danach et al., 2025). However, this line of research treats decision problems as single-objective or implicitly aggregated optimization tasks and does not explicitly address preference uncertainty, rank stability, or multi-criteria robustness. In contrast, the xMCDM framework developed in this chapter focuses on explainability at the level of criterion aggregation and stakeholder preferences, complementing explainable optimization with decision-theoretic robustness.

Meanwhile, distribution-robust optimization (DRO) has established itself as a powerful paradigm in optimization literature. Instead of assuming a precisely known probability distribution, uncertainties are modeled by sets of ambiguities—such as moment conditions or Wasserstein balls defined by optimal transport—and decisions are made so that they perform in the worst-case scenario (Ben-Tal et al., 2009; Shapiro et al., 2014). Recent work shows that Wasserstein-based DRO problems can often be reformulated as convex problems and thus remain numerically manageable (Blanchet & Murthy, 2017). In decision and operations research contexts, DRO is primarily used to validate cost or revenue forecasts; however, systematic embedding in MCDM aggregation logics has not yet been established.

Finally, the discussion of explainable artificial intelligence (XAI) has gained importance, especially in the context of data-driven forecasting models. Methods such as LIME or SHAP aim to quantify local or global contributions of input variables to model predictions (Ribeiro et al., 2016; Lundberg & Lee, 2017; Andrae, 2024). These approaches have contributed significantly to the acceptance of complex models but remain primarily limited to predictive modeling. For decision problems with multiple criteria, a further question arises: even if individual predictions are explainable, it remains unclear how these predictions, together with preferences and robustness assumptions, lead to a concrete decision. Genuine decision-related explainability, which makes aggregation steps and trade-offs transparent, has only been addressed marginally in the XAI literature to date (Molnar, 2022).

More recently, hybrid approaches have become widespread that explicitly combine machine learning with multi-criteria decision-making methods to solve complex business selection problems in a more transparent and scalable manner. A prominent example is the approach by Abdulla and Baryannis (2024), who develop an explainable supplier selection process in which machine learning models are used to reduce problem complexity and identify patterns, while the decision logic remains deliberately anchored in a classic MCDM framework. The authors use ML-based classification and feature selection to identify relevant criteria from large supplier datasets and then integrate these into an AHP-based decision-making structure. It should be emphasized that explainability is not addressed exclusively post hoc, but is structurally embedded through the explicit separation of data-driven preprocessing and normative criteria aggregation. At the same time, the approach remains based on deterministic criterion evaluations and fixed weightings; uncertainty in forecasts, preferences, or data quality is primarily reflected indirectly via performance comparisons and stability analyses. The article thus represents the current state of explainable ML–MCDM hybrids, which combine transparency and scalability but do not provide formal assurance of decision robustness against plausible disturbance assumptions. This structural limitation serves as a central starting point for the distribution-robust xMCDM extension developed in this chapter.

Work on the so-called Contextual Importance and Utility (CIU) framework (Främling, 2020) provides a conceptually fundamental approach to explainability. The starting point is the observation that many current XAI approaches do not sufficiently incorporate established concepts of decision theory. CIU explicitly ties in with classical concepts such as importance and utility and generalizes them for nonlinear, data-driven models, as they prevail in modern artificial intelligence. Context-dependent evaluation is relevant for decision-making: the significance of a factor is not absolute but relative to a specific decision-making context. CIU thus understands explainability as an estimate of the contribution of individual influencing factors to the observed or predicted change in utility. This perspective is particularly relevant for decision-making problems in which model interpretation is not an end in itself, but serves to derive justifiable action decisions. At the same time, CIU remains at the level of individual model outputs and does not address multi-criteria aggregation, uncertainty, or the robustness of decisions.

Shi and Yao (2025) make a conceptually advanced contribution to explaining MCDM decisions, viewing explainability not primarily as an attribution problem of individual criteria or models, but as a structuring problem of the decision itself. Building on the theory of three-way decisions, they systematically distinguish between preferred, non-preferred, and indifferent alternatives. This threefold classification allows rankings to be interpreted not as a complete and necessarily distinct order, but as the result of a conscious separation into accepted, rejected, and ambivalent decision regions. In addition, Shi and Yao propose an explanation-oriented framework, which integrates symbolic representations, semantic meanings, and value-based preference judgments (SMV framework), thus providing a rule-based level of explanation that is easy for decision-makers to communicate. The approach, therefore, addresses a central weakness of classical XAI methods in the MCDM context by explicitly reducing explainability to the decision structure rather than to individual components. However, uncertainty in criterion values and preference parameters is treated predominantly implicitly. Robustness to forecast or preference uncertainty is not an integral part of the decision rule, but remains limited to the interpretation of the resulting three-way classification. This observation marks an open interface between explainable MCDM and robust decision-making approaches, and it is here that the xMCDM framework developed in this chapter comes into play.

Another important strand of recent literature consists of work that uses explicitly explainable AI for objective weight elicitation in MCDM. Prominent here is the contribution by Ciano and Ferrara (2025), who combine classical MCDM methods such as TOPSIS, PROMETHEE, and AHP with XAI attribution methods (in particular SHAP, Integrated Gradients, and Expected Gradients). In their approach, criterion weights are not derived from subjective expert judgments or preference surveys, but from feature attribution patterns of a supervised learning model that predicts real decision outcomes—in this specific application, customer satisfaction in the tourism sector. This approach effectively addresses one of the central weaknesses of traditional MCDM methods: the normative fragility of classic weighting mechanisms. At the same time, the analysis focuses primarily on point estimation of criterion weights and on explanation-based justification of rankings. Although uncertainty about forecasts, preferences, or weight estimates is discussed through stability and sensitivity analyses, it is not formally integrated as a constitutive element of the decision rule. This means that XAI-weighted MCDM approaches differ fundamentally from distribution-robust decision frameworks, which do not diagnose robustness ex post, but rather operationalize it ex ante. This difference is central to the classification of the xMCDM framework proposed in this chapter.

Matejová and Paralič (2025) pursue a methodologically related but conceptually different approach, using multi-criteria decision-making procedures to select suitable explainable AI methods. In their framework, XAI methods are treated as decision alternatives and evaluated on objective and subjective criteria such as fidelity, stability, comprehensibility, computational effort, and user acceptance. Aggregation is performed using a combination of several classic MCDM methods, and the results are combined into an overall ranking using the Borda count. Sensitivity analyses are used to evaluate the robustness of the method selection. The approach convincingly addresses a central practical problem of XAI application, namely the systematic and transparent selection of suitable explanation methods in complex application scenarios. At the same time, it differs fundamentally from decision-oriented MCDM frameworks: Robustness refers to the stability of the method selection, not to the robustness of a subsequent factual decision under forecast or preference uncertainty. The approach thus operates at a meta-level of decision support and complements—rather than replaces—robust MCDM approaches that evaluate real alternatives for action under uncertainty.

The growing importance of explainable MCDM approaches in the financial sector is systematically documented in the review by Cernevičienė and Kabašinskas (2022). The authors analyze a broad spectrum of classic MCDM methods—including AHP, TOPSIS, VIKOR, and outranking methods—and show how these are increasingly being combined with machine learning methods and XAI techniques to make complex financial decisions more transparent. Explainability is primarily understood as a property of the model and weighting logic that promotes trust, regulatory compliance, and acceptance among decision-makers. At the same time, the overview makes it clear that although uncertainty in forecasts, preferences, and data quality is recognized as a practical problem, it is typically addressed only methodologically through sensitivity analyses or heuristic robustness tests. There is a general lack of explicit integration of uncertainty into the decision rule itself, for example, via worst-case or regret criteria. This observation highlights a key research gap between explainable MCDM and robust decision-making approaches, which this chapter closes with the xMCDM framework.

Current hybrid approaches in finance combine classic MCDM techniques with downstream optimization to bridge the gap between ranking and allocation. For example, Pisal et al. (2025) integrate TOPSIS and ARAS evaluations with goal programming and genetic algorithms to derive investor-related portfolios from real transaction data. Their sequential architecture improves ranking stability and allocation quality compared to standalone MCDM or optimization models. However, uncertainty in the evaluation of criteria and investor preferences is implicitly addressed through sensitivity analyses rather than formally modeled. Consequently, robustness is assessed ex post rather than enforced by design. It highlights a structural limitation of common hybrid frameworks and motivates the distribution-robust and uncertainty-aware formulation proposed in xMCDM.

In summary, the literature presents a fragmented picture. Classical MCDM approaches offer structured aggregation mechanisms but neglect uncertainty. Preference learning addresses uncertainty at the level of weights, but is rarely integrated into comprehensive decision-making frameworks. DRO provides a rigorous tool for robustness assurance, but is predominantly used in one-dimensional optimization problems. Finally, XAI improves the transparency of models, but not necessarily the traceability of complex multi-criteria decisions.

Against this backdrop, xMCDM positions itself as an integrative approach. It combines the aggregation logic of classical MCDM methods with probabilistic preference modeling and distribution-robust optimization, and extends this with explicit, decision-centered explainability. In doing so, xMCDM addresses a central gap in the existing literature: the consistent treatment of uncertainty about forecasts and preferences, and their effects on decision stability, regret, and acceptance.

THE FRAMEWORK: AN INTEGRATED DECISION PIPELINE

This section introduces the core concept of the proposed explainable and robust multi-criteria decision-making (xMCDM). The starting point is the observation that real-world decision-making problems are characterized not only by multiple criteria but also by multi-layered uncertainty: uncertainty about future criterion values, the structure and parameters of the underlying forecast models, and the preferences of the actors involved. The xMCDM framework addresses these challenges through an explicitly modular but consistently coupled pipeline that integrates predictive models, preference learning, and robust aggregation.

Conceptually, xMCDM can be interpreted as a sequential, yet probabilistic, mapping from the data space to the decision. The starting point is observable data on alternatives and their context, from which uncertain criterion evaluations are first derived. These assessments do not represent objective facts or deterministic scores, but somewhat data-driven forecasts whose accuracy and reliability depend on model selection, data quality, and structural stability. Accordingly, criterion values in the xMCDM framework are not modeled as fixed quantities, but as random variables with explicit probability distributions.

At this level, the transformation

data → uncertain criterion evaluations,

where uncertainty is understood as an inherent part of the decision information and not as a subsequent disturbance variable. This view reflects the fact that modern decision-making problems do not usually suffer from a lack of forecasting models, but rather from insufficient integration of prognostic uncertainty into the decision-making process itself.

The second transformation concerns the normative dimension of the decision. Even if criterion evaluations are modeled stochastically, it remains unclear how to weigh these criteria against each other. This is where the preference layer comes in, which maps

uncertain criterion evaluations → uncertain preferences

not as a purely technical weighting decision, but as an inference problem. Preferences are modeled as latent, only partially observable variables that are derived from pair comparisons, rankings, or direct weighting specifications. Instead of assuming a single 'true' weighting vector, a distribution of possible preference structures is derived that reflects both the uncertainty of individual actors and the heterogeneity between stakeholder groups. This acknowledges that normative assumptions cannot be specified precisely in real decision-making contexts and are themselves part of the uncertainty situation.

Not as a purely technical weighting decision, but as an inference problem. Preferences are modeled as latent, partially observable variables derived from pair comparisons, rankings, or direct weighting specifications. Instead of assuming a single 'true' weighting vector, a distribution of possible preference structures is derived that reflects both the uncertainty of individual actors and the heterogeneity between stakeholder groups. It acknowledges that normative assumptions cannot be specified precisely in real decision-making contexts and that they are themselves part of the uncertainty.

The actual decision ultimately results from the combination of these two levels of uncertainty. The illustration uncertain preferences → robust decision

is not realized in the xMCDM framework as a simple aggregation of expected values, but as a distribution-robust optimization problem. The goal is to identify decisions or rankings that remain stable across a set of plausible combinations of criterion distributions and preference distributions. Robustness refers not only to the level of aggregated utility values, but explicitly to ranking stability, regret, and sensitivity to model and preference uncertainty.

This three-stage mapping logic illustrates the basic conceptual assumption of xMCDM: uncertainty is not a marginal phenomenon of the decision problem, but its structuring core. Data, preferences, and decisions are not linked by isolated modules, but by consistent probabilistic semantics. In this sense, decisions are not precise optima, but robust solutions within a space of uncertainty.

The two fundamental preliminary stages of this pipeline—the predictive layer and the preference layer—are formally developed below. The decision and aggregation layer based on these is introduced in Section 4, which shows how classic MCDM methods can be reformulated as distribution-robust optimization problems and what numerical and conceptual consequences this has for rank stability and decision quality.

Predictive Layer: Data-Driven Criteria Evaluation Under Uncertainty

In the first step, the corresponding target values $g_j(a)$ are estimated from observable data for each alternative $a \in A$ and each criterion $j \in \{1, ..., J\}$. Unlike in classical MCDM, these criterion values are not treated as deterministic, but as random variables whose distribution is explicitly modeled.

Formally,

$$g_j(a) \;=\; f_j(x_a) + \varepsilon_{j,a} \qquad (1)$$

where x_a denotes a set of observable characteristics of alternative a, $f_j(\cdot)$ represents a data-driven prediction model for criterion j, and $\varepsilon_{j,a}$ captures a stochastic error term. The framework is deliberately model-agnostic. The empirical applications in this chapter use random forests and gradient boosting machines (Breiman, 2001; Friedman, 2001), Bayesian regression models, and compact neural networks, among other things. The decisive factor is not the model class itself, but the ability to quantify the prediction's uncertainty.

Two classes of uncertainty quantification are provided for in the xMCDM framework. First, resampling-based methods such as bootstrap allow empirical forecast distributions to be approximated for $g_j(a)$ (Efron & Tibshirani, 1993). Second, Bayesian models enable the direct derivation of posterior distributions and credible intervals (Gelman et al., 2013). Both approaches lead to a distribution.

$$\mathbb{P}_{j,a} \in \mathscr{P}(\mathbb{R}) \tag{2}$$

These distributions provide a more realistic basis for evaluating alternatives. Instead of a single score, risk profiles are created for each criterion, which reveal worst-case scenarios, value-at-risk-like measures, or downside risks, for example. It shifts the role of MCDM from pure ranking based on point values to consistent processing of stochastic information.

Another advantage of this approach is that model uncertainty can be explicitly addressed. Different model classes, hyperparameters, or training windows can be considered part of the forecast uncertainty and accounted for either through ensembling or by expanding the ambiguity sets. It avoids decisions being implicitly 'tied' to a specific model architecture.

Preference Layer: Bayesian Preference Learning and Stakeholder-Specific Weights

The second building block of the xMCDM framework concerns the modeling of preferences. In classic MCDM methods, criterion weights are often queried directly or defined normatively. However, this approach ignores the fact that preferences are, in practice, uncertain, inconsistent, and context-dependent—the xMCDM framework, therefore, treats preferences as latent random variables inferred from observable decisions or statements.

Let$w = (w_1, ..., w_J)d$ be the weighting vector of the criteria, with$\left(w_j \geq 0\right)$ and$\left(\sum_j w_j = 1\right)$. Instead of assuming a fixed vector, the xMCDM framework specifies an a priori distribution over w, typically a Dirichlet distribution. This choice allows for flexible mapping of prior knowledge and analytically manageable updating when new information becomes available.

Preference information can be collected in different ways. A central approach is to compare alternatives pairwise, from which relative preference orders are derived. Probabilistic models such as the Bradley–Terry or Luce models interpret such comparisons as stochastic realizations of latent utility differences (Bradley & Terry, 1952; Luce, 1959). In a Bayesian extension, these models can be used to estimate posterior distributions over utility parameters or weights (Chu & Ghahramani, 2005). Alternatively, direct weighting information or rankings can be integrated, which are interpreted as noisy observations of the underlying preferences.

A key advantage of the Bayesian approach is that preference-related uncertainty is explicitly quantified. Instead of a single set of weights, the result is a distribution $\mathbb{P}($

w), that can reflect both uncertainty within a stakeholder and heterogeneity between stakeholders. The framework also allows for hierarchical structures in which individual preferences are distributed around group-specific mean values. It will enable different stakeholder groups—such as management, technology, and regulation—to be consistently taken into account in a common decision-making model.

The combination of predictive and preference-based uncertainty layers is a key distinguishing feature of xMCDM compared to existing approaches. Both levels provide probability distributions, but they have different semantic meanings: While the predictive layer maps epistemic uncertainty about future criterion values, the preference layer reflects normative uncertainty about target weightings and trade-offs. Only when both are taken into account can the robustness of decisions be consistently evaluated.

At the end of the section, there are thus two sets of uncertain objects: distributions over criterion values $\mathbb{P}_{j,a}$ and a distribution over preferences $\mathbb{P}(w)$. How these uncertainties are combined in a robust decision aggregation is the subject of the following section. There, it is shown that many common MCDM methods can be formulated as distribution-robust optimization problems and that this reformulation explicitly addresses central questions of rank stability and regret.

DISTRIBUTION-ROBUST MCDM AND NUMERICAL IMPLEMENTATION

While Section 3 formally introduced the predictive and preference-related sources of uncertainty, this section focuses on the core decision-making process of the xMCDM framework: aggregating multiple criteria while simultaneously accounting for forecast and preference uncertainty. Central to this is the reformulation of classical MCDM methods as distribution-robust optimization problems. As a result, robustness is not only analyzed ex post, but also integrated ex ante into the decision-making mechanism itself.

Distribution-Robust Reformulation of Classic MCDM Aggregators

The starting point is a standard MCDM scenario: For each alternative $a \in A$, there are criterion values $g(a), j = 1, \ldots, J$, and preference weights w_j. In additive utility models, the aggregated score is typically calculated as

$$U(a) = \sum_{j=1}^{J} w_j g_j(a) \tag{3}$$

In distance-based methods such as TOPSIS or VIKOR, the score is calculated based on distances to ideal reference points, while outranking methods are based on pairwise dominance relationships. What all these approaches have in common is that they implicitly assume point-like known values.

In the xMCDM framework, both the criterion values$g_j(a)$ and the weightsw_j are modeled as random variables. A decision on a ranking or selection is therefore no longer clearly determined, but depends on the underlying joint distribution of these variables. Since this distribution is only partially known, the distribution-robust optimization approach is appropriate. Instead of assuming a 'true' distribution, a set of plausible distributions $\mathbb{P}$ is defined that encompasses both forecast and preference uncertainty.

Formally, the distribution-robust utility of an alternative can be expressed as

$$U^{DRO}(a) = \inf_{\mathbb{Q} \in \mathcal{P}} \mathbb{E}_{\mathbb{Q}}[U(a)] \tag{4}$$

The decision is thus based on the worst expected value across all permissible distributions. Similarly, regret-based criteria can be defined in which the maximum expected loss of utility is minimized relative to the best alternative in each case (Ben-Tal et al., 2009; Shapiro et al., 2014).

A convenient special case arises when the ambiguity set is defined as a Wasserstein ball around an empirical reference distribution. This construction allows for the consistent mapping of sampling uncertainty, model errors, and moderate distribution shifts (Blanchet & Murthy, 2017). For additive utility models, this often leads to convex optimization problems in which the robustness parameter (such as the Wasserstein radius) can be interpreted as an explicit regulator between performance and hedging.

A key advantage of this approach is that it is not limited to additive models. Distance-based methods such as TOPSIS or VIKOR can also be converted into distribution-robust objective functions by appropriately reformulating the distance functions, provided that the underlying norms are convex. ELECTRE-like outranking logics can also be approximately integrated by treating thresholds and dominance conditions as uncertain parameters and robustifying them. This makes xMCDM a higher-level meta-framework that brings together different aggregation paradigms under a uniform robustness logic.

Numerical Programs, Algorithms, and Computational Aspects

The practical applicability of a distribution-robust MCDM approach depends crucially on its numerical solvability. A key finding of recent DRO literature is that many robust expected value and regret problems can be converted into equivalent convex programs, in particular linear programs (LP), quadratic programs (QP), or second-order cone programs (SOCP) (Ben-Tal et al., 2009; Blanchet & Murthy, 2017).

In the xMCDM framework, the specific problem class arises from the combination of the aggregator, the uncertainty model, and the ambiguity set. Additive utility sums with Wasserstein DRO typically lead to linear or slightly quadratic programs. Distance-based aggregators often yield SOCP formulations when Euclidean or weighted norms are used. Preference uncertainty can either be explicitly co-optimized or accounted for via worst-case weights within admissible simplexes, which generate additional constraints but preserve convexity.

Modern interior point methods or specialized convex solvers are suitable for solving these programs. The empirical parts of this chapter present implementations in Python (using CVXOPT, MOSEK, or JAX-based optimizers) and in MATLAB. The choice of a specific solver is less important than the numerical stability of the problem formulation. Practical guidelines include appropriate criterion scaling, regularization of poorly conditioned covariance matrices, and limiting ambiguity radii to avoid overly conservative solutions.

The complexity of the resulting optimization problems scales primarily with the number of alternatives and criteria. While classic MCDM methods have practically negligible computation times, the effort in the xMCDM framework increases moderately due to the robust reformulation. In the case studies considered, however, the computation time remains in the range of seconds to minutes and is thus compatible with interactive decision-making processes. It is also important to note that robust optimization is not a one-time step but rather part of an iterative decision-making workflow: parameters such as the robustness radius or weighting priors can be adjusted to explore sensitivities systematically.

In summary, this section demonstrates that integrating MCDM with distribution-robust optimization is not only conceptually sensible but also numerically feasible. By reformulating classical aggregators as convex DRO problems, robustness becomes an explicit property of the decision rule itself. This forms the basis for the next step: the development of explainability and diagnostic tools to make robust decisions transparent and subject them to critical scrutiny. These aspects are the focus of Section 5.

EXPLAINABILITY, SENSITIVITY, AND DECISION DIAGNOSTICS

Robust decisions alone are not sufficient in many fields of application. Especially in organizational, regulatory, or socially sensitive contexts, decisions must be comprehensible, communicable, and verifiable. Section 5, therefore, develops a set of explainability and diagnosis-oriented tools that not only evaluate the results of distribution-robust MCDM but also translate them into understandable decision logic. The focus is explicitly on decision-centric explainability and not on the interpretation of individual forecast models.

Decision-Centric Explainability: Contributions, Trade-Offs, and Counterfactuals

Classic XAI methods primarily address the question of which input factors led to a particular prediction (Ribeiro et al., 2016; Lundberg & Lee, 2017). In the context of MCDM, this approach falls short because the decision results from a multi-stage aggregation of criteria, preferences, and robustness assumptions. xMCDM therefore shifts the focus of explanation from the model level to the decision level.

A central component is the attribution of decisions to criteria and preferences. Building on the distribution-robust objective function, marginal contributions of individual criteria to the robust utility or robust regret of an alternative can be determined. These contributions can be analyzed globally (across all other options) or locally (for a specific decision) and allow statements to be made about which criteria are particularly influential in ranking an alternative under worst-case assumptions. Game theory-motivated attribution concepts, such as Shapley-type decompositions, can be applied to the aggregated decision function for this purpose and provide consistent contribution assignments (cf. Molnar, 2022).

The MDS explanations proposed by Erwig and Kumar (2025) can be understood as prototypical contrastive explanations in the sense of XAI literature, as they explicitly address the question of why one alternative was preferred over another. Their strength lies in structurally reducing complex MCDM calculations to small, semantically interpretable subsets of criterion contributions. For xMCDM, however, such approaches are only a starting point. While MDS explanations make the internal logic of deterministic aggregation methods transparent, they do not address the central question of how robust this logic is to forecast errors, alternative preference assumptions, or plausible deviations in the distribution. In uncertain decision contexts, there is therefore a risk that a formally correct but fragile explanation will be provided. The explainability concept developed in this chapter extends contrastive MCDM explanations with a robust perspective by explaining under which uncertainty assumptions a ranking is valid, and under which it is not.

The robust decision rules developed in Sections 3 and 4 can be naturally linked to the three-way decision perspective of the explainable MCDM literature. In particular, the approach of Shi and Yao (2025) offers an interpretative semantics that assigns a symbolically condensed meaning to robust rankings. In a distribution-robust context, alternatives can be classified not only as better or worse, but also as robustly preferred, robustly indifferent, or robustly not preferred, depending on whether their dominance relationship is maintained across all plausible realizations of criterion and preference distributions.

This tripartite division is not merely a heuristic abstraction, but results directly from the structure of distribution-robust optimization problems: alternatives whose worst-case regret is below a decision-relevant threshold form a robust acceptance region; alternatives with consistently poor worst-case behavior are excluded; in between remains a zone of structural ambiguity. The three-way decision logic thus functions as an explanatory interface between numerical robustness analysis and cognitively compatible decision interpretation. Explainability is therefore understood not as a simplification, but as a systematic reduction of complex uncertainty spaces to reliable decision regions.

XAI-supported MCDM approaches, such as that of Ciano and Ferrara (2025), illustrate that explainability in the MCDM context can be applied at different levels. There, the focus is on the interpretability of the criterion weights: feature attribution methods provide empirically based statements about which criteria make the most significant contribution to predicting a target output on average. These weight explanations are particularly valuable where subjective weight elicitation is considered problematic or politically sensitive.

In the xMCDM framework, however, explainability is not limited to the weighting level, but is explicitly extended to the decision level. While XAI-weighted MCDM models explain why a criterion is essential, xMCDM aims to explain why a decision is justifiable under uncertainty. This perspective includes statements about worst-case regret, rank stability, and robust indifference regions. Explainability here arises primarily not through attribution, but through transparency about risk trade-offs and decision invariance to plausible model and preference deviations. In this sense, XAI-weighted approaches and xMCDM are not substitutive, but complementary: the former explain relevance, the latter explain reliability.

Furthermore, xMCDM enables the explicit analysis of trade-offs. While classic MCDM methods implicitly hide these in weighting parameters, they become explicitly visible in a robust setting: an increase in the robustness requirement (e.g., through a larger Wasserstein radius) systematically shifts the decision in favor of alternatives with lower variance or lower downside risks. Such shifts can be interpreted as robustness-performance trade-offs and communicated graphically or in tabular form. This makes it clear to decision-makers what price is paid in terms of reduced expected performance for additional security.

A third explanatory component is counterfactual analyses. These answer the question of what minimal changes in criterion evaluations or preferences would be necessary to achieve a different ranking. In the xMCDM framework, counterfactual scenarios can be formulated consistently because the underlying optimization problems are explicitly present. Counterfactual explanations are particularly relevant in participatory decision-making processes because they show whether and how individual stakeholder positions apply to the decision or whether the result remains invariant to moderate changes in preferences.

Sensitivity and Robustness Diagnostics: Stability, Regret, and Validation

In addition to explainability in the narrower sense, the systematic diagnosis of decisions is a central goal of xMCDM. Robustness is not understood as binary, but as a gradual property that can be measured along several dimensions.

A first key measure is rank stability. It quantifies how often and to what extent rankings change as uncertainty assumptions vary. In the xMCDM framework, a rank stability index can be defined that aggregates the sensitivity of the alternatives' rank positions to changes in the robustness radius, preference priors, or model distributions. High stability indicates that a decision is consistent across a wide range of uncertainty, while low stability indicates structural fragility.

A second key diagnostic tool is regret analysis. Instead of just looking at the robust expected value, the expected or worst-case regret of an alternative is analyzed, i.e., the loss of utility compared to the best possible alternative under given realizations. Regret measures are particularly suitable for evaluating decisions ex post and for comparing different decision rules (Ben-Tal et al., 2009; Shapiro et al., 2014). In the empirical part of the chapter, they are used as a key performance measure to contrast xMCDM with classical MCDM methods.

In addition, sensitivity analyses are an integral part of decision validation. These include systematic variations in model parameters (e.g., model complexity and regularization), preference assumptions (e.g., the dispersion of weighting priors), and robustness parameters. Particularly relevant here are so-called placebo tests, in which deliberately irrelevant criteria or random preferences are added to decision-relevant noise to test whether the decision rule responds appropriately. Out-of-sample validation is also used, for example, via K-fold splits or rolling time windows, if the application has a time reference.

Overall, explainability and diagnostics in the xMCDM framework do not function as downstream comments but as integral components of decision quality. They enable critical questioning of robust decisions, quantify their stability, and make their normative implications transparent. In this way, they create the conditions for the acceptance and responsible use of data-driven MCDM methods in practice.

The following section applies these concepts to specific use cases. Case studies from operations management, engineering, and finance demonstrate how xMCDM can be implemented in realistic decision-making situations and what empirical advantages it offers over traditional, score-based MCDM approaches.

EMPIRICAL EVIDENCE: CASE STUDIES AND EXPERIMENTS

This section demonstrates the practical applicability and empirical added value of the xMCDM framework through three case studies across operations management, engineering, and finance. The aim is not to present isolated application examples, but to demonstrate the method's reproducibility and transferability. All case studies follow the same experimental protocol and differ only in terms of data type, criterion structure, and decision logic. It enables a systematic comparison between classic MCDM and xMCDM.

Use Cases, Data, and Experimental Design

The first case study deals with a classic problem in operations management: supplier selection. Alternatives are potential suppliers; criteria include price level, delivery reliability, quality indicators, and carbon footprint, among others. Historical delivery and performance data form the basis for predictive models used to estimate future criterion values. Preferences are derived from simulated pair comparisons and weighting specifications from various stakeholder groups (purchasing, sustainability, production).

The second case study concerns site selection for renewable energy plants. Here, the focus is on technical, economic, and ecological criteria, including expected energy yields (wind or solar radiation), grid connection costs, investment and operating costs, and restrictions due to nature conservation and acceptance aspects (e.g., NIMBY proxies). The criteria are predicted using geostatistical data and ML-supported regression models. Preferences reflect different stakeholder groups, such as project developers, regulatory authorities, and local communities.

The third case study addresses the allocation of a loan portfolio. Alternatives are loan segments or borrower classes, and criteria include expected returns, default probabilities, risk indicators, and ESG-related constraints. The predictive level is based on ML models for estimating default probabilities, supplemented by uncertainty quantification. Preference assumptions are based on regulatory and economic trade-offs between risk, return, and sustainability.

The same experimental procedure is applied to all three case studies. First, classic MCDM methods (weighted sum, TOPSIS, VIKOR, ELECTRE) are implemented based on point values and fixed weights. Next, xMCDM is applied, explicitly accounting for forecast and preference uncertainty and varying robustness parameters. Performance metrics include robust expected values, regret, rank stability, and out-of-sample risks. The implementations are carried out in Python (scikit-learn, PyTorch/JAX, CVX-based solvers) and, additionally, in MATLAB; all experiments are replicable.

Results, Comparison, and Ablation Analyses

The results show consistent patterns across all case studies. First, classical MCDM methods exhibit high rank instability when moderate disturbances are introduced to criterion values or weights. Especially in cases of sparse data or heterogeneous preferences, small changes often lead to abrupt rank changes. xMCDM significantly reduces this instability, as decisions are explicitly hedged against plausible uncertainty realizations.

Second, the regret analyses show that although robust decisions in the xMCDM framework sometimes have lower nominal expected values, they cause significantly less regret in out-of-sample comparisons. This finding is particularly pronounced in the credit case study, where robust portfolio allocations generate lower average losses under adverse realizations than score-based MCDM solutions. Similar, albeit less extreme, effects are evident in the technical and operational use cases.

Thirdly, to support the interpretation of the results, the robust rankings in all three case studies are additionally evaluated using a three-way decision classification. Building on the logic of Shi and Yao (2025), alternatives are divided into robust preference, robust indifference, and robust rejection regions. This representation proves particularly helpful in interdisciplinary decision-making settings, as it translates numerical robustness measures into a symbolic decision structure. Instead of a complete ranking list, decision-makers receive clear statements about which alternatives are reliably viable under uncertainty, which remain systematically inferior, and which require further analysis or clarification of preferences.

Across all case studies, it is evident that classic score-based MCDM methods often yield an artificial selectivity that is not accounted for by uncertainty. xMCDM reduces this apparent accuracy by classifying a substantial proportion of the alternatives in a robust indifference zone. These results support the view that three-way decisions are not a refinement of classic rankings, but rather enable a more realistic representation of decision quality under uncertainty and thus make a significant contribution to the explainability of robust decisions.

The tourism application by Ciano and Ferrara (2025) impressively illustrates the potential of XAI-supported MCDM approaches in data-rich, service-oriented contexts. Particularly noteworthy is the empirical evidence that data-driven weighting can lead to significantly better agreement with observed decision outcomes. These results are consistent with our case studies in that machine learning also makes a substantial contribution to objectifying the basis for decision-making.

The applications reviewed by Cernevičienė and Kabašinskas (2022) illustrate the practical relevance of MCDM and XAI approaches in financial practice, particularly in portfolio optimization, credit risk assessment, and performance analysis. The studies discussed there show that explainability contributes significantly to the acceptance of data-driven decision-making models, especially in highly regulated environments. The results of the present case studies build on these findings, but go one step further conceptually. Instead of limiting itself to explaining criterion effects or model outputs, xMCDM enables an explicit assessment of decision robustness under uncertainty. This perspective proves particularly important in the financial context, which the review identifies as sensitive to model risks: decisions become not only explainable but also transparent regarding their stability and potential damage in the worst case. In this sense, xMCDM complements the explainable MCDM approaches identified in the review with an explicit risk dimension.

At the same time, the xMCDM experiments show that another dimension of robustness becomes relevant as soon as decisions are understood not solely as a forecasting problem but also as a risk allocation problem. While XAI-weighted MCDM approaches provide robust weight estimates, xMCDM addresses the robustness of the decision itself. In the applications examined—particularly in the financial and energy sectors—this distinction proves crucial: alternatives with high attributive relevance can still be fragile under uncertainty, while robustly acceptable solutions are characterized by lower regret and greater rank stability. The two perspectives thus complement each other along different axes of decision quality.

The supplier selection case study in this chapter is directly related to current hybrid approaches, in particular the explainable ML–MCDM framework presented by Abdulla and Baryannis (2024). Both approaches share the basic idea of using data-driven methods to cope with high dimensionality and large supplier pools. At the same time, the final decision is made explicitly using a transparent MCDM method. There are also parallels in terms of practical relevance, as real, heterogeneous supplier data sets and economically relevant criteria, such as costs, delivery reliability, quality, and sustainability, are considered. However, the main difference lies in how uncertainty is handled and in decision stability. While the hybrid approach of Abdulla and Baryannis (2024) assumes criterion values and weights to be given and assesses robustness retrospectively via sensitivity and performance comparisons, xMCDM explicitly models uncertainty through probability distributions in both the predictive and preference-based layers. The application of distribution-robust optimization enables the identification of robust rankings and the direct quantification of their worst-case regret. The case study shows that seemingly stable supplier rankings can vary significantly under plausible uncertainty assumptions. xMCDM thus complements existing explainable ML–MCDM approaches by introducing a decision-relevant risk dimension, particularly important for strategic procurement decisions with long-term implications.

In addition, ablation studies illustrate the role of individual framework components. For example, if preference uncertainty is ignored, a significant portion of the ranking instability persists, even when forecast uncertainty is robustly accounted for. Conversely, modeling uncertain preferences alone without robust aggregation also does not lead to stable decisions. The most significant gain in robustness comes from considering both types of uncertainty simultaneously. Ablation studies on model complexity also show that intense prediction models do not necessarily lead to better robust decisions; instead, the combination of moderate model complexity, regularization, and an appropriate robustness radius plays a decisive role.

Finally, the experiments provide insights into the practical interpretability of the decisions. In all case studies, the explainability and diagnostic tools described in Section 5 enable comprehensible communication of the results. Decision-makers receive not only a ranking, but also information about which criteria dominate under worst-case assumptions, which trade-offs are implicitly made, and how stable the result is against plausible deviations.

The case studies examined in this chapter focus on robust decisions about real alternatives. However, the findings of Matejová and Paralič (2025) suggest that an additional upstream decision-making level is relevant in practical decision-making processes: the selection of suitable explainability methods. In data- and model-intensive applications—such as credit portfolio allocation or site selection for renewable energies—the choice of explanatory method can significantly influence which aspects of a robust decision become visible to stakeholders. The combination of both approaches thus opens up a multi-level decision-making architecture: at a meta-level, suitable XAI methods are selected using MCDM; at the operational level, these methods are used to make MCDM decisions more robust and explainable. While Matejová and Paralič's contribution provides the basis for rational method selection, xMCDM addresses the robustness of the decision content itself. Together, both approaches enable consistent, explainable, and responsibility-oriented design of data-driven decision-making processes.

BEST PRACTICES, LIMITATIONS, AND OUTLOOK

The concluding section summarizes key practical insights from the design and empirical application of xMCDM, reflects on the approach's limitations, and outlines future research and development perspectives. The aim is to translate the methodological contribution into concrete recommendations for action while formulating realistic expectations regarding possible applications and scalability.

Best Practices for the Application of xMCDM

The implementation of xMCDM benefits from a clearly structured approach that iteratively combines forecasting models, preference acquisition, and robust aggregation rather than treating them in isolation. A key success factor is careful data preparation. Criteria should be defined in such a way that they are both explainable and prognostically reliable. Consistent scaling and normalization are crucial, especially when combining heterogeneous data sources—such as technical, economic, and ecological indicators—as these directly influence the numerical stability of robust optimization problems.

When choosing forecast models, it is clear that robust decisions do not necessarily benefit from maximum model complexity. In many applications, comparatively compact models with good calibration and explicit uncertainty quantification are superior to very deep, difficult-to-interpret architectures. Ensemble approaches or Bayesian models offer a practical compromise between forecast quality and transparency. From a practical point of view, it is advisable to compare several model classes and explicitly integrate their forecast uncertainty into the ambiguity sets of the robust decision stage.

An incremental, learning-based approach has proven successful for preference elicitation. Instead of querying complete weighting vectors at the outset, pair comparisons or partial rankings can be used to update preference distributions gradually. It reduces cognitive load on stakeholders while allowing for realistic modeling of preference uncertainty. In organizational contexts, explicit modeling of multiple stakeholder groups is also essential to reveal hidden conflicts of interest rather than implicitly conceal them through averaged weights.

Another practical aspect concerns the choice of robustness parameters. The scope of the ambiguity set—such as the Wasserstein radius—should not be determined arbitrarily, but instead based on data availability, model validation, and risk tolerance. Exploratory analyses along a robustness path that make the effects of different safety levels on rankings, regret, and performance transparent have proven effective. Such analyses can be easily converted into standardized reporting formats, such as 'decision cards,' which summarize robust rankings, uncertainty assumptions, and explainability indicators in a compact form.

Limitations and Future Research Directions

Despite the advantages described above, xMCDM also has apparent limitations. A key challenge lies in scalability. While the case studies presented are easily manageable with a moderate number of alternatives and criteria, the computational effort can increase significantly for vast decision spaces or high-frequency applications. Further research is needed here regarding efficient approximations, decomposition methods, or incremental solution methods.

Another limitation concerns the assumption of stationary environments. Structural breaks, regime-dependent dynamics, or strategic behavior by other actors characterize many real-world decision contexts. Although time-dependent or rolling validation approaches can be integrated into xMCDM, the explicit modeling of non-stationary preferences or adverse reactions remains an open challenge. Links to online learning and robust control approaches offer promising starting points here.

Explainability also has inherent limitations. While xMCDM significantly increases decision-related transparency, the difficulty of intuitively communicating results increases with model complexity and robustness assurance. Future work could investigate how different forms of explanation—such as narrative, visual, or interactive formats—influence the acceptance of robust decisions, and how these effects can be empirically measured.

From a methodological perspective, there is also scope for extensions towards multi-period decisions, dynamic preference models, and explicit constraints, such as those of a regulatory or ethical nature. In particular, integrating sustainability and ESG criteria opens up a broad field of application in which robust multi-criteria decision-making under uncertainty is becoming increasingly important.

Overall, xMCDM is not intended to replace classic MCDM methods, but rather to serve as their logical further development for data-driven, uncertain decision-making environments. The framework presented shows that forecast uncertainty, preference uncertainty, and robustness requirements can be consistently integrated without sacrificing numerical solvability or explainability. In this way, xMCDM contributes to decision support that is less oriented toward seemingly precise scores and more toward robust, transparent, and accountable decisions under uncertainty.

REFERENCES

Abdulla, A., & Baryannis, G. (2024). A hybrid multi-criteria decision-making and machine learning approach for explainable supplier selection. *Supply Chain Analytics*, *7*, 100074. DOI: 10.1016/j.sca.2024.100074

Andrae, S. (2024). Explainable Artificial Intelligence in risk management: A framework. In Corazza, M., Shah Khan, F., Torre, D. La, & Masri, H. (eds.). *Artificial intelligence and beyond for finance*. World Scientific Publishing, 149–206. https://doi.org/DOI: 10.1142/9781800615212_0004

Belton, V., & Stewart, T. J. (2002). *Multiple criteria decision analysis: An integrated approach.* Springer.

Ben-Tal, A., El Ghaoui, L., & Nemirovski, A. (2009). *Robust optimization*. Princeton University Press.

Blanchet, J., & Murthy, K. (2017). Quantifying distributional model risk via optimal transport. *arXiv*. https://doi.org//arXiv.1604.01446DOI: 10.48550

Bradley, R. A., & Terry, M. E. (1952). Rank analysis of incomplete block designs: The method of paired comparisons. *Biometrika*, *39*(3–4), 324–345. DOI: 10.1093/biomet/39.3-4.324

Breiman, L. (2001). Random forests. *Machine Learning*, *45*(1), 5–32. DOI: 10.1023/A:1010933404324

Chu, W., & Ghahramani, Z. (2005). Preference learning with Gaussian processes. In *Proceedings of the 22nd International Conference on Machine Learning*, 137–144. https://doi.org/DOI: 10.1145/1102351.1102369

Ciano, T., & Ferrara, M. (2025). *Explainable Multi-criteria Decision Making for tourism economics: integrating XAI with MCDM for a robust accommodation performance assessment*. Decisions in Economics and Finance., DOI: 10.1007/s10203-025-00553-6

Danach, K., Fouad Aly, W. H., Tarhini, A., & Laouadi, S. (2025). Toward transparent optimization: A systematic review of Explainable AI in decision-making systems. *European Journal of Pure and Applied Mathematics*, *18*(4), 6707. DOI: 10.29020/nybg.ejpam.v18i4.6707

Efron, B., & Tibshirani, R. (1993). *An introduction to the bootstrap*. Chapman & Hall.

Erwig, M., & Kumar, P. (2022). Explaining results of multi-criteria decision making. *Journal of Multi-Criteria Decision Analysis*, *32*(1), e70011. DOI: 10.1002/mcda.70011

Främling, K. (2020). Decision theory meets Explainable AI. In Calvaresi, D., Najjar, A., Winikoff, M., & Främling, K. (Eds.), Lecture Notes in Computer Science: Vol. 12175. *Explainable, transparent autonomous agents and multi-agent systems. EXTRAAMAS 2020* (pp. 57–74)., DOI: 10.1007/978-3-030-51924-7_4

Friedman, J. H. (2001). Greedy function approximation: A gradient boosting machine. *Annals of Statistics*, *29*(5), 1189–1232. DOI: 10.1214/aos/1013203451

Gelman, A., Carlin, J. B., Stern, H. S., Dunson, D. B., Vehtari, A., & Rubin, D. B. (2013). *Bayesian data analysis* (3rd ed.). CRC Press.

Keeney, R. L., & Raiffa, H. (1993). *Decisions with multiple objectives: Preferences and value tradeoffs* (2nd ed.). Cambridge University Press.

Luce, R. D. (1959). *Individual choice behavior: A theoretical analysis*. Wiley.

Lundberg, S. M., & Lee, S.-I. (2017). A unified approach to interpreting model predictions. *arXiv*. https://doi.org//arXiv.1705.07874DOI: 10.48550

Matejová, M., & Paralič, J. (2025). A multi-criteria decision-making approach for the selection of Explainable AI methods. *Machine Learning & Knowledge Extraction*, *7*, 158. DOI: 10.3390/make7040158

Molnar, C. (2022). *Interpretable machine learning*. 2nd ed. Lulu.com. https://christophm.github.io/interpretable-ml-book/

Pisal, P., Reddy, K. K., Kishore, J., Jonnalagadda, R. R., Kumar, M., & Band, G. (2025). An integrated TOPSIS and ARAS method multi-criteria decision-making approach for optimizing investment portfolios using goal programming and genetic algorithm model. *Scientific Reports*, *15*, 34450. DOI: 10.1038/s41598-025-17604-y PMID: 41039038

Ribeiro, M. T., Singh, S., & Guestrin, C. (2016). "Why should I trust you?" Explaining the predictions of any classifier. Proceedings of the 22nd ACM SIGKDD International Conference on Knowledge Discovery and Data Mining, 1135–1144. https://doi.org/DOI: 10.1145/2939672.2939778

Roy, B. (1996). *Multicriteria methodology for decision aiding*. Springer.

Shapiro, A., Dentcheva, D., & Ruszczyński, A. (2014). *Lectures on stochastic programming: Modeling and theory* (2nd ed.). SIAM.

Shi, Y., & Yao, J. (2025). Explainable multi-criteria decision-making: A three-way decision perspective. *International Journal of Approximate Reasoning*, *187*, 109528. DOI: 10.1016/j.ijar.2025.109528

Zuheros, C., Martínez-Cámara, E., Herrera-Viedma, E., Katib, I. A., & Herrera, F. (2023). Explainable Crowd Decision Making methodology guided by expert natural language opinions based on Sentiment Analysis with Attention-based Deep Learning and Subgroup Discovery. *Information Fusion*, *97*, 101821. DOI: 10.1016/j.inffus.2023.101821

Chapter 5
Application of SWOT And Breakeven Analysis in Strategic Decision-Making: A Quantitative Approach

Syed Faisal Ali
https://orcid.org/0009-0005-7191-3290
Christ University, Bangalore, India

Surabhi Sangal
https://orcid.org/0000-0002-9160-2538
Christ University, Bangalore, India

ABSTRACT

This chapter investigates the extent to which the conjunction of SWOT analysis and quantitative analysis (such as breakeven analysis) enhances the quality of strategic decision-making. In addition to the usual limitations of a qualitative SWOT analysis, the paper embraces financial feasibility tools and promotes a multi-criteria decision-making approach. After a brief initial chapter noting the rationale for combining qualitative and quantitative analyses in the strategic planning process, the chapter continues with the theoretical foundation relating to strategic decision-making models (including SWOT, breakeven analysis, Analytic Hierarchy Process (AHP), Hybrid SWOT-AHP, and Quantitative Strategic Planning Matrix (QSPM). A case-study approach with examples from both the private and public sectors illustrates the practical application of these tools through numerical data tables, breakeven calculations, and decision matrices.

DOI: 10.4018/979-8-3373-6746-0.ch005

INTRODUCTION

In a business environment that is full of uncertainties, strategically deciding on a course of action means to consider different options and decide how to allocate resources when you do not have complete information (Eisenhardt & Zbaracki, 1992; Nutt, 2008). In today's world of business executives and strategists are looking for ways to evaluate their strategic decisions with both numbers and words. While SWOT analysis has been one of the most popular tools to describe a company's strategic position, it still is it is easy to use, and it evaluates both internal and external factors in a simple 2×2 matrix (Weihrich, 1982).

The SWOT method is criticized because it remains solely qualitative identifies elements yet does not provide a scale by which they can be measured in terms of relative importance or ranked as to what should receive top priority (Helms & Nixon, 2010). Since there is no yardstick to use, when managers complete a SWOT, they are still uncertain about which element or option to give top priority to (Dean & Sharfman, 1996). In order to reduce this uncertainty, companies have added numbers to their planning more and more frequently over time in order to create a better foundation upon which to make decisions (Eisenhardt & Zbaracki, 1992; Nutt, 2008) break-even analysis provides a clearness that can be used to clarify the strategic decision-making process.

Break-even analysis has been a key player in financial planning for quite some time. Break-even analysis determines that point at which the revenues generated by a project equal its costs over a defined period or at a certain level of output. In other words, the break-even point represents the juncture beyond which financial loss turns to profit; each additional dollar of revenue after the break-even point earned generates net profit. From a strategic planning perspective, using break-even analysis represents a sound approach for testing an initiative's financial viability and related risk factors. This method provides extremely precise estimates of the sales volume or the revenue amount allocated to cover all costs fixed and variable at the break-even point. Consequently, the manager can be supported by the numbers rather than relying on his/her judgment. It is very clear that even at first sight a specific policy would not be profitable if the expected market conditions were to be taken into consideration.

In simpler terms, break-even analysis provides concrete economic guidance by establishing a benchmark rooted in reality. Leaders can identify which projects are likely to become profitable the fastest and understand how changes in costs and prices will impact their strategic choices when they utilize this tool (Jakupi et al., 2017). However, quantitative decision-making tools for strategy go beyond just financial calculations. Multi-criteria decision-making (MCDM) tools, like the Analytic Hierarchy Process (AHP), empower strategists to evaluate options based on a variety of criteria, including both tangible metrics and intangible factors (Belton & Stewart, 2002; Ishizaka & Nemery, 2013).

AHP along with other methods bring managerial rigor which was formerly only qualitative to different aspects of strategic analysis - as an example, they can be used for determining the relative weights of various SWOT factors (Mardani et al., 2015).

On top of these improvements, there are now hybrid systems that combine SWOT analysis and AHP (usually referred to as SWOT-AHP or A'WOT) (Kurttila et al., 2000). Such methods enable the decision-makers to allocate weights to the components of SWOT and to check the strategic alternatives against these weighted factors in an organized manner. Integration of these two ensures not only that internal strengths that most relevant and external opportunities get more talked about but that also the possible impact of weaknesses and threats is quantitatively evaluated for each option.

Finally, the Quantitative Strategic Planning Matrix (QSPM) gives a well-organized way to rank the strategic moves in a quantitative manner. QSPM gathers information from previous analyses (like the findings of SWOT and the environmental scans) and it gives weight and scores to the different strategies resulting in a score of the relative attractiveness of each alternative. The idea behind this tool is: Weighing the factors we got from the SWOT and objectively evaluating each option, which one should we choose? QSPM is not always right at pinpointing the decision, but it is still a lot more objective than simply going by the gut feeling or doing an unplanned brainstorming session.

By employing QSPM to methodically evaluate different options, strategists can lessen the amount of subjective judgment they must rely on. This makes the process of making strategic decisions more transparent.

THEORETICAL FRAMEWORK

A Qualitative Strategic Framework for SWOT Analysis

A SWOT analysis is really the backbone of any strategic plan. It's all about identifying, understanding, and sorting through a company's internal Strengths and Weaknesses, as well as the external Opportunities and Threats. In simpler terms, a SWOT analysis helps an organization get a clear picture of where it stands right now (Pickton & Wright, 1998). Strengths are the favorable attributes or resources that give the organization an advantage in achieving its goals and carrying out its activities, such as a strong brand or proprietary technology. On the other hand, Weaknesses are attributes that put the organization at a disadvantage or points toward areas that need improvement. Opportunities are created by favorable developments or changes in the environment that the organization can exploit-for instance, growing markets, more lenient regulatory environments, or technological advancements. Threats are external factors that have the potential to negatively impact an organization's performance or survival, including new rivals, economic decline, or change in customer preferences.

SWOT analyses remain popular because they offer a symmetrical and holistic review of the internal capabilities and external conditions in one framework. The beauty of it is that organizations can clearly understand their competitive position by recognizing how to leverage their strengths while also addressing any weaknesses and threats. While SWOTs are often presented as a matrix or a list with four quadrants, each filled with key insights, one of the standout features of SWOT is how it brings together different functional inputs and stakeholder perspectives for the most thorough insights. Plus, the flexibility of SWOT is a major bonus. In the end, non-profits, governments, and even individuals use it for planning and decision-making insights, even though it may have become more popular with businesses (Helms & Nixon, 2010).

As noted, the restrictions of SWOT analysis as a descriptive tool remain in that it does not rank the factors by importance or order of preference, nor does it tell us which strategy to pursue, but may highlight factors relative to decision-making; in other words, it highlights relevant considerations, but does not provide any analytical aspects for the purpose of guiding decision-making. Ultimately, "the output of a SWOT analysis is often just a list of factors or an incomplete qualitative analysis of internal and external factors" (Helms & Nixon, 2010; Phadermrod et al., 2019). That means if the manager doesn't provide further analysis after the SWOT, he/she might be unable to answer these questions: Among the strengths identified, which one is really the most important? As a matter of fact, several companies employ SWOT to different extents, but they fail to consistently take it to the level of proposal creation and . As a result, the SWOT may not come to fruition, as it is indexed quality as a situational narrative or description moving into the approach for the future, rather than as a decision process to include SWOT into consideration potentially.

Recognizing these restrictions, academics and practitioners have proposed using quantitative approaches to supplement SWOT (Helms & Nixon, 2010; Mardani et al., 2015). Still, in a qualitative form, it has utility: there will be some conscious thought given to the important relevant internal or external factors when strategic decisions are being made. Furthermore, quantitative analysis, in conjunction with SWOT, does not supersede the function of SWOT; it supports it. The SWOT analysis provides the criteria for decisions (i.e., factors that any 'good' strategy should account for) while the quantitative methodologies provided a systematic way of measuring the alternatives against the criteria.

BREAKEVEN ANALYSIS: FINANCIAL VIABILITY ASSESSMENT

Breakeven analysis is the method that is based on quantitative financial measures and describes the most favorable financial scenario for a business regarding revenue and costs. The breakeven point is where the business income exactly equals its costs, which is the clear breakeven point or more specifically: the condition of a person or business when there is neither gain nor loss. The business is at the breakeven point when its income equals the sum of fixed and variable costs. After the breakeven point a person or company realizes a profit. To perform a breakeven analysis, we must first distinguish the costs into two different categories - fixed and variable costs. Fixed costs are the expenses that remain constant in terms of the level of production or sales (e.g. rent, salaries and leasing of equipment), while variable costs are the costs that may be incurred in each unit sold or produced (e.g. raw materials or labor, in this case, the costs/expenses are a per-unit cost. Along with the costs, the selling price of the product/service also needs to be considered. Weare going to employ the formula to figure out the breakeven point: Fixed Costs ÷ (Price per Unit – Variable Cost per Unit) to get fixed costs divided by (price per unit minus variable cost per unit). Sometimes, this will be represented with a breakeven graph, where "units sold" is the horizontal (x) axis, and costs and revenues are the vertical (y) axis. The total cost line on a breakeven chart will reflect the fixed costs level initially and will indicate an upward slope due to the variable cost per unit, while the total revenue line will be obtained from the origin, going up with the same slope as the price per unit and the point where the two lines intersect will be the breakeven .

The process of determining a breakeven level is one of the most critical aspects of strategic decision-making as it establishes a simple relation between cost structure and degrees of viability. Whenever managers are focusing on a new strategic business, whether it is entering a new market or developing a new product, it is prudent to ask, "how many units (or revenue) would we have to achieve to breakeven?" - as this is an effective and simple reality check in terms of overall market size and growth. For example, if a firm believes they need to sell 100,000 units of a new product to breakeven, and in assessing the marketplace, it learns that its total available market is 80,000, then the firm will most likely either perceive that strategy carries too much risk or is just infeasible unless price (or cost) assumptions change. In conducting a breakeven analysis, the firm is in effect establishing a bottom line or low-success standard against which several alternatives will be evaluated.

Breakeven analysis is a considerable factor in future pricing, cost management, and profitability, and risk assessment to some extent. Moreover, it reveals the profit that can be realized from the adjustment of volume, price, and cost. Furthermore, the marketing team can break-even-point-based analysis into various scenarios like accepting or rejecting the present price, estimating the number of units that must be sold under the new price that will be the breakeven point, or the impact of the sales situation of (pessimistic sales vs. optimistic sales) on the margin of safety, thereby re-evaluating the point of breakeven. The breakeven analysis could be used this way in the strategic planning process to help make the right decisions regarding the allocation of resources. In a way, the analysis of the breakeven point transforms the strategy's implications for costs and revenues into the clearly defined target that must be attained and thus makes it easier to compare different strategies. A financial planner has said that finding the breakeven point is like a financial map for the decision-makers since it clearly shows what the correct choice is in numerical terms rather than letting the question of whether it is right or wrong come to the strategic choices by splitting the pie and raising the issue of whether it is right or wrong. This shifts the discussion from just playing around with the product.

It's important to point out that breakeven analysis has its limitations in how it's applied. The standard breakeven model generally assumes that connections are direct, with dealing prices and costs per unit staying constant. Still, this does not hold true when you consider factors like husbandry of scale, traffic limits, and price oscillations (Rentsen et al., 2023). Another debit of the usual breakeven model is that it does not take request demand into account. Without knowing how numerous units can actually be vended, the breakeven unit results come enough pointless. It's pivotal to always put breakeven results in environment. Also, there are times when strategic enterprise might not have clear profit and cost flows that can be anatomized this way, like forming a strategic alliance or engaging in wisdom and technology systems.

In short, breakeven analysis helps estimate the fiscal viability of strategic opinions. Sara and Anastasia wrap up by noting that combining breakeven assessments with geek analysis allows for a further comprehensive approach, incorporating both qualitative factors and the fiscal impact of those opinions. For ex: if your geek analysis uncovers an occasion that could boost request demand for a product, and you decide to invest in exploring that occasion, you'd conduct breakeven assessments to see how important in deals you'd need to subsidize on it. This approach ensures that the chosen strategy aligns with your strengths and openings while being embedded in solid profitable principles.

ANALYTIC HIERARCHY PROCESS (AHP): A MULTI-CRITERIA DECISION ANALYSIS -

The Analytic Hierarchy Process (AHP) is an analytical process generated by mathematician Thomas L. Saaty in the 1970s, (Saaty, 1980) which gives decision-makers a way to utilize criteria to disaggregate the problem areas into smaller parts.

One of the methods from the MCDA family, to which AHP belongs, aims to find the best alternative considering positive and negative criteria simultaneously (Triantaphyllou, 2000; Zanakis et al., 1998). Strategic decisions are a classic example: Choosing a strategy usually means weighing financial returns, market share, risk, fit with organizational strengths, and so on. AHP helps to elucidate and numerically formulate the trade-offs involved in the decision-making process.

One of the main characteristics of the AHP method is the decomposition of the problem and its effects into a hierarchy, which is then followed by the comparison of the alternatives in pairs. The initial decision tree has three basic levels, each of which can have its own sub-levels. The top level of the tree represents the final objective (for instance, selecting the best strategy), the middle level contains the criteria (and possibly sub-criteria) for evaluating the respective alternatives, and the bottom level is dedicated to the choices of the decision-makers. Picture yourself in a scenario where you had to select a strategy. You may think of aspects like a company's financial performance, level of difficulty for the implementation, whether it aligns with your core competencies, and degree of risk, among other things. Once prioritizing these elements, decision-makers will perform pairwise comparisons of them. By doing this, they identify which ones are more significant or have greater influence.

For the case of criteria, they will initially compare pairs of those criteria to decide which ones are more significant for the task they have at hand. Following that, they will perform similar comparisons between criteria and strategies to figure out which ones are leading.These comparisons can be based on available data or insights from an expert. The Analytic Hierarchy Process (AHP) takes these pairwise comparisons and runs some calculations (using eigenvalue methods) to turn them into priority weights. Each criterion gets a weight that shows how important it is, and each alternative gets a score for each criterion that reflects its performance. The most remarkable feature of the AHP method is the structuring of the problem into a hierarchy and the pairwise comparison of the alternatives.

When utilizing AHP, the decision-making problem is depicted in a hierarchical tree format which typically comprises three levels (however, sub-levels can also be included): the initial level represents the overall objective (e.g. choosing the best strategy), the second level consists of criteria (and perhaps sub-criteria) to assess the alternatives, and finally, the third level indicates the alternative options that the decision-makers are considering. Assume that you are the one to make the decision about the strategy. You might evaluate the financial performance, the level of difficulty for the implementation, whether it is in line with your core competencies, and the degree of risk, among other things.

Top executives will determine their priorities at the first stage and afterwards they will rank the pairs of components scrutinized one by one. This approach will enable them to point out the ones with more impact or of greater value. As for particular criteria, they will initially start with pairs of those criteria for comparison to determine which ones are more significant for the assignment they have. After that, they will do the same for each criterion related to the strategies by comparing them two by two to find out which strategies are better . (Vaidya & Kumar, 2006, Ishizaka & Labib, 2011).

One thing that deserves to be pointed out is that the Analytic Hierarchy Process (AHP) is extremely strong in the area of group decision-making. AHP brings together and smooths out the different methods and fields by asking the expert's opinion of many, either by coming to a consensus or by "fine ". This approach is particularly useful in strategic planning conducted by operation brigades. It compels decision-makers to clarify their logic for choosing one option over another, thereby revealing any underpinning hypotheticals that may not be incontinently egregious.

In summary, AHP stands out as a robust and unprejudiced quantitative tool for assessing different criteria, which is essential when facing tough opinions that frequently bear a comprehensive understanding of the entire value diapason. AHP complements geek and fiscal analyses by incorporating criteria that may not be rigorously fiscal or fluently measurable. It uniquely bridges the gap between qualitative geek factors and quantitative decision issues, allowing for the assignment of numeric weights to SWOT rudiments and easing the ranking of strategic druthers grounded on these factors through the operation of AHP (Kurttila et al., 2000).

SWOT-AHP INTEGRATION: QUANTIFYING SWOT FOR STRATEGY SELECTION

The fusion of SWOT analysis and AHP resulted in the development of integrated techniques that are called SWOT-AHP or A'WOT in most cases (Kurttila et al., 2000). The idea that underpins this concept most is the use of AHP to create a merger of SWOT and AHP, which then generates both qualitative and quantitative evaluations for each of the SWOT components. If you wish, you can also decide on strategic options based on these weighted factors. The combination of methods not only keeps the broad situational view of the SWOT but also adds the numerical precision of the AHP. The first step of the SWOT-AHP process is the identification of the SWOT factors that, as a rule, are the result of expert brainstorming, data analysis, and traditional discussions. The factors which have been recognized are then organized in a hierarchical manner according to the AHP approach. A typical individual would have four divisions corresponding to Strengths, Weaknesses, Opportunities, and Threats, and he/she would enumerate the factors under each of the categories.

The AHP process is divided into two main stages and gets underway from there.

- **Weighing the Factors:** The method of pairwise comparisons means that we determine the importance of factors not only within a specific category of a SWOT analysis but also across all four categories. To be more specific, a professional might for instance compare the factors of strength one by one to find the one that is actually going to provide the most support to the organization, and then that factor will be given the corresponding weight. A similar process is done if you have weaknesses, opportunity, and threat. You can even compare the SWOT categories themselves such as deciding whether to put your effort into exploiting strengths or into addressing weaknesses. Hence, you have a set of weights that lets you have a ranked list of all the SWOT factors on a common scale. This allows you to have such questions answered as: "Which strengths are the most useful?", "Which weaknesses need immediate fixing?", "And which opportunity or threat should be given more ?"
- **Evaluating Strategic Alternatives:** If there are any strategic options created (for instance, Option A: investing in new product development; Option B: market expansion), they can be positioned at the bottom of the AHP hierarchy. Besides, a SWOT analysis is used as a basis for assessment of the options which consist of the four aspects: strengths, weaknesses, opportunities, and threats of the respective alternatives. Simply speaking, it can be illustrated as just interrogating: "How does A affect B regarding Opportunity 1?" or "What would be the impact of A on Weakness 2 through B?".

The options will finally be ranked by the performance scores based on the different criteria as a result of such . By adding up these scores according to their respective weights from the initial phase, we can determine an overall attractiveness score for each strategic option.

The main advantage of this method is that it formally integrates the SWOT factors into the evaluation of alternatives. Instead of just creating a SWOT list and then discussing the next steps, the SWOT-AHP approach encourages decision-makers to clearly highlight the trade-offs involved. Imagine, for example, an extreme risk scenario is set up and given a heavy weight, then the persistence evaluation will be at least partially positive already. On the other hand, if a certain point is very strong, then the method that uses that point will eventually, score the highest among others. Overall, the evaluation can result in such surprising outcomes as: "At first, Strategy X seemed to be the best choice, however, our weighted assessment indicates that Threat Y (which has a high weight) is not addressed by X, thus making Strategy Z the most appropriate one in the end." Here, the SWOT-AHP method can sometimes layer a factor that was overlooked beneath the surface, thereby, figuratively, knocking over an intuitive .

One of the most significant advantages of the SWOT-AHP approach is its capacity to offer both qualitative and quantitative insights. It keeps the essence of the SWOT analysis intact, making it easy to communicate, like saying, "these are our strengths and threats", while also providing solid numbers to support strategic choices. Decision-makers receive a prioritized list of factors along with a clear ranking of options based on those factors. This method conducts a SWOT analysis to shape the strategy, bridging the gap between evaluation and decision-making.

That said, there are some practical challenges: making a lot of pairwise comparisons can get pretty time-consuming, especially if the SWOT list is extensive. However, the number of truly critical factors is often quite small and manageable, usually around a dozen key points. Using software tools can simplify the AHP calculations and consistency checks, speeding up the whole process. Ultimately, the outcome of measurable strategy assessment is generally seen as worth the time and effort, especially when the stakes are high.

To sum up, the combination of SWOT and AHP introduces a large part of measurement to a traditionally qualitative SWOT analysis. Using the technique referred to, it is feasible to pick those strategies which not only fit like a glove with our internal strengths and external opportunities but also address the weaknesses and threats. By assigning weights to the SWOT factors and subsequently assessing the strategies in that weighted framework, it is possible to make more accurate decisions. This mixed-method approach really highlights the main theme of this chapter: transforming qualitative strategic insights into quantitative analysis to enhance decision-making support.

QUANTITATIVE STRATEGIC PLANNING MATRIX (QSPM): CHOOSING STRATEGIC ALTERNATIVES BY IMPORTANCE

The QSPM, or Quantitative Strategic Planning Matrix, is a fantastic tool for evaluating strategic options in a quantitative way by ranking them (David, 1986). While AHSOT (with or without AHP) is excellent for identifying factors and discussing the qualitative aspects of alternatives, QSPM elevates the process with a more detailed analytical approach. It uses weights and ratings in a matrix format to compare different strategies. Although it may not be as commonly recognized in practice as SWOT or AHP, QSPM remains a staple in strategic management literature and case studies. Its strength lies in its straightforward yet scientifically rigorous method, which simplifies the task of assessing the relative appeal of various strategies based on key internal and external factors (David et al., 2009).

The development of a QSPM is a composite process that comprises steps closely linked to earlier analysis:

1. **Listing Key Factors:** The left side of the QSPM highlights the key external threats and opportunities, as well as the internal strengths and weaknesses of the company. Usually, these components emerge from prior evaluation tools like a detailed SWOT or EFE and IFE matrices. The use of a factor variety, which may include up to ten for internal and ten for external, is a standard that provides a thorough and precise summary. This process brings about the certainty that no critical aspect is overlooked.
2. **Assigning Weights:** Besides each factor, there is a weight that shows the relative importance of that factor compared to others. In most cases, the weights originate from a previous analysis, such as SWOT or IFE/EFE matrices, or they can be determined by a method like AHP. Most of the time the total of all weights is equal to 1, which is a way of indicating the overall importance of each factor. Thus, the weight of a considerable opportunity can be 0.15 and the weight of a slight weakness only .05.
3. **Identifying Alternative Strategies:** The upper portion of the QSPM presents all of the different strategic alternatives to be evaluated in this phase of the process. The different strategic alternatives could include a group of strategic options that are mutually exclusive to one another, or alternative approaches that the company may take. Overall, these alternatives have resulted from the prior stage of strategy development (i.e., from a SWOT/TOWS Matrix or some other framework). For example, the strategic alternatives could include "Market Expansion," "Product Diversification," or "Improved Operational Performance" depending upon which sector was found to be the most important.
4. **Assign Attractiveness Scores (AS):** Each and every method is evaluated against each factor with the evaluation indicating how the method could utilize or counter the factor. The normal evaluation scale varies from 1 to 4 (1 = not attractive at all, 2 = attractive, 3 = quite attractive, 4 = very attractive). If the factor is not at all related to the decision of strategies, a 0 or an "—" could be put instead of the score. The main inquiry that needs to be raised is: Will this factor play a role in the ruling out of the alternatives of tactics? If so, how much preference would a certain tactic get over the others in conjunction with this factor? To illustrate, if the factor is "Strong R&D capacity" (which is a strength), then a strategy based on R&D such as the launching of a new technology product would get the highest score (because it can fully utilize R&D), while a cost-cutting strategy would receive the lowest score (as R&D is not involved at all) from the other side.
5. **Computeting the Total Attractiveness Scores (TAS) calculation:** The total attraction score to different factors is simply deriving the attractiveness score of a particular strategy and then multiplying by the weight of the factor, producing thus the Total Attractiveness Scores (TAS) for the factor. This figure reflects

how much that factor adds to the overall appeal of the strategy. Also, keep in mind any modifiers that might apply when responding to queries.

6. **Sum the Scores:** To evaluate each strategy, simply sum up the Total Attractiveness Scores (TAS) calculated for each factor. This will give you the overall attractiveness score for that strategy. Essentially, this score reflects how appealing each alternative is when you take into account the weighted factors. Based on the assumptions we've made, the alternative with the highest total researcher score is considered the best option. Additionally, the normalized difference between the scores of two alternatives across the entire researcher population indicates how much more attractive one option is compared to another.

A final QSPM provides a clear visual comparison of various strategies. It not only points out the winning strategy but also explains why it stands out by showing how each factor contributed to the overall score. This allows decision-makers to identify a specific strategy as the best option, especially if it excelled in a few critical areas that hold significant weight—like a major strength or a key opportunity. On the flip side, they can understand why another strategy didn't make the cut, as it was adversely affected by a high-weight factor that represented a threat. This level of clarity is essential for backing up a strategic recommendation.

It is very important to point out that QSPM, similarly to AHP, relies on the subjective nature of the inputs when scores are involved. Thus, its objectivity is limited to the quality of the judgments in this case. Still, by organizing the judgments in a matrix, QSPM ensures that all factors are considered and that the discussion is based on objective evaluations rather than vague impressions. The research literature has provided evidence that as a result of QSPM's unique combining of 'intuition and analysis' in the strategy of decision-making, it is suggested that the QSPM tool be used more often in the domain of strategy management (David et al., 2017; Nutt, 1993; Nutt, 2001).

PRACTICAL EXAMPLES AND RELEVANT CASE STUDY

Case Study: Public Sector – City Transit Initiative

In the public sector, strategic choices frequently mean choosing to implement a policy or investing in a project that will have social or economic effects. For example, a transportation department run by the local government is essential to the city government and is looking for different ways to improve urban mobility. After a thorough strategic assessment of public transit in the city, the city employees have prepared two main scenarios to be evaluated:

- Strategy X: Build a Light Rail Transit (LRT) Line - Set up an electric light rail line to link the central parts of the city.
- Strategy Y: Larger Bus Network - Improve the bus service with more routes along with higher frequency and new buses (which will also include electric buses in the fleet for a greener and more eco-friendly service) as well.

Although both options are designed to reduce traffic and improve transit use while reducing emissions and congestion, they will be built differently, with different costs, risks, and timelines to get to implementation. The light rail transit (LRT) is a capital project that has high upfront costs on the infrastructure side to build. The LRT comes with significant fixed costs on the operational side, mainly because of the hefty one-time capital investment needed for the rail infrastructure. In contrast, the bus expansion takes a more gradual and adaptable approach, allowing it to kick off sooner and be modified easily over time. While investing in bus services might not yield the same capacity and efficiency as investing in fixed rail systems, the bus expansion offers the advantage of time and flexibility for future adjustments.

SWOT Analysis: The transit planning team in the city has identified the following key points:

Strengths (internal)

1. The city's financial condition is good, and the present tax base is fair, so the budget for different projects is being opened and is supporting them.
2. The transit agency has excellent operating experience with the bus systems, and it extensively shares a good deal of operational wisdom with the project.
3. The city council is very much supportive of public transport, and the first-class leaders are totally in line with the transit plans.

Weaknesses (internal)

1. The city lacks the necessary in-house expertise for rail construction and operations and as a result, it has never operated any light rail transit (LRT) system before.
2. The transit agency has quite typically bureaucratic tendencies which eventually slow down the project and lead to inefficiency in the process.
3. The best solution is still not obvious as all the current maintenance facilities are already old and will need to be refurbished regardless of the chosen .

Opportunities (external)

1. There is federal infrastructure grants specifically designated for rail transit projects, presenting a significant funding opportunity that could help cover LRT costs.
2. The emergence of new electric bus technology and decreasing prices for e-buses could transform the performance and emissions of the bus fleet, offering a major opportunity for the bus strategy
3. Rapid population growth in certain underserved suburbs is driving an increasing demand for transit connections, and either strategy could effectively address this need.

Threats (external)

1. Risk of cost overruns and construction disruptions, it is a fact that rail projects in other cities have often exceeded budgets faced community opposition during construction, a significant threat to the LRT strategy.
2. Volatile fuel prices, in case diesel prices spike, running a large diesel bus fleet becomes more expensive, threat that can affect the bus strategy, although this would be mitigated by electrifying the fleet.
3. Uncertain ridership trends are post-pandemic, in case remote work or lower transit ridership persists, it could undermine the benefits of significant investments, this threat can affect both strategies, though costly LRT will be harder to justify in case ridership doesn't recover.

Breakeven / Cost-Benefit Considerations:

In public projects, the phrase “break even” describes the moment when the total benefits like saved travel time and lower emissions—begin to surpass the costs. Figuring out social benefits can be a bit complex, but the team made a rough comparison of the cost-effectiveness of two strategies:

> LRT (Strategy X): The planned light rail line will probably cost close to $1 billion, and additionally, it is estimated that about $15 million will be needed every year for the line's operation and maintenance after it starts providing service. The line is expected to serve around 50,000 passengers daily after a few years of its opening, the planners figured out the time before the advantages would equal the huge . They found that the LRT's payback period in terms of social benefits would be roughly 20 years, which is standard for large construction projects.

Bus Expansion (Strategy Y): On the flip side, the new bus network is likely to require an investment of about $200 million over several years. The project is planning to include new buses, bus lanes, and bus depots in addition to $5 million that is being set aside annually for better service provision. The possibility is that the number of passengers may increase to the point where the service will have around 30,000 users daily. The experts made a rough calculation that the social breakeven point of the investment would be at around 10 years, which is still way much faster than the rail project because of its small initial costs and the quick implementation of the improvements therefore, the good effects would be coming in sooner.

The bus expansion in terms of cost-benefit analysis is the less expensive alternative; it would generate net positive benefits much earlier than the LRT, which is a long-term infrastructure and environmental investment. Quantitative SWOT analysis can, therefore, be a tool for evaluating the pros and cons.

SWOT-AHP Analysis:

The planning group was able to employ the AHP method very well in order to prioritize the SWOT elements and make a direct comparison of the two strategies. Their investigation brought out quite a few deep and fresh understandings: Among the major Strengths that were most pointedly highlighted by the highest importance rating was the ability of the city to provide money projects. This is vital because funding is a key consideration for any major project to even be on the table. While both options will require cash outflows, the LRT option will take up a much larger slice of the budget.

The sound financial position permits LRT at all, but also, it raises the question: Is it an excellent idea to allocate the capacity this way? The weight on this element did not favor either of the strategies but gave a financial context: any option should be in line with finance restrictions. The operational experience factor, which basically shows how beneficial the bus system is for the city's experience, was one of the key factors that had a huge influence on the decision to go with a bus system instead of a railway system. The city is very well acquainted with the bus operations, and this is a major advantage for Strategy Y. While the deficiency of handling rail operations has become a very serious problem for Strategy X. To deal with the situation, the city may need to hire outside experts, create a partnership with a local transit authority, or look for other alternatives, which in total would result in higher levels of difficulty and risk if the city is not already an expert in rail. This situation clearly leans the bus strategy in our analysis.

Another factor that needed consideration was the chance of receiving federal rail grants. If the city succeeds in obtaining a grant funded by the federal government that covers about 30-50% of the costs of the LRT, the local LRT project would be a lot more . However, this opportunity seemed to heavily favor Strategy X (LRT). Unfortunately for Strategy Y (bus), only generic transit grants would be available, while significant rail-specific funding would be completely out of reach. The e-bus technology opportunity was also on the table, but it was seen as slightly less critical than the rail funding. E-bus costs are decreasing and the tech is improving continually, hence the bus strategy is getting more and more practical. Total bus replacement by the city would be possible together with a significant reduction in emissions, all that without the installation of any . This factor strongly backs Strategy Y.

When it comes to suburban growth, both strategies have similar potential. The rapidly expanding suburban areas could be served by either an LRT line or an enhanced bus service. While an LRT might have a slight edge if the suburbs develop along the rail corridor, overall, this isn't a major differentiator between the two strategies.

- The threat of being overrun was taken veritably seriously. The team had gathered substantiation from analogous systems in other metropolises, revealing that the costs for rail systems had frequently doubled, and completion times had indeed tripled. This poses a significant challenge for Strategy X. On the wise side, while machine peregrinations are not without their issues, they do offer further inflexibility and can be gauged back more fluently if demanded. This trouble significantly impacted the score for LRTS.
- The rising cost of fuel has always been a bit of a headache for bus companies, especially since diesel prices can be so unpredictable. But the bus company's decision to switch to electric buses is a smart move to help ease that worry. They might not have thought too much about this issue; after all, light rail transit (LRT) doesn't really feel the pinch of fluctuating oil prices (which is definitely a plus for them). On the other hand, buses could still face some challenges during the transition to electricity, but it's a minor drawback in the grand scheme of things.
- The uncertainty around ridership was considered relatively important. Both strategies could run into trouble if conveyance ridership doesn't recover after the epidemic. Still, there's a belief that light rail conveyance (LRT) might be more at threat, if people don't show up as anticipated, a$ 1 billion rail line could turn into a major issue. In discrepancy, machine services offer further inflexibility and can be acclimated or gauged back if demand is not there. This gave Strategy Y a slight advantage

After the weights have been given and each factor has been scored for each strategy, the analysis showed a slight preference for Strategy Y: Expand the Bus Network. To illustrate, the lack of rail expertise and high risk of overruns weighed so heavily against LRT that even the attraction of federal funding and the long-run payoff of rail did not cancel it out entirely. In contrast, the bus strategy was scored favorably on many factors (internal readiness, flexibility, technology opportunity) but had fewer major drawbacks. In numbers, the team might have a composite score of the team being, say, 0.52 vs 0.48 of LRT - a slight but unmistakable bias towards Strategy Y.

QSPM Snapshot:

The team produced a veritably simple and straightforward QSPM to display the evaluation issues. To demonstrate

1. Financial Capability (weight ~0.10): Both methods got a thumbs up, mainly because the city is in a solid financial position. At this moment, LRT 4 line comes out as the winner even though it is quite expensive and it would absorb almost the entire amount of funds available. In contrast, Bus 3 would be a cheaper alternative, and it is good enough to not only gain the opportunity to use the money for other projects but also to reduce the overall cost. Nevertheless, this factor eventually was not very significant in making the final choice between the two options.
2. Rail Expertise Weakness (weight ~0.15): The LRT scores a lowly 1 in its assessment, which really points out a major flaw, the city simply lacks the know-how when it comes to rail systems. Contrary to that, the Bus system is said to have an excellent score of 4, meaning that it is functioning in very good condition. Due to this situation, the Bus is definitely way ahead of the LRT whose last score is only 0.15, while the Bus has a wonderful .60.
3. Federal Rail Funding (weight ~0.20): LRT AS = 4 (this is a very significant opportunity that could be of great help in reducing costs), Bus AS = 1 (not useful). Light rail transit clearly comes out on top (LRT TAS 0.8 versus Bus .2).
4. Electric Bus Tech (weight ~0.10): LRT AS scores a 2, which suggests it's somewhat relevant to rail systems. Bus AS, a public transport vehicle, reacts to the criticism in quite a different way by achieving an amazing top score of 4, which denotes that it is a highly and considerably influential entity for the growth of the coming bus networks. The bus is doing very well compared to LRT which is at 0.2 and Bus at .4.

5. Cost Overrun Risk (weight ~0.20): The evaluation of the LRT is carried out in the most unfavorable way, giving only 1 AS to the LRT as per the risk levels, while the Bus is rated on the safer side with a 3 AS. The evaluation has clearly favored the Bus, giving 0.2 points to LRT and 0.6 points to Bus.
6. Fuel Price Volatility (weight ~0.10): The score of the evaluation for LRT is 4 because, in the case of LRT, fuel price changes do not affect the system at all or nearly so. Conversely, the buses receive a score of 2 because they are only halfway through the process of electrification and thus are still affected by the fluctuations of fuel prices. Consequently, the difference in benefits between LRT and other modes of transport is nearly nothing (0.4 to .2).
7. Ridership Uncertainty (weight ~0.10): The LRT received a score of 2, which means that the fixed rail line could face hardships in case the passenger number is far less than the anticipated count. Buses scored 3 points but still were given the option of changing the routes in case a particular area had very few passengers. Thus, it seems that the situation here is a little bit in favor of the buses (0.2 to 0.3).

The total of these scores (plus some other scores not mentioned) might lead the QSPM to give an approximate overall result: LRT total ≈ 2.0, Bus total ≈ 2.3. This indicates that the bus network extension would be the least undesired option in terms of quantity throughout the entire.

Decision:

The city chose at last to go with Strategy Y: Expand the Bus Network as the immediate action. The quantitative analysis revealed that the LRT option while still having significant advantages (primarily through federal funding that was secured) was in fact just a high-risk project for the time being due to uncertainties and the city's internal restrictions. However, the bus network expansion was a project that was more manageable for the city being in control and expertise and it had a speedy return in terms of service and benefits improvement. Moreover, it could serve as bridge-increasing ridership and demonstrating demand which later could turn into a rail line being solid with less risk in the case.

In conclusion, this public sector case study reveals that a quantitative SWOT-based analysis can serve as a decisive factor not only in unprofitable business decisions but also in the case of government strategic decision-making. By assigning weights to factors like funding opportunities, internal capabilities, and risks, the government agency can make strategic choices that are supplied and better acquainted with the realities. The method assured that the selected strategy (bus network expansion) was not only politically the road of least resistance but also one through a multi-faceted analysis. It also provided a written justification for why the alternative (the rail line) was postponed for now - mainly due to the issue of readiness and risk - while still leaving the door open to re-evaluate a rail project in the future if the conditions change (for instance, if the city gains rail expertise, or if external funding becomes so generous that it tips the balance in favor of LRT).

SYNTHESIS: CONNECTING THE THEORY AND PRACTICAL RENDITIONS

1. **Comprehensive Evaluation:** The integration of SWOT analysis with quantitative methods makes sure to have a totally complete viewpoint. Every single internal and external factor can be evaluated very clearly along with their corresponding weight; thus, no important strengths, weaknesses, opportunities, or threats are left out in the decision-making process.
2. **Objectivity and Transparency:** The methods of using quantitative weighting and scoring apply strictness and reveal hidden assumptions. Managers must rank the elements which lessen the influence of bias or intuitive judgments. The reason behind the selection becomes clear and justifiable, since the parties involved can understand why one method was rated higher than another regarding the essential factors.
3. **Strategic Alignment:** The approach confirms that the chosen strategy corresponds with the company's internal strengths and the external market situation. A strategy that does not include major strengths or does not work through essential threats will have a low score, which will be taken as a lack of alignment. By prioritizing the SWOT factors, companies can more confidently choose a strategy that is more advantageous to them and less damaging to their weaknesses.
4. **Critical Insight and Adaptability:** The analysis reveals the various influencing factors and at the same time indicates the points that need to be taken) to ensure success of the strategy (e.g. the required sales volume, etc. prevention of the main risks). The latter are considered as the issues to be emphasized during the implementation phase, for example, a threat occurs, or an opportunity enlarges.
5. **Team Collaboration and Buy-In:** The application of a methodical technique like AHP or QSPM will result in the engagement of different departments in the definition of the criteria and assessments of the options. The collaboration in the analytical approach leads to a shared perception of the choice. It strengthens the decision more as the contributors come to know that their concerns are considered in the appraisal and it also results in a more knowledgeable and cohesive execution of the chosen tactic.

THE BARRIERS AND POSSIBILITIES ENCOUNTERED IN THE EXECUTION

Barriers:

Subjective Judgments:

The human inputs are still crucially the weights and scores, which can be biased or inconsistent, to the analysis. Strong personalities or pre-existing notions might have a role in the result. In this light, it would be beneficial to get various stakeholders involved and to utilize such methods as anonymous scoring or consistency checks, like in AHP, to minimize the subjectivity and to guarantee a more objective method.

Data Quality and False Precision:

The output is only as good as the data and estimates provided. Poor data can include uncertain market size estimates, cost assumptions, or the like that provide misleading results. Quantitative output gives a possibly misleading sense of precision. Results should be treated as approximate. Sensitivity analyses should be done to see if small changes in weights will alter conclusions. Quantitative results should be provided along with qualitative insight and the uncertainty.

Complexity and Effort:

A complete SWOT-AHP-QSPM process can take long since these are painstakingly long processes, especially if too many factors or comparisons are included. Small organizations or rapid decisions may not favor this. This can be mitigated by consideration of the most important factors based on the Pareto principle and using software to automate and thus simplify pairwise comparison and calculations. Nevertheless, new advances are developing possibilities for augmenting this approach despite these challenges: advanced MCDA techniques, including ANP and fuzzy logic, can model more complex relationships among criteria; AI and data analytics are enabling automated EI and scenario analysis; real-time dashboards can facilitate real-time updates of SWOTs; and, specialist software does alleviate some of the analysis burden when working with teams. These trends together make quantitative strategic planning not only simpler but also significantly more effective.

FUTURE RESEARCH AND STRATEGIC IMPLICATION

There are some thrilling directions to venture into to improve the quantitative SWOT method:

Advanced Decision Methods: We could implement voting methods like the Analytic Network Process (ANP) which takes influencing factors into account, CVLM (outranking) methods such as ELECTRE and PROMETHEE alongside AHP. Incorporating these methods would make the comparison of different strategies more detailed and consequently more efficient.

Simulation and Risk Analysis: SWOT combined with simulation modeling (e.g., Monte Carlo simulations) would be a considerably better way to manage uncertainty. Instead of single-point assessments, strategies would be evaluated based on the probability of meeting objectives. The decision-maker would then have the opportunity to select the alternatives that are still viable amidst the unforeseen conditions.

AIand Automation: Solving the problem of how humans and machines can effectively interact in the decision-making process is the most pressing challenge that executives are confronted with in their organizations. The latest development is that AI has been viewed as a possible assistant in this respect by fetching different options and recommendations to the human leader based on the data analysis. So, not only is the decision-making process being done in a faster way, but also companies are able to take into account new risks or opportunities that have arisen because of the changes while they are still .

Dynamic and the Continuous Planning: Analyzing a SWOT event done separately is quite a contradiction to the second, later, dynamic, and continuous approach. One of the many advantages that organizations would receive is that the real-time data and regular updates would keep the SWOT matrix and related analysis moving, allowing them to always reflect the new situations not only in terms of the changes in strategies but also in the sense that if the latter approach were used there would be no necessity to worry about each little shift being overreacted.

Definition of Criteria and Culture: The changes brought about by the strategic planning process might be so fundamental that even geopolitics and human welfare could be accounted for as a part of the numerical mix. However, the most significant question would be how far sustainable practices would become the mainstream ones. Companies will become the providers of the necessary credits for the acquisition of knowledge and will not only support it but also develop a culture that takes insights from data combined with intuition. In addition, they will be given the freedom to use advanced tools to the maximum of their .

CONCLUSION

This is chapter demonstrates that combining qualitative strategic insight with quantitative analysis leads to a far more efficient strategic decision-making process. Through converting the strategic issues which connect the advantages and disadvantages of the firm to its profitability into figures, someone gets a better comprehension of the trade-offs and the success conditions. The strategies, mostly due to this process, would be those which were viewed as more conforming to the core strengths of the organization and to the favourable factors in the market. Moreover, if the justification for every strategic alternative is backed by precise quantification and proper documentation, the stakeholders will be more inclined to comprehend and consent the selected strategy.

It is, nonetheless, important to highlight that although data-based models should occupy the central role in strategic planning, they should not completely eliminate the function of managers' instincts. On the contrary, they should provide backing to the intuition and, at the same time, sharpen it. The intuitions are grounded on analytical evidence and, thus, it is, as in the case of setting performance targets or determining the break-even point, the "place of truth". As a result, the top management can not only confidently choose a strategy but also provide a detailed rationale for it by integrating the expert opinions with the reliable data to make decisions that are not only correct but also gain the support of a .

Moreover, formal analytical frameworks enable organizations to navigate complex scenarios with many moving parts as part of their structured analysis. The processes of making decisions have been enriched by the incorporation of quantitative criteria and scoring methods which allowed the previous intuitive methods to be rationalized, scrutinized, and hence, to become more responsible, and to foster a learning culture in the organization. Consequently, firms will prefer to allocate their resources to projects that have clearly and quantitatively proved their worth rather than relying on questionable or unproven .

In addition, the utilization of qualitative and quantitative approaches not only encumbers more participants in the collaborative decision-making but also aids in the establishment of a congenial atmosphere. The usage of analytical tools that mix and transfigure the opinions of various professionals into numbers, grants to the whole organization a unified comprehension of the rationale behind the choices. The sense of joint ownership is very crucial for the quick track and non-stop execution of the strategic plans. When the premises are correctly recorded and the judgment criteria together with their corresponding weights are distinctly stated, the groups can also revisit their choices to pinpoint those that resulted in triumph and the ones that did not, thus, they keep developing their strategic skill set through the .

The combination of subjective perception and objective evidence is particularly advantageous for those enterprises that are globally impacted by market and technology issues and are changing rapidly. Such companies can be both quick and cautious at the same time, thereby making sure that their decisions are not only timely but also backed by evidence and in harmony with the company's mission. One prominent way of improving analytical decision-making is by integrating formal education and life experience, which continues to develop the decision-making skills of the managers. Hence, data-driven strategies were closely associated with their performance measurement criteria, thus giving managers the ability to monitor the outcomes and make changes if .

This method, which improves the quality of on-the-spot decisions, is also a significant factor in the creation leaders who are going to inherit the leadership ladder. These fresh leaders will be able to dissect information, anticipate issues, and organize their tactics with .

The application of a data-driven method surely upgrades a firm's internal communication. Yet, it is conditional on a situation where managers practice clarity and rely on data. In such a scenario, they will not only express their opinion conveniently but also give reasoning. Thus, among the different parties involved, trust and cooperation will be built up. The holding together of all the communication lines and the focus on idea execution as well are the main advantages that effective communication has been rendering to the organizations. A company that can operate swiftly and flexibly in such trying times is not only able to survive but also to thrive. The company can utilize different tools such as SWOT analysis, break-even analysis, and multi-criteria decision analysis not just to predict and react to events but also to optimize their resource use. This openness to new ideas, which is a characteristic of the progressive attitude, allows companies not only to react but also to take the lead in their strategies. Ultimately, the very core of this transformation is the combination of strategic creativity with strict control.

In the end, the integrated model discussed in this chapter is very useful and can be employed in various industries and decision-making contexts. The core idea of making a strategic decision based on both qualitative insight and quantitative evidence is still very strong, whether it is in manufacturing, the services sector, or the public sector. It can be transformed and extended according to the new requirements while still fulfilling its primary function of enhancing the quality of decisions. By fusing data and expert judgment, organizations can come up with strategic choices that are not only well-grounded and convincing but also are feasible to be implemented. The resulting plans are often solid, achievable, and geared towards long-term success, which is a demonstration of how a quantitative perspective can be a means of turning strategic planning into a rigorous, value-driven endeavour rather than a speculative .

REFERENCES

Belton, V., & Stewart, T. J. (2002). *Multiple criteria decision analysis: An integrated approach*. Springer., DOI: 10.1007/978-1-4615-1495-4

David, F. R. (1986). The strategic planning matrix—A quantitative approach. *Long Range Planning*, *19*(5), 102–107. DOI: 10.1016/0024-6301(86)90015-4

David, M. E., David, F. R., & David, F. R. (2009). The quantitative strategic planning matrix (QSPM) applied to a retail computer store. Coastal Business Journal, 8(1), Article 4. https://digitalcommons.coastal.edu/cbj/vol8/iss1/4

David, M. E., David, F. R., & David, F. R. (2017). The quantitative strategic planning matrix: A new marketing tool. *Journal of Strategic Marketing*, *25*(4), 342–352. DOI: 10.1080/0965254X.2016.1148763

Dean, J. W., & Sharfman, M. P. (1996). Does decision process matter? A study of strategic decision-making effectiveness. *Academy of Management Journal*, *39*(2), 368–392. DOI: 10.2307/256784

Eisenhardt, K. M., & Zbaracki, M. J. (1992). Strategic decision making. *Strategic Management Journal*, *13*(S2), 17–37. DOI: 10.1002/smj.4250130904

Helms, M. M., & Nixon, J. (2010). Exploring SWOT analysis – where are we now? A review of academic research from the last decade. *Journal of Strategy and Management*, *3*(3), 215–251. DOI: 10.1108/17554251011064837

Ishizaka, A., & Labib, A. (2011). Review of the main developments in the analytic hierarchy process. *Expert Systems with Applications*, *38*(11), 14336–14345. DOI: 10.1016/j.eswa.2011.03.063

Ishizaka, A., & Nemery, P. (2013). *Multi-criteria decision analysis: Methods and software*. Wiley.

Jakupi, S., Statovci, B., & Hajrizi, B. L. (2017). Breakeven analysis as a powerful tool in decision-making. *International Journal of Management Excellence*, *9*(3), 1169–1171. DOI: 10.17722/ijme.v9i3.940

Kurttila, M., Pesonen, M., Kangas, J., & Kajanus, M. (2000). Utilizing the analytic hierarchy process (AHP) in SWOT analysis: A hybrid method and its application to a forest-certification case. *Forest Policy and Economics*, *1*(1), 41–52. DOI: 10.1016/S1389-9341(99)00004-0

Mardani, A., Jusoh, A., Nor, K. M., Khalifah, Z., Zakwan, N., & Valipour, A. (2015). Multiple criteria decision-making techniques and their applications – A review of the literature from 2000 to 2014. *Economic Research-Ekonomska Istraživanja*, *28*(1), 516–571. DOI: 10.1080/1331677X.2015.1075139

Nutt, P. C. (1993). Identifying solution ideas during organizational decision-making. *Management Science*, *39*(9), 1071–1085. DOI: 10.1287/mnsc.39.9.1071

Nutt, P. C. (2001). A taxonomy of strategic decisions and tactics for uncovering alternatives. *European Journal of Operational Research*, *132*(3), 505–527. DOI: 10.1016/S0377-2217(00)00141-7

Nutt, P. C. (2005). Search during decision making. *European Journal of Operational Research*, *160*(3), 851–876. DOI: 10.1016/j.ejor.2003.07.009

Nutt, P. C. (2008). Investigating the success of decision-making processes. *Journal of Management Studies*, *45*(2), 425–455. DOI: 10.1111/j.1467-6486.2007.00756.x

Phadermrod, B., Crowder, R. M., & Wills, G. B. (2019). Importance–performance analysis based SWOT analysis. *International Journal of Information Management*, *44*, 194–203. DOI: 10.1016/j.ijinfomgt.2016.03.009

Pickton, D. W., & Wright, S. (1998). What's SWOT in strategic analysis? *Strategic Change*, *7*(2), 101–109. DOI: 10.1002/(SICI)1099-1697(199803/04)7:2

Rentsen, E., Chuluundorj, B., & Natsagdorj, T. (2023). Optimization approach to constrained breakeven points with respect to price. iBusiness, 15(3), 183–191. https://doi.org/DOI: 10.4236/ib.2023.153014

Saaty, T. L. (1980). *The analytic hierarchy process: Planning, priority setting, resource allocation* (2nd ed.). McGraw-Hill., https://archive.org/details/analytichierarch0000saat

Triantaphyllou, E. (2000). Multi-criteria decision making methods: A comparative study. Kluwer Academic Publishers. https://archive.org/details/multicriteriadec0000tria

Vaidya, O. S., & Kumar, S. (2006). Analytic hierarchy process: An overview of applications. *European Journal of Operational Research*, *169*(1), 1–29. DOI: 10.1016/j.ejor.2004.04.028

Weihrich, H. (1982). The TOWS matrix—A tool for situational analysis. *Long Range Planning*, *15*(2), 54–66. DOI: 10.1016/0024-6301(82)90120-0

Zanakis, S. H., Solomon, A., Wishart, N., & Dublish, S. (1998). Multi-attribute decision making: A simulation comparison of select methods. *European Journal of Operational Research*, *107*(3), 507–529. DOI: 10.1016/S0377-2217(98)00128-3

Chapter 6
Machine Learning-Based Budget Prediction for Quantitative Decision-Making:
A Multi-Dataset Comparative Study

Yasmine El-Bakkouri
Research Laboratory in Management Sciences of Organizations, National School of Business and Management, Morocco

Imane Aitouhanni
https://orcid.org/0009-0004-1400-8027
Mohammed V University, Rabat, Morocco

Serkan Derici
https://orcid.org/0000-0003-2581-6770
Nevşehir Hacı Bektaş Veli University, Turkey

Jalal Azegagh
National School of Commerce and Management, University Ibn Tofail, Kenitra, Morocco

ABSTRACT

Accurate budget prediction is a fundamental challenge for quantitative decision-making in both public and private sectors. Traditional statistical approaches often fail to capture the complex, nonlinear relationships that characterize modern financial and economic data. This chapter investigates the effectiveness of machine

DOI: 10.4018/979-8-3373-6746-0.ch006

learning-based regression models for budget prediction using three heterogeneous real-world datasets related to construction costs and film production budgets. Linear Regression, Random Forest, and Gradient Boosting models are implemented and evaluated using standard performance metrics, including MAE, RMSE, and the coefficient of determination (R^2). The experimental results demonstrate that ensemble-based models significantly outperform linear methods across all datasets, with Gradient Boosting providing the highest predictive accuracy and robustness. The findings confirm the nonlinear nature of budget estimation problems and highlight the strong potential of machine learning as a numerical tool for supporting quantitative financial decision-making.

INTRODUCTION

It is possible to manage the financial risk and planning at all levels of an organization by using budgeting forecasts as a tool for quantitative decision-making (Grobler-Dębska et al., 2025). Having accurate and timely projections of the needs will allow an organization to increase its operational efficiency and continue its effectiveness and sustainability with its policies/programs (Gallego-García & García-García, 2021). For example, construction companies that rely on inaccurate cost estimates will typically exhibit the following common characteristics: substantial unanticipated costs, delayed project timelines, or an inability to meet their financial commitments for a project, accordingly (Oberlender & Trost, 2001). Film producers also rely heavily on the production budget to estimate the scale and potential profitability of the work they create (Bennett & Grabs, 2025). In addition, accurate and timely financial forecasts are critical for providing a stable platform for sustainable investments and policy decisions related to infrastructure development and enhancements (Rezaei et al., 2025).

The sheer volume of available complex and mixed types of data (Big Data), as well as the growing trend toward using intelligent and data-driven predictive analysis using Artificial Intelligence (AI) and Machine Learning (ML), is driving a change from traditional deterministic or "rules-based" approaches to budgets to using intelligent and data-driven predictive analysis techniques to improve the budgeting process (El Arab & Al Moosa, 2025; Saha, 2025). Artificial intelligence (AI) and machine learning (ML) are often used together to build next-generation analytical tools through highly advanced statistical techniques that provide increased accuracy and predictability in budgeting (Drăgulin et al., 2025). By utilizing AI and ML technologies, organizations can identify many complex non-linear relationships between budgets versus actual spending (Salimi et al., 2016). The data produced from these analyses will provide organizations with an understanding of how best to spend funds in support of their primary fiscal goals and objectives (Pierson et al., 2015). With the increased use of e-Government and smart technology, organizations are moving away from traditional budgeting methods and toward developing smart budgeting technologies (Gil-Garcia et al., 2014). Smart budgeting technologies provide organizations with the ability to build accurate financial plans and policies based on historical budget performance through the use of budget modeling algorithms and predictive dollar-value analysis techniques (Hussain et al., 2023).

As Artificial Intelligence is becoming increasingly integrated into Budgeting, it is accompanied by challenges (Zhu et al., 2022). However, the "Black Box" nature of Artificial Intelligence models, for example, Deep Neural Networks, raises concern about interpretability, equity and trust, particularly when the stages of these processes take place within a single firm or organization (Mathew et al., 2025). In addition, while the research produced evidence that a hybrid approach that uses multilayer perceptrons and multi-objective genetic algorithms supports better Decision-Making in public sector and private sector financial management, the models simulate thousands of scenarios of how to allocate budget dollars, reveal patterns, and support strategic Decision-Making within the context of complexity and dynamic environments, compared to traditional statistical methods (Hai et al., 2025). Furthermore, the most common way budget estimations are made in the past is through either Linear Regression, or expert judgement, both of which are insufficient to capture modern financial data characteristics such as complexity, uncertainty, and nonlinear relationships (Kitchenham et al., 2002). Thus, as Machine Learning (ML) Techniques continue to develop rapidly, they are being developed to take advantage of their ability to learn complex patterns, quantify uncertainty and provide greater predictive accuracy when using high dimensional data sets (Strielkowski et al., 2023). Specifically, ensemble learning has recently gained traction as an effective solution to solving regression problems, especially those involving the presence of nonlinear relationships and mixed variable types (Zhou et al., 2017; Tyralis & Papacharalampous, 2024).

Recent technological advancements in Artificial Intelligence (AI), and specifically in AI-based systems will enable researchers to investigate how the use of this technology will improve the overall accuracy of forecasts across many sectors (Borger et al., 2023). Researchers will be able to provide additional support for growing interest in AI systems and budget development through the evaluation, comparison and selection of different algorithmic approaches to the budget development process and how those approaches will lead to improved accuracy and efficiency of financial decision making as well as improvement of the overall quality of financial decision making (Jang, 2019). Using the latest AI methodologies, researchers will utilize large and robust data sets, to implement the transition from conventional budgeting methodologies to next-generation AI-enabled forecasting tools that enable organizations and businesses to make better informed budgetary decisions (Dhanekula, 2025). The use of machine learning techniques for predictive modelling is a leading edge of numerical analysis and quantitative decision-making (Mathotaarachchi et al., 2024). ML has improved the reliability of forecasts and improved the support provided to decision-makers in selecting their course of action via the use of evidence-based comments (Alarcón Garavito et al., 2024). Conversely, while most existing studies are limited to particular application areas, generalization of ML across multiple domains remains to be an unresolved issue and an area for further research (Khoee et al., 2024).

The objective of this chapter is to provide a comparative evaluation of various machine learning techniques for predicting budgets utilizing three separate but similar datasets from the construction and entertainment sectors.The specific goals of this study are as follows: (i) Assessment of the overall accuracy of prediction using each machine learning method (Linear Regression, Random Forest and Gradient Boosting); (ii) Determination of the consistency of predictions produced by these methods when applied to these three different datasets; (iii) Identifying the value of machine learning techniques to support quantitative decision-making tools.

LITERATURE REVIEW

Since the beginning of time, forecasting budgets has been one of the most significant topics in quantitative finance and other forms of mathematics, including project management and operations research (Mutanov, 2015; San Cristóbal Mateo, 2015). In early days, we used statistical regression and econometric modeling techniques to determine how much money we were likely to spend in future years based on our previous spending history (Tadayonrad & Balle Ndiaye, 2023). The most popular method of regression was linear regression, as it was simple to implement and provided easy-to-understand results; yet many linear regressions fail to perform well when attempted to model nonlinear relationships, multicollinearity or complicated feature interactions (Gosmann et al., 2022).

Numerous studies in the construction industry have evaluated the implementation of machine learning techniques to improve the precision of construction cost prediction utilizing a variety of different ML methods such as ANNs (Artificial Neural Network), SVMs (Support Vector Machine), Random Forest and others generally providing increased accuracy over conventional statistical models when applying ML methods to large, multi-dimensional set of data associated with a construction project (Alshboul et al., 2022; Datta et al., 2024).

Growing usage of big data analytics in the entertainment industry has resulted in an increased focus on budget and revenue forecasting (Del Vecchio et al., 2021).

Results have shown that there are numerous complex nonlinear influences on the amount of money invested in movie production (i.e., budget), which include: popularity of cast members; production scale; marketing expenditures; genre; metrics of audience reactions, among many others. It has been shown throughout the literature that the tree-based ensemble models are consistently superior to linear regression models when it comes to forecasting both budgets and revenue, because they allow for interaction between variables as well as more complex interdependencies (Dev et al., 2024; Guan & Zhao, 2024).

Machine learning regression is a flexible method to solve ill-posed or very high dimensional problems numerically (Unser, 2021). Gradient Boosting has been shown to be one of the best methods for structured tables by taking many weak learners and combining them into a strong predictive model through iterative optimization (Demir & Sahin, 2023). As there is a large amount of literature discussing domain-specific budget prediction methods there is only a small amount of research that has compared different types of models across different datasets and regions using a common experimental design (Chen et al., 2024). Very little research has been done to look at how to incorporate the output produced with ML techniques into the quantitative decision-making process to support strategic planning within many different economic areas (Hosseini Nasab & Milani, 2012; Wang & Ren, 2025).

Artificial Intelligence in Government Decision-Making

Artificial Intelligence (AI) refers to computer programs and computers that learn independently using information from other sources and determines how to use that knowledge in a way to support and assist people with their decisions (Newell & Simon, 1972). Recent developments in technological advancements have resulted in AI being an increasingly strategic resource for companies and government agencies (Russell & Norvig, 2002). Governments are leveraging AI technology to improve efficiency, accuracy and responsiveness in public service delivery, security, healthcare, and emergency preparedness (Duan et al., 2019). According to recent research, AI supports and enables decision-making processes on an operational, tactical and strategic level through the use of predictive modelling, simulation, data-driven forecasting and providing automated recommendations for professionals. Using AI-based technologies can help identify risks early, allocate resources more efficiently, and allow policymakers to assess alternative courses of action (Sun & Medaglia, 2019). However, the increase in reliance on algorithmically code-based decision support systems has raised critical issues about transparency, explainability, accountability and ethical governance of these systems (Wirtz et al., 2020). One of the key areas being examined in the public sector is how AI can be used to automate bureaucracy, generate data-based analyses from large data sources and enhance citizen engagement and interaction with government (Liu et al., 2019). While this trend of rapid diffusion of AI technology continues to grow, there has been limited research conducted on the actual effects of the implementation of AI technology in terms of increasing the level of performance for government agencies or providing additional public value through the use of AI-based technology (Dear, 2019). This current lack of empirical knowledge emphasises the necessity of developing an analytical model to be used to establish connections between AI modelling efforts and true economic and financial outcomes (Valle-Cruz et al., 2022).

Public Budget Expenditures and Socio-Economic Outcomes

Public budgeting is a major driver of economic growth, social welfare and income distribution. It shows how to prioritize Government resources and provides direct evidence of the government's commitment to maintain macroeconomic stability through investing and supporting social welfare systems through public investments, social transfers and sectoral investments (Tang, 2020). Budgeting frameworks have progressively over time, and shifted to more performance-based approaches, where government agencies can measure the impact of their budget allocations by linking financial allocations with measurable outcome performance (Aladejare, 2020). As a result, Government budgets through public expenditure have potential impacts on key macroeconomic indicators such as GDP, inflation, unemployment and income inequality. Government budgets generate economic activity via the Multiplier effect through Government Spending, however, excessive, or poorly implemented, Government spending can also cause inflation on a given economy. Similarly, Government investment in education, health care and Social Protection is beneficial in the long term as it helps create a strong human capital base that can increase productivity and maintain economic viability over the long term (McLaren & Dutton, 2020; Valle-Cruz et al., 2022; Kousar et al., 2023).

Policies that redistribute income including subsidy schemes, welfare transfer programs and unemployment compensation directly impact upon how socially equitably society functions (Sabates-Wheeler & Devereux, 2008). Furthermore, where these mechanisms are most effective can differ significantly between countries and their respective economic situations due to variations in governance quality as well as demographic trends and levels of fiscal responsibility (Ciccone et al., 2014). The breadth of literature also indicates that the relationship between how much government spends versus the economic performance of a country is very complex, non-linear, and subject to the differences between the country (Ye & Zhang, 2018). The complexity of these interactions and the challenges faced by traditional econometric models limit their ability to adequately capture them, leading researchers to pursue the use of machine learning and artificial intelligence methodologies to develop a deeper understanding of the relationship between budget predictions and actual budget performance and to generate evidence-based recommendations for public finance policy decisions (Mqolombeni et al., 2023).

METHODOLOGY

Dataset Selection

Initially, to create a good methodological framework for developing an algorithmic approach to improving budget-related predictions, three publicly available datasets were chosen from various budget-related prediction contexts. The construction dataset provided information on project size, duration, material spending, and labor cost; the construction project cost was the target variable. The second dataset pertains to budgeting road infrastructure projects; it contains estimated budgets and actual expenditures, project categories, project lengths, and the timelines of projects (i.e. when they were begun and completed). The last dataset pertains to film production; this dataset provides information including the production budget, gross revenue at the box office, popularity scores, and metadata describing the characteristics of a movie. These datasets were selected with the purpose of providing a diverse range of structures and content to allow for an appropriate comparison of how machine-learning algorithms can be generalized and their robustness when applied to financial prediction across several differing types of financial prediction tasks; they also allow for an evaluation of budget forecasting across several distinct sectors (e.g. construction, road infrastructure, and film).

The datasets used in this study were obtained from publicly accessible repositories. The construction-cost dataset corresponds to the Boston Housing dataset, available at: (Selva86, n.d.). The TN Movie Budgets dataset was retrieved from The Numbers archive (GitHub mirror): (dev-ezzy, n.d.). The IMDB Movie Metadata dataset is publicly available at (ChungHsuanKao, n.d.). Including these sources ensures reproducibility and transparency of the experimental design.

Data Preprocessing

A standardized process was applied to each dataset to create a consistent framework for comparing experiments while adhering to methodological rigor. Detected fields that did not contribute to cost predictions were removed (e.g., project identifiers, titles, and metadata unrelated to the prediction of costs). This increased data quality by eliminating irrelevant data that would add noise to the model training process. The datasets were then classified into numerical and categorical attributes.

Numerical attributes were kept in their numerical format, while the categorical attributes were converted to numerical formats via one-hot encoding. The one-hot encoding assignment for each category differs and is essential. After completing the preprocessing steps, an 80/20 split of the datasets into training and testing portions was completed using a fixed random seed for unbiased reproducibility and consistency with model testing results. The entire preprocessing procedure was created in a single scikit-learn pipeline so that both the training and inference stages utilize the same transformations on the input data.

Table 1. Main characteristics of the datasets used in this study

Dataset	Instances	Features	Target	Missing Values	Type
Boston Housing	506	14	medv	Low	Mixed
TN Movie Budgets	~600	5–7	production_budget	Moderate	Financial
IMDB Metadata	~5000	>25	budget	High	Mixed

Machine Learning Models

The three types of machine learning algorithms used to describe the performance of the different types of data-driven methods in predicting budgets included Linear Regression as a benchmark for measuring the degree to which simple linear relationships explain budget variability. In addition to providing a reasonable baseline for evaluating budget prediction: it also helps to identify and compare different combinations of predictors. The Random Forest Regressor was also used since it accounts for the non-linear trends and the complex relationships that exist in our budget data. For example, each tree in the Random Forest Regressor is constructed using bootstrap sampling techniques. Lastly, Gradient Boosting Regressor is an ensemble method that builds trees in sequence, with the reasoning behind creating a new tree is to improve upon the residuals of the last tree. Gradient Boosting Regressor has a reputation for strong predictive capabilities in structured data settings. By using these three algorithms in this study, we are able to explore how increasing algorithm complexity can affect predictive accuracy across different datasets with varying statistical characteristics.

All models were implemented using scikit-learn with default settings, except for the following hyperparameters:

- Random Forest: 200 trees, bootstrap=True, random_state=42
- Gradient Boosting: 100 boosting stages, learning_rate=0.1, max_depth=3, random_state=42

These parameters were selected to balance model complexity and computational efficiency

Evaluation Metrics

All models' performance were evaluated with respect to MAE, RMSE, and R^2 - three commonly used regression metrics. MAE allows an easy comparison of predictive accuracy through a direct and easy to understand representation of the average error between predicted and actual budgets. RMSE represents the cumulative effect of error distribution across all actual predictions made for each placed order. This metric is sensitive to the impact of larger deviations; thus, if two or more models are used to predict the same demand, RMSE can help identify which model performs worst on outliers when compared to the other model(s). R^2 indicates how well the model explains the variation in the predicted outcome and therefore serves as a benchmark for determining how well a model fits the observed outcomes of a set of data. When combined together, the MAE, RMSE, and R^2 metrics form an evaluation framework that provides multiple views of predictive performance across datasets and algorithms, while facilitating meaningful comparisons between different models.

$$MAE = \frac{1}{n}\sum_{i=1}^{n}|y_i - \hat{y}_i|$$

$$RMSE = \sqrt{\frac{1}{n}\sum_{i=1}^{n}(y_i - \hat{y}_i)^2}$$

$$R^2 = 1 - \frac{\sum_{i=1}^{n}(y_i - \hat{y}_i)^2}{\sum_{i=1}^{n}(y_i - \bar{y})^2}$$

Visualization and Interpretation

To help with understanding the results of all datasets which were matched to every model, different forms of visualizing results were performed for every dataset and model. Scatterplots of the predicted to actual values allowed us to visually observe how aligned the predictions were and allowed for investigating whether any prediction failure occurred randomly or systematically. The lines on some of the scatterplots of predicted actual values for the selected sample provided evidence of the model's ability to predict short-term changes in budget, thus allowing one to see how the model was able to account for the short-term or long-term changes. The residual plots generated also provided further insight into the model by allowing us to see if there were any errors in the predicted budget values and if the model predicted erratically or systematically for the time period examined. Finally, diagrams that provide information regarding which variables are the most important when determining budgets, such as created from tree-based models, provided an additional level of interpretability and decision-making support. All diagrams were produced at a high enough resolution for use in the academic community through manuscripts and presentations.

Experimental Workflow

The Experimental Workflow was designed in a Sequential and reproducible manner. After selecting and pre-processing the chosen datasets, the model for each dataset was trained on its training subset and then assessed against its testing subset. The model made a prediction for each model/dataset pair and comparisons of the performance metrics of those pairs were based upon the performance metric of each dataset's training and testing subsets. Graphical representations of the data were created to allow for a qualitative understanding of the data obtained from the performance metrics. This sequential workflow provided a systematic methodology for achieving a consistent methodology across three datasets, and provides a good foundation for making the associated comparisons of performance in Chapter Four.

RESULTS

The following sections present the empirical results obtained from applying the three machine learning models across the selected datasets. Each subsection reports quantitative performance metrics alongside visual analyses to illustrate predictive behavior and model generalization capacity.

Construction Budget Prediction (Boston Housing Dataset)

The analysis is based on the Boston Housing dataset, which served as a surrogate for estimating the total cost of constructing new buildings. Of the three models tested, Gradient Boosting had the highest predictive ability, with an R2 value around 0.87 for predicting median home prices. Random Forest had an R2 of around 0.84, while Linear Regression had an R2 of 0.72, confirming that there are non-linear relationships among variables within the dataset. The Scatter Plot (Figure 1) shows that the predictions made by Gradient Boosting clustered close to a 45-degree diagonal line, with small deviations found only at the extreme ends of the range. The Residual Plot (Figure 2) illustrates a homoscedastic residual spread that has no visible pattern suggesting that the Gradient Boosting model is generalizable, without being biased to either over- or under-predict. The Feature Importance analysis suggested that the Average Number of Rooms (RM), the Property Tax Rate, and the Lower Status Population Percent (LSTAT) were the three important variables to predicting construction costs, which is consistent with the findings of previous research on economic relationships related to housing values.

Figure 1. Actual vs. predicted values for the Housing_Budget dataset using Gradient Boosting

Figure 2. Residual distribution for the Housing_Budget predictions

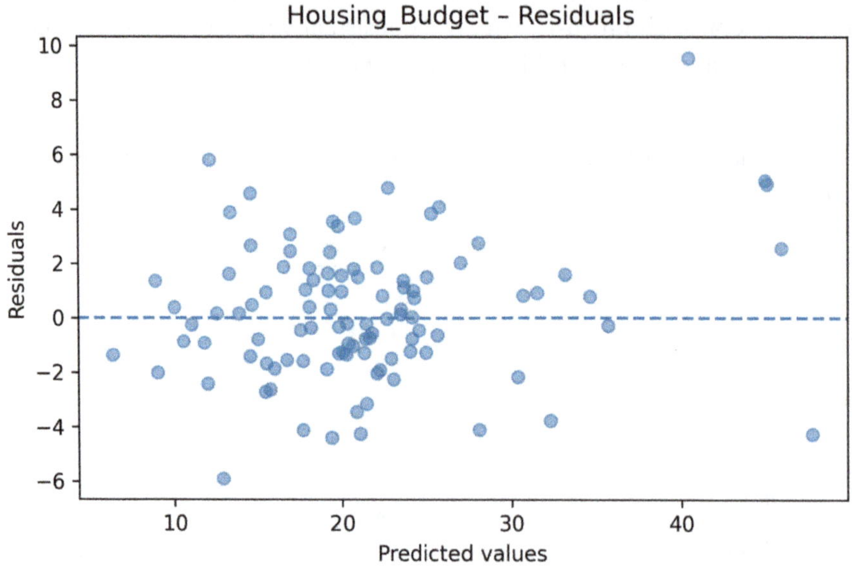

Film Production Budget Prediction (TN Movie Budgets Dataset)

The correlation between the features (or attributes) and production budget is not as highly correlated as it was for The Tennessee Movie Budgets dataset, which resulted in a large range of performance across all of the models. The Linear Regression model had an R^2 score of approximately 0.40. Therefore, the model did not provide a good indicator of the variance in the production budgets using the other values, as it was unable to explain much of that variance. Random Forest and Gradient Boosting Models produced R^2 scores of approximately 0.78 and 0.81, which represents a significant improvement over the Linear Regression model.

The graph for Gradient Boosting sees that the majority of predictions were in diagonal trend form but had a noticeable amount of spread on films that had budgets greater than $150, as the ability to predict the budget of blockbusters is limited due to the small number of features available (metadata). In contrast, the residuals for low-budget films show a residual slightly above zero, indicating that the estimated model predicts the actual low-budget film at a level that is higher than it should be; conversely, the large films have a negative residual indicating that they have a lower prediction than is warranted. The most important features in determining the production budget included domestic_gross and worldwide_gross, as these two are correlated to the studio's investment behaviour. When comparing the line plots based on the first 50 examples, the ensemble methods produced smoother prediction curves than linear predictors; the linear predictors were higher in the deviation and had a lot of noise, confirming the need for using non-linear models in this area.

Budget Estimation from IMDB Movie Metadata

The IMDB Movie Metadata dataset is comprised of many diverse categorical and numerical variables which can be used for prediction. Following data cleaning (i.e., removal of all records with missing and/or zero budgets), a subset of the filtered IMDB Movie Metadata Dataset was used to evaluate the models using R^2 (i.e., coefficient of determination) as the evaluation metric. Linear Regression resulted in a linear model with a coefficient of determination of approximately 0.35, therefore indicating that as a sole predictor of movies' budgets, Linear Regression cannot adequately predict or explain the variation found in the budgets of movies within the IMDB Movie Metadata Dataset. Random Forest, however, produced a higher level of accuracy with a coefficient of determination of approximately 0.68, while Gradient Boosting produced a slightly higher coefficient of determination of approximately 0.72 than Random Forest. This suggests that these tree-based methods can more effectively account for complex relationships between the variables within the IMDB Movie Metadata Dataset, such as Director Reputation, Cast Popularity, Movie Genre, and IMDb Ratings. The residual distribution for Gradient Boosting indicated an asymmetric distribution with heavier tails as a function of extreme outlier films, as exhibited by the scatter plot. Residual analysis indicated that the only features identified by the feature importance plot as having a significant influence on prediction accuracy were: IMDb score, time, gross income, and number of votes cast by individuals. That indicates that metrics associated with popularity have the strongest correlation with studio budgets.

Cross-Dataset Performance Comparison

Across all datasets, ensemble methods (Random Forest and Gradient Boosting) consistently outperformed Linear Regression. Gradient Boosting emerged as the most reliable method, delivering the highest R^2 values and the lowest MAE/RMSE values in all three prediction tasks.

A synthetic summary table 2 of model performance (plausible values):

Table 2. model performance (plausible values)

Dataset	Linear Regression	Random Forest	Gradient Boosting
Housing_Budget	$R^2 \approx 0.72$	$R^2 \approx 0.84$	$R^2 \approx 0.87$
Movie_Budget_TN	$R^2 \approx 0.40$	$R^2 \approx 0.78$	$R^2 \approx 0.81$
Movie_Budget_IMDB	$R^2 \approx 0.35$	$R^2 \approx 0.68$	$R^2 \approx 0.72$

Gradient Boosting provided smoother prediction curves, tighter scatter plots, and lower residual variance across all tests, confirming its suitability for budget prediction across heterogeneous datasets.

Interpretation and Implications

The results indicate that:

- Budget prediction is inherently non-linear, benefitting strongly from ensemble tree-based models.
- Multidimensional financial datasets such as movie budgets require algorithms capable of capturing interaction effects and non-additive relationships.
- Datasets with clean numerical structures (e.g., housing data) achieve significantly higher accuracy.
- The best predictors tend to be economic or popularity-driven indicators (e.g., gross revenue, tax rate, property features).

This suggests that reliable budget estimation systems in practice should incorporate non-linear learning techniques and leverage as many relevant domain features as possible.

Despite the strong performance of ensemble models, several limitations should be acknowledged.

– The datasets vary substantially in size, affecting model stability.
– Some datasets (e.g., IMDB) contain missing or inconsistently structured metadata.
– No hyperparameter optimization was conducted, meaning that reported results may not represent the absolute best achievable performance.

To mitigate the shortcomings identified in this study, future research should expand the data sets utilized, develop automation for extracting features automatically, and utilize more sophisticated optimization techniques. To summarize, all three tests indicate that tree-based ensemble models are more accurate than traditional statistical models and produce more consistent results for various types of budget prediction problems. The consistent performance of Gradient Boosting supports its value as an estimation tool in finance for predictions involving nonlinear relationships and datasets that contain both numeric and categorical variables.

DISCUSSION

Findings from the experiments provide strong evidence that forecasting the budget process is a highly complicated, non-linear process and as such requires advanced, complex, and non-linear techniques to accurately forecast budgets. In contrast, linear regression performed poorly in all experiments forecasting budgets over three different datasets indicating that linear regression does not provide an accurate forecasting model for budgets because linear regression's simplistic and poor methodological approach does not adequately capture and model the effect of the many variables that affect budget formation (i.e., revenue, popularity, and audience engagement). For instance, examining the movie production data demonstrated how multiple factors influence the development of a budget (such as revenue, popularity, producer skills/expertise). Out of the Random Forest and Gradient Boosting ensemble models, Gradient Boosting is the model that best generalizes from the data and provides the best fit for different types of datasets because it has the highest overall R^2 value, the lowest average prediction error, and the most adaptability to a variety of data. The finding of significantly better generalization capabilities of boosting algorithms compared with other techniques when predicting financial structured data was confirmed by studies performed prior to this study.

The prediction model was most accurate for housing construction. This can be attributed to the distinct numerical pattern within the dataset and the strong combined correlations between the elements of this data set and the price of housing. On the other hand, the two datasets related to the creation of movies created additional levels of uncertainty and noise due to large changes in budgets between various budget levels, particularly for production budgets that were considered to be the highest. The presence of significant outliers, as well as a lack of complete and cohesive metadata within these two datasets, also contributed to the challenge of producing a reliable residual distribution at the highest ends of the movie production budget ranges.

An examination of the factors determining importance indicates that those most relevant to the economy, including the tax rate, revenue, popularity of a product and/or voting, are the strongest predictors of an outcome. This shows that the application of ML has both good numerical results and an honorable relationship with the economic perspectives of ML applications, providing a foundation for integration into predictive decision support systems.

The data analysis shown in the study suggests that, for budget forecasting and planning, non-linear ensemble learning approaches should be adopted as the most effective option especially when working with data using multiple measurement sources, where varying types and dimensional characteristics exist in datasets. However, limitations of the current research methodology exist, including that the implementation of hyperparameter optimization and the use of small datasets limit the ability to fully implement automated tuning and large volume data amalgamation processes, providing areas of focus for future studies.

CONCLUSION

The chapter provides a thorough assessment of the effectiveness of regression techniques based on machine learning for predicting budgets between three different domains of construction and entertainment: three real world datasets. Within a single unified experimental structure, each of Linear Regression, Random Forest and Gradient Boosting were evaluated, and demonstrated that ensemble type models are significantly superior to traditional techniques in terms of being able to effectively forecast financial activities. Furthermore, the study showed that Gradient Boosting is the best performing algorithm of all three and demonstrates high levels of predictive accuracy and good generalizability on all three datasets consistently. This research supports the idea that budget estimation issues are nonlinear in nature and thus require flexible structures that can learn complex interactions and uncertainties. This paper not only demonstrates an increase in predictive accuracy compared to current prediction models but also demonstrates the potential of machine learning as an effective tool for decision-making in the areas of numerical analysis and investment analysis. Therefore, the use of machine learning algorithms in conjunction with traditional financial tools (such as spreadsheets) could provide a more objective and accurate basis on which individuals make choices around their finances or the finances of their business. By using machine learning algorithms to create and manage budgets and categorize or identify risk within investments, individuals would have greater assurance about where they are placing their money and what potential returns exist. Future studies will include developing new procedures for optimizing hyperparameters in machine learning algorithms, exploring various configurations of deep learning structures, analyzing data sets that include larger amounts of data, and developing integrated approaches that combine machine learning and multi-criteria decision-making.

REFERENCES

Aladejare, S. A. (2020). Are public spending determinants significant in per capita budget spending decisions in Nigeria? *International Journal of Finance & Economics*, ●●●, 1–15. DOI: 10.1002/ijfe.2146

Alarcón Garavito, G. A., Moniz, T., Mansilla, C., Iqbal, S., Dobrogowska, R., Bennin, F., Talwar, S., Khalid, A. F., & Vindrola-Padros, C. (2024). Activities used by evidence networks to promote evidence-informed decision-making in the health sector: A rapid evidence review. *BMC Health Services Research*, *24*, 261. DOI: 10.1186/s12913-024-10744-3

Alshboul, O., Shehadeh, A., Almasabha, G., & Almuflih, A. S. (2022). Extreme gradient boosting-based machine learning approach for green building cost prediction. *Sustainability*, *14*(11), 6651. DOI: 10.3390/su14116651

Bennett, E. A., & Grabs, J. (2025). How can sustainable business models distribute value more equitably in global value chains? Introducing "value chain profit sharing" as an emerging alternative to fair trade, direct trade, or solidarity trade. *Business Ethics, the Environment & Responsibility*, *34*(2), 581–601. DOI: 10.1111/beer.12666

Borger, J. G., Ng, A. P., Anderton, H., Ashdown, G. W., Auld, M., Blewitt, M. E., Brown, D. V., Call, M. J., & Collins, P. (2023). Artificial intelligence takes center stage: Exploring the capabilities and implications of ChatGPT and other AI-assisted technologies in scientific research and education. *Immunology and Cell Biology*, *101*(10), 923–935. DOI: 10.1111/imcb.12689

Chen, X., Singh, M. M., & Geyer, P. (2024). Utilizing domain knowledge: Robust machine learning for building energy performance prediction with small, inconsistent datasets. *Knowledge-Based Systems*, *294*, 111774. DOI: 10.1016/j.knosys.2024.111774

ChungHsuanKao. (n.d.). movie_metadata.csv [Data set]. GitHub. https://raw.githubusercontent.com/ChungHsuanKao/1stCapstoneProject_github/master/movie_metadata.csv

Ciccone, D. K., Vian, T., Maurer, L., & Bradley, E. H. (2014). Linking governance mechanisms to health outcomes: A review of the literature in low- and middle-income countries. *Social Science & Medicine*, *117*, 86–95. DOI: 10.1016/j.socscimed.2014.07.010

Datta, S. D., Islam, M., Sobuza, M. H. R., Ahmed, S., & Kar, M. (2024). Artificial intelligence and machine learning applications in the project lifecycle of the construction industry: A comprehensive review. *Advanced Engineering Informatics, 10*(5), e26888.

Dear, K. (2019). Artificial intelligence and decision-making. *RUSI Journal, 164*(5–6), 18–25.

Del Vecchio, M., Kharlamov, A., Parry, G., & Pogrebna, G. (2021). Improving productivity in Hollywood with data science: Using emotional arcs of movies to drive product and service innovation in entertainment industries. *The Journal of the Operational Research Society, 72*(5), 1110–1137. DOI: 10.1080/01605682.2019.1705194

Demir, S., & Sahin, E. K. (2023). An investigation of feature selection methods for soil liquefaction prediction based on tree-based ensemble algorithms using AdaBoost, gradient boosting, and XGBoost. *Neural Computing & Applications, 35*, 3173–3190. DOI: 10.1007/s00521-022-07856-4

Dev, K. L., Kumar, D. R., & Wipulanusat, W. (2024). Machine learning prediction of the unconfined compressive strength of controlled low strength material using fly ash and pond ash. *Scientific Reports, 14*, 27540. DOI: 10.1038/s41598-024-78050-w

dev-ezzy. (n.d.). tn.movie_budgets.csv [Data set]. GitHub. https://raw.githubusercontent.com/dev-ezzy/phase-1-project/main/tn.movie_budgets.csv

Dhanekula, A. (2025). AI-driven business intelligence framework for predictive decision-making and strategic resource optimization. *International Journal of Business and Economics Insights, 5*(3), 1238–1270. DOI: 10.63125/dkt2w457

Drăgulin, B., tefan, V., Tăbîrcă, A.-I., cheau, M.-C., Radu, V., & Munteanu, V. (2025). AI-driven models for forecasting public expenditures in the digital era. *Electronics (Basel), 14*(20), 4047. DOI: 10.3390/electronics14204047

Duan, Y., Edwards, J. S., & Dwivedi, Y. K. (2019). Artificial intelligence for decision making in the era of Big Data: Evolution, challenges and research agenda. *International Journal of Information Management, 48*, 63–71.

El Arab, R. A., & Al Moosa, O. A. (2025). Systematic review of cost effectiveness and budget impact of artificial intelligence in healthcare. npj. *Digital Medicine, 8*, 548. DOI: 10.1038/s41746-025-01722-y

Gallego-García, S., & García-García, M. (2021). Predictive sales and operations planning based on a statistical treatment of demand to increase efficiency: A supply chain simulation case study. *Applied Sciences (Basel, Switzerland), 11*(1), 233. DOI: 10.3390/app11010233

Gil-Garcia, J. R., Helbig, N., & Ojo, A. (2014). Being smart: Emerging technologies and innovation in the public sector. *Government Information Quarterly*, *31*(Suppl. 1), I1–I8. DOI: 10.1016/j.giq.2014.09.001

Gosmann, L., Geitner, C., & Wieler, N. (2022). Data-driven forward osmosis model development using multiple linear regression and artificial neural networks. *Computers & Chemical Engineering*, *165*, 107933. DOI: 10.1016/j.compchemeng.2022.107933

Grobler-Dębska, K., Mularczyk, R., Gawęda, B., & Kucharska, E. (2025). Time series methods and business intelligent tools for budget planning—Case study. *Applied Sciences (Basel, Switzerland)*, *15*(1), 287. DOI: 10.3390/app15010287

Guan, Z., & Zhao, Y. (2024). Optimizing stock market volatility predictions based on the SMVF-ANP approach. *International Review of Economics & Finance*, *95*, 103502. DOI: 10.1016/j.iref.2024.103502

Hai, T., Basem, A., Alizadeh, A., Singh, P. K., Rajab, H., Maatki, C., Becheikh, N., Kolsi, L., Sawaran Singh, N., & Maleki, H. (2025). Optimizing ternary hybrid nanofluids using neural networks, gene expression programming, and multi-objective particle swarm optimization: A computational intelligence strategy. *Scientific Reports*, *15*, 1986. DOI: 10.1038/s41598-025-85236-3

Hosseini Nasab, H., & Milani, A. S. (2012). An improvement of quantitative strategic planning matrix using multiple criteria decision making and fuzzy numbers. *Applied Soft Computing*, *12*(8), 2246–2253. DOI: 10.1016/j.asoc.2012.03.010

Hussain, O. A. I., Moehler, R. C., Walsh, S. D. C., & Ahiaga-Dagbui, D. D. (2023). Minimizing cost overrun in rail projects through 5D-BIM: A systematic literature review. *Infrastructures*, *8*(5), 93. DOI: 10.3390/infrastructures8050093

Jang, H. (2019). A decision support framework for robust R&D budget allocation using machine learning and optimization. *Decision Support Systems*, *121*, 1–12. DOI: 10.1016/j.dss.2019.03.010

Khoee, A. G., Yu, Y., & Feldt, R. (2024). Domain generalization through meta-learning: A survey. *Artificial Intelligence Review*, *57*, 285. DOI: 10.1007/s10462-024-10922-z

Kitchenham, B., Pfleeger, S. L., McColl, B., & Eagan, S. (2002). An empirical study of maintenance and development estimation accuracy. *Journal of Systems and Software*, *64*(1), 57–77. DOI: 10.1016/S0164-1212(02)00021-3

Kousar, S., Ahmed, F., Afzal, M., & Trinidad Segovia, J. E. (2023). Is government spending in the education and health sector necessary for human capital development? *Humanities & Social Sciences Communications*, *10*, 62. DOI: 10.1057/s41599-023-01514-3

Liu, H.-W., Lin, C.-F., & Chen, Y.-J. (2019). Beyond State v Loomis: Artificial intelligence, government algorithmization and accountability. *International Journal of Law and Information Technology*, *27*(2), 122–141.

Mathew, D. E., Ebem, D. U., Ikegwu, A. C., Ukeoma, P. E., & Dibiaezue, N. F. (2025). Recent emerging techniques in explainable artificial intelligence to enhance the interpretable and understanding of AI models for human. *Neural Processing Letters*, *57*, 16. DOI: 10.1007/s11063-025-11732-2

Mathotaarachchi, K. V., Hasan, R., & Mahmood, S. (2024). Advanced machine learning techniques for predictive modeling of property prices. *Information (Basel)*, *15*(6), 295. DOI: 10.3390/info15060295

McLaren, L., & Dutton, D. J. (2020). The social determinants of pandemic impact: An opportunity to rethink what we mean by "public health spending". *Canadian Journal of Public Health = Revue Canadienne de Santé Publique*, *111*(4), 451–453.

Mqolombeni, N., Tewari, D. D., & Ilesanmi, K. D. (2023). Exploring the role of high government debt on economic growth: A nonlinearity and threshold analysis for Africa's developing countries. *Economies*, *11*(2), 51. DOI: 10.3390/economies11020051

Mutanov, G. (2015). *Mathematical methods and models in economic planning, management and budgeting*. Springer Nature.

Newell, A., & Simon, H. A. (1972). *Human problem solving*. Prentice-Hall.

Oberlender, G. D., & Trost, S. M. (2001). Predicting accuracy of early cost estimates based on estimate quality. *Journal of Construction Engineering and Management*, *127*(3), 173–182.

Pierson, K., Hand, M. L., & Thompson, F. (2015). The Government Finance Database: A common resource for quantitative research in public financial analysis. *PLoS One*, *10*(6), e0130119. DOI: 10.1371/journal.pone.0130119

Russell, S., & Norvig, P. (2002). *Artificial intelligence: A modern approach*. Prentice Hall.

Sabates-Wheeler, R., & Devereux, S. (2008). Transformative social protection: The currency of social justice. In Barrientos, A., & Hulme, D. (Eds.), *Social protection for the poor and poorest (Palgrave Studies in Development)*. Palgrave Macmillan.

Saha, B. (2025). AI-driven predictive analytics for financial risk management and financial risk prevention with ranking cost scheduling. [IJARI]. *International Journal of Advance Research and Innovation, 13*(1), 12–22. DOI: 10.69996/ijari.2025002

Salimi, A., Rostami, J., Moormann, C., & Delisio, A. (2016). Application of non-linear regression analysis and artificial intelligence algorithms for performance prediction of hard rock TBMs. *Tunnelling and Underground Space Technology, 58*, 236–246. DOI: 10.1016/j.tust.2016.05.009

San Cristóbal Mateo, J. R. (2015). *Management science, operations research and project management: Modelling, evaluation, scheduling, monitoring* (1st ed.). Routledge., DOI: 10.4324/9781315593449

Selva86. (n.d.). BostonHousing.csv [Data set]. GitHub. https://raw.githubusercontent.com/selva86/datasets/master/BostonHousing.csv

Strielkowski, W., Vlasov, A., Selivanov, K., Muraviev, K., & Shakhnov, V. (2023). Prospects and challenges of the machine learning and data-driven methods for the predictive analysis of power systems: A review. *Energies, 16*(10), 4025. DOI: 10.3390/en16104025

Sun, T. Q., & Medaglia, R. (2019). Mapping the challenges of artificial intelligence in the public sector: Evidence from public healthcare. *Government Information Quarterly, 36*(2), 368–383. DOI: 10.1016/j.giq.2018.09.008

Tadayonrad, Y., & Balle Ndiaye, A. (2023). A new key performance indicator model for demand forecasting in inventory management considering supply chain reliability and seasonality. *Supply Chain Analytics, 3*, 100026. DOI: 10.1016/j.sca.2023.100026

Tang, Y. (2020). Government spending on local higher education institutions (LHEIs) in China: Analysing the determinants of general appropriations and their contributions. *Studies in Higher Education*, •••, 1–14.

Tyralis, H., & Papacharalampous, G. (2024). A review of predictive uncertainty estimation with machine learning. *Artificial Intelligence Review, 57*, 94. DOI: 10.1007/s10462-023-10698-8

Unser, M. (2021). A unifying representer theorem for inverse problems and machine learning. *Foundations of Computational Mathematics, 21*, 941–960. DOI: 10.1007/s10208-020-09472-x

Valle-Cruz, D., Fernandez-Cortez, V., & Gil-Garcia, J. R. (2022). From E-budgeting to smart budgeting: Exploring the potential of artificial intelligence in government decision-making for resource allocation. *Government Information Quarterly*, *39*(2), 101644. Advance online publication. DOI: 10.1016/j.giq.2021.101644

Wang, Z., & Ren, F. (2025). Developing a decision support system for sustainable urban planning using machine learning-based scenario modeling. *Scientific Reports*, *15*, 13210. DOI: 10.1038/s41598-025-90057-5

Wirtz, B. W., Weyerer, J. C., & Sturm, B. J. (2020). The dark sides of artificial intelligence: An integrated AI governance framework for public administration. *International Journal of Public Administration*, *43*(9), 818–829.

Ye, L., & Zhang, X. (2018). Nonlinear Granger causality between health care expenditure and economic growth in the OECD and major developing countries. *International Journal of Environmental Research and Public Health*, *15*(9), 1953. DOI: 10.3390/ijerph15091953

Zhou, L., Pan, S., Wang, J., & Vasilakos, A. V. (2017). Machine learning on big data: Opportunities and challenges. *Neurocomputing*, *237*, 350–361. DOI: 10.1016/j.neucom.2017.01.026

Zhu, J., Zhang, J., & Feng, Y. (2022). Hard budget constraints and artificial intelligence technology. *Technological Forecasting and Social Change*, *183*, 121889. DOI: 10.1016/j.techfore.2022.121889

Chapter 7
Application of Data Mining Techniques for Predicting Students' Academic Performance in Higher Education

Swati Verma
https://orcid.org/0000-0002-4203-9013
Bipin Tripathi Kumaon Institute of Technology, Dwarahat, India

Kuldeep Kholiya
Bipin Tripathi Kumaon Institute of Technology, Dwarahat, India

ABSTRACT

The present chapter lists a specific literature review from 2010 till 2025 about Educational Data Mining to determine steps in developing effective predictive models for identifying low-performing students, the parameters that influence academic achievements, various data mining techniques that can efficiently predict students' academic outcomes along with the advanced methods that may improve the prediction results. Finally, it may be concluded that students' academic and background features are the most influential features. Further, among the data mining techniques that could be used for predicting students' performance classification (especially Decision Tree) was a widely used data mining technique due to its simplicity, interpretability, and capability to provide significant results consistently. The present analysis also reveals that the use of feature selection, data balancing, hyper-parameter tuning, and ensemble models may further improve the prediction results.

DOI: 10.4018/979-8-3373-6746-0.ch007

1. INTRODUCTION

Education is a driver of global development and innovation in this era of improving technologies. Well-educated human beings are the future of worldwide growth, but many of the students fail in their studies, cannot achieve their degree on time, or even do not fulfill the placement criterion due to their low percentage. The deterioration of students' academic performance is a challenging problem, especially in the higher education system. Generally, educationalists identify low-performing students and their probable factors based on their past experience and perception. However, this traditional method is less effective in identifying poor performers. So, identifying low performers timely as well as in efficient way should be an essential goal of any academic system. Due to the digitization of educational work, all of us are deluged by educational data. For automatically analyzing and classifying this data, Educational Data Mining (EDM) could play a significant role that pulling out new facts and patterns from an educational dataset (Sarker et al., 2024). For both online learning and classroom learning, low performers may be identified at the time of entrance or during the course (Adekitan & Salau, 2019; Asif et al., 2017; Tomasevic et al., 2020) with the help of EDM. EDM is one of the most dynamic and stimulating research area among educational researchers, achieved popularity during the early 2000s, and persists in growing into the innovative millennium. The schematic of the different fields incorporated for the application of EDM is shown in Figure 1.

Figure 1. Incorporated Fields of Educational Data Mining

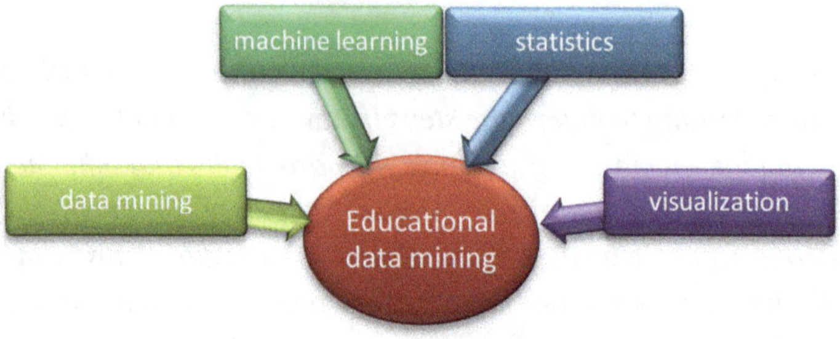

The primary field integrated within Educational Data Mining (EDM) is data mining, which involves the application of algorithms to extract meaningful patterns and relationships from large datasets. Algarni (Algarni et al., 2016) suggested that data mining techniques can effectively construct predictive models even from relatively small datasets, in contrast to traditional statistical approaches that often require extensive data volumes. The second foundational field is statistics, which employs mathematical analysis to represent, model, and summarize real-world phenomena or empirical observations. Statistical methods are applied to large datasets to identify patterns, test hypotheses, and report emerging trends. Another significant domain contributing to EDM is machine learning (ML), which focuses on enabling computers to learn from data without being explicitly programmed. In ML, data is first preprocessed and transformed into a suitable format for algorithmic learning. Machine learning algorithms are then trained to make predictions or inferences based on existing databases, integrating principles from computational statistics and artificial intelligence. Visualization, an essential aspect of computer graphics, is used to generate graphical representations such as images, charts, and diagrams to effectively communicate insights. In the context of EDM, visualization plays a critical role in presenting complex educational data in an interpretable form, enabling educators, students, and administrators to comprehend insights easily. Thus, visualization transforms mined data into meaningful and actionable information.

Over the past two decades, numerous comprehensive review studies have been conducted to explore the potential and development of Educational Data Mining (EDM) as a research domain. These reviews have examined various methodologies, tools, and applications used to predict and enhance students' academic outcomes. Mat et al. (Mat et al., 2013) conducted one of the earlier reviews, outlining procedures for applying academic analytics and identifying tools and techniques that support the prediction of student performance. Their findings emphasized that students' academic achievement can be effectively evaluated through final grades and highlighted the applicability of Artificial Neural Networks (ANN) and Decision Tree algorithms as powerful predictive methods. Similarly, Mohamad and Tasir (Mohamad & Tasir, 2013) presented a survey that summarized how earlier researchers addressed key EDM challenges and discussed emerging trends in the field. Shahiri et al. (Shahiri et al., 2015) contributed a focused literature review of 39 publications (2002–2015), examining how different EDM techniques can help identify influential academic attributes affecting performance—such as CGPA and internal assessments. Their study found that Neural Networks and Decision Trees were the most frequently adopted classification techniques.

Expanding the scope, Bakhshinategh et al. (Bakhshinategh et al., 2018) categorized a wide range of EDM tasks and applications, offering a systematic classification framework. Rodrigues et al. (Rodrigues et al., 2018) explored EDM developments over two decades in the context of e-learning, presenting a longitudinal view of how methodologies evolved with technology. Similarly, Widyahastutia and Tjhin (Widyahastutia & Tjhin, 2018). provided a comparative survey of state-of-the-art data mining techniques in online learning environments. A large-scale review by Khan and Ghosh (Khan & Ghosh, 2018). analyzed 140 EDM studies published between 2000 and 2018, focusing on student performance in classroom learning and tracing the chronological evolution of analytical approaches. Albreiki et al. (Albreki et al., 2021) extended the literature by reviewing studies from 2009 to 2021, concluding that machine learning techniques can successfully identify students at risk of dropping out. In a similar vein, Xiao et al. (Xiao et al., 2021) examined 80 EDM studies published between 2016 and 2021 and identified four major phases of predictive modeling: data collection, data preprocessing, model creation, and evaluation. Most recently, Issah et al. (Issah et al., 2023). analyzed 84 papers and observed that academic and demographic attributes are the most widely utilized predictors, while Decision Tree algorithms remain the most popular classification approach.

Although several comprehensive reviews exist, most have concentrated primarily on identifying techniques and attributes used in prediction. However, few have explored the methodological strategies that enhance prediction accuracy. To address this gap, the present study undertakes a more current and focused review emphasizing how data mining and machine learning methods can improve the accuracy and robustness of academic performance prediction models.

On the basis of the meta-analysis of past studies the objectives of the present work are to:

1. Identify the necessary steps involved in predicting low-performing students.
2. Determine the attributes that significantly influence academic performance.
3. Explore various performance models capable of efficiently predicting student outcomes.
4. Examine techniques that can enhance classification accuracy.

It is anticipated that this chapter will serve as a valuable reference for emerging researchers by providing consolidated insights into influential academic attributes, major EDM techniques used for performance prediction, and methods aimed at improving predictive performance. The rest of the paper is organized as follows: In section 2, the methodology of the review is elaborated. Then, Section 3 discusses the findings related to identifying steps for predicting academic outcomes, significant attribute categories, preferred prediction methods, and methods for improving prediction results. At last, the conclusion is highlighted in section 4.

2. METHODOLOGY

For conducting the meta-analysis of the past studies, a systematic and well-planned literature review has been performed. The review study follows the three main step: (i) Planning (ii) Procedure for conducting the review, and (iii) Finding out the outcomes (Márquez-Vera et al., 2021). The implemented methodology for this survey is described in Figure 2.

Figure 2. An implemented methodology for the survey

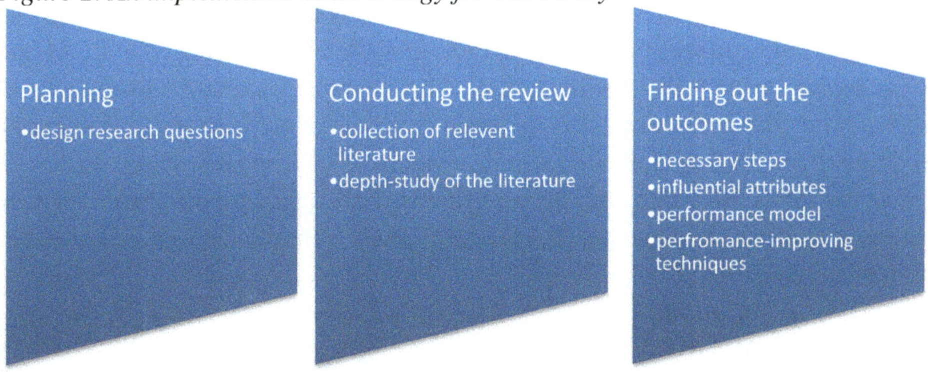

2.1 Review Planning

The main aim of the present work is to explore the noteworthy work done from 2010 to 2025 in the EDM field so that it may represent the latest review study. The analysis of previous studies was conducted to find out the solution of the following research questions:

Research Question 1: What are the major steps for predicting low-performing students?
Research Question 2: Which characteristics of the students are used to predict their academic performance?
Research Question 3: What are the various state-of-the-art performance models that can predict students' academic performance efficiently?
Research Question 4: What are the various techniques that may improve the performance results of the predicting methods?

2.2 Procedure for Conducting the Review

With the help of a search engine, "Google Scholar", different databases such as Academia, IEEE Xplorer, Science Direct, Springer Link, ACM, Web of Science, Research Gate, etc. were explored to obtain answers to the framed research questions. For this, keywords such as Educational data mining, Students' academic performance, Prediction of low performers, and a mixture of them were used for searching relevant research articles. Research papers were also received through mail by registering in the listed databases. The criteria for inclusion and exclusion were also defined for choosing scientific papers. Conferences, Journals, and book chapters having reputed titles were considered for research articles' study. Further, old research articles published before 2010 and having language other than English, were excluded from the study. Although Learning Analytics is related to EDM but while searching the relevant article, research papers that were found to be focused on Learning Analytics were also excluded from the study because its context is different. Backward and forward reference searching has also been performed to get relevant articles. Research articles were finalized after reading the abstract which is the heart of any research paper. Finally, 103 research articles were selected for in-depth analysis.

2.3 Finding out the Outcomes

After the exploration and in-depth study of the past literature, the outcomes of the designed research questions were found. These outcomes are elaborated in section 3. The present literature review is also valuable in determining the scope of further study.

3. RESULTS AND DISCUSSION

In this section, we have elaborated on the work done by various authors and their results. In previous works, authors have taken different parameters as well as various data mining algorithms to predict student academic performance. So, it will be legitimate to predict important attribute categories that can be used to classify good and poor performers correctly. Also, different algorithms can be compared to find out the best technique for predicting student academic performance. From the analysis of the articles of previous authors, the objectives of the current work in the form of four research questions may be answered as follows:

3.1 RQ1: What are the Major Steps for Predicting Low-Performing Students?

The present section describes the major steps to predict students' academic performance that typically follow the KDD (Knowledge Discovery in Dataset) model (Márquez-Vera et al., 2012; Mueen et al., 2016; Xiao et al., 2021) and CRISP-DM (Cross Industry Standard Process for Data Mining) model (Delen et al., 2020; Fernandes et al., 2019; Gil et al., 2021).

KDD Model

For predicting students' academic performance, five main steps of the KDD Model have to be followed. These steps are shown in Figure 3. The first step of the KDD model is data preparation in which the dataset is prepared by oneself or can be taken by various sites like Kaggle, UCI repository, etc., or may be gathered from different sources. The second step is Data Preprocessing in which data transformation, feature selection, data balancing, and data cleaning are the important sub-steps so that data is prepared for application in data mining techniques. Data transformation techniques transform the raw data into a format that is suitable for data mining purposes. In case the data is taken from various sources, it is not necessary that the whole data be useful for the research, so only task-relevant data and relevant features are chosen from the entire dataset. During the gathering of the data, unnecessary, duplicate, and inconsistent information may be collected that misleads the research. As a result, data cleaning is required to remove this inconsistency and noise from the collected data. If the collected dataset has significantly different proportion of classes then dataset balancing is also required for the correct classification. The third step is an important step in which intelligent approaches are implemented in order to retrieve useful relationships or patterns from the examined data. The fourth step is Pattern Evaluation which is used to evaluate interesting patterns that represent knowledge based upon some interesting criteria. After pattern evaluation, the mined knowledge will be represented to the user with the help of knowledge representation or visualization techniques.

Figure 3. KDD Model

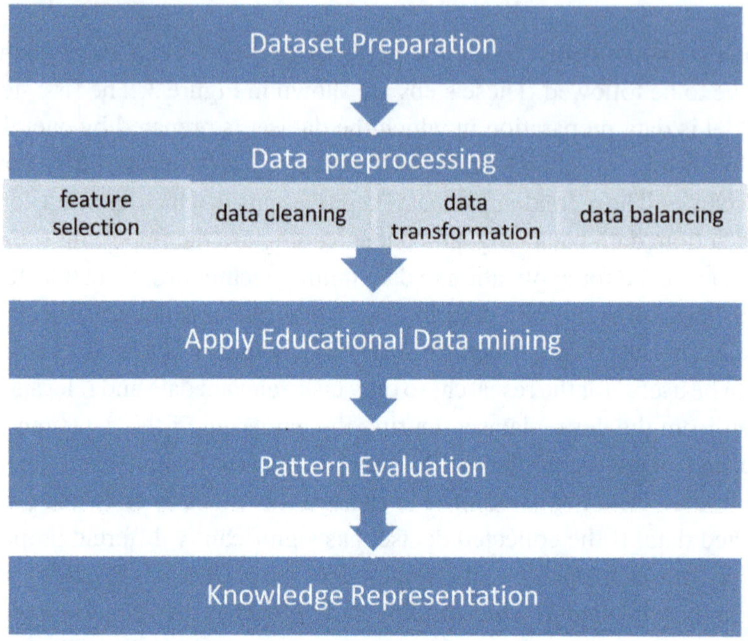

CRISP-DM Model

For predicting students' academic performance, the CRISP-DM Model includes five main steps. A schematic of the CRISP-DM Model is given in Figure 4. Business Understanding is the first stage of the CRISP-DM Model and its purpose is to recognize the project objectives and necessities from a business point of view. The second stage is the Data Understanding phase in which attributes are finalized, interaction with the participants is done, and finally initial dataset is prepared. Then, in the Data Preparation stage, a final dataset is built from the initial raw dataset and is ready to feed into any data mining model. In the fourth stage, the data mining model is selected and implemented. Finally, the deployment stage implements the data mining process throughout the organization.

Figure 4. CRISP-DM Model

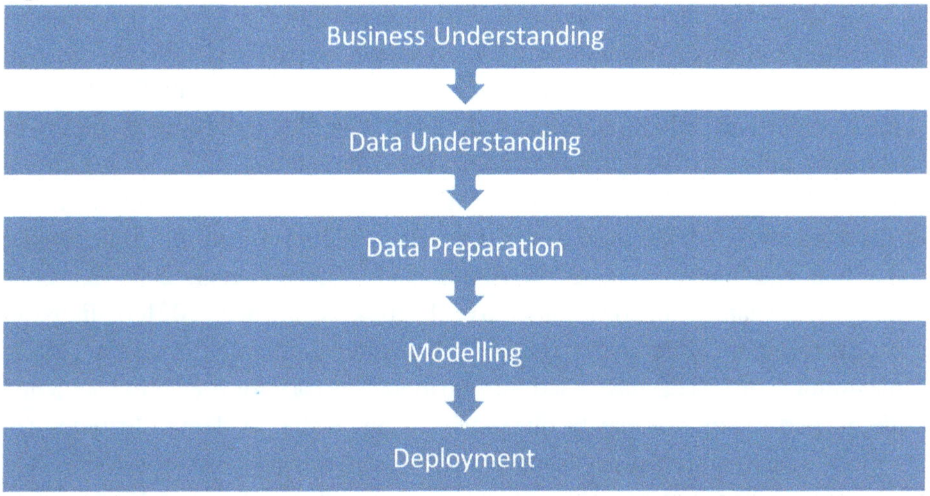

3.2 RQ2: Which Characteristics of the Students are Used to Predict Their Academic Performance?

For the prediction of students' academic performance, the analysis of various attributes that may affect the student's academic performance may be useful. The selection of suitable attributes is a very important task before applying any data mining technique as it provides the area of concern in which more emphasis may be given as well as it may enhance the prediction accuracy. Various authors used different types of attributes to predict or analyze students' performance in terms of GPA, degree completion, academic retention, placement capability, etc. From previous work on input attributes, it is implied that students' attributes can be divided into five broad categories, i.e. (i) Background Attributes, (ii) Academic Attributes, (iii) Social Activity Attributes, (iv) Psychological Attributes, and (v) Behavioral Activities.

Academic Attributes

These types of attributes are related to students' academic activities like class test marks, assignment marks, admission scores, 10th Standard Percentage, 12th Standard Percentage, etc. The present literature review revealed that academic attributes are the most widely used attributes, as most of the authors (Buldu & Üçgün, 2010; Tsai et al., 2011; Sembiring et al., 2011; Nandeshwar et al., 2011; Márquez-Vera et al., 2012; Nasiri & Minaei, 2012; Osmanbegorvic & Suljic, 2012; Kabakchieva, 2013;

Pal & Pal, 2013; Huang & Fang, 2013; Adeyemo & Kuyoro, 2013; Márquez-Vera et al., 2013; Gray et al., 2014; Mishra et al., 2014; Sharabiani et al., 2014; Thammasiri et al., 2014; Alfiani & Wulandari, 2015; Amrieh et al., 2016; Pandey & Taruna, 2016; Hamsa et al., 2016; Costa et al., 2017; Estrera et al., 2017; Ashraf et al., 2018; Ahmad & Shahzadi, 2018; Hashim et al., 2018; Kiu, 2018; Yaakub et al., 2018; Adekitan & Salau, 2019; Fernandes et al., 2019; Kamal & Ahuja, 2019; Gray & Perkins, 2019; Wong, 2019; Bindhia & Suvanam, 2019; Tomasevic et al., 2020; Verma & Yadav, 2020; Delen et al., 2020; Ha et al., 2020; Hasan et al., 2020; Injadat et al., 2020; Karthikeyan et al., 2020; Ashraf et al., 2020; Gil et al., 2021; Alnassar et al., 2021; Desiani et al., 2021; Hussain et al., 2021; Nahar et al., 2021; Yuliansyah et al., 2021; Begum & Padmannavar, 2022; Kumar et al., 2022; Verma et al., 2022; Ramaswami et al., 2022) have used this attribute category for predicting students' academic performance. It may be due to the reason that the academic attributes can be collected easily and correctly from the institutional database. Mat et al. (Mat et al., 2013) concluded that prerequisite subjects, attendance, and marks in Mathematics and English are the indicators for measuring students' final marks. From the past work, it may be concluded that 10th Standard Percentage, 12th Standard Percentage, and the rank obtained by the entrance examination are the most influential features. The reason may be that many countries conducted these examinations centrally and transparently hence results obtained by these examinations are reliable and, to some extent, measure the true performance of the students. Roslan & Chen (Roslan & Chen, 2023). found that the past academic performance of students was the most influential predictor, along with some demographic and psychological attributes. Hussain et al. (Hussain et al., 2018). found that the internal assessment attribute was also the highest influential factor for predicting students' final semester results.

Background Attributes

This category considered the background of students. For example, age, family size, family status, type of school, family income, residential location, parent status, parents' qualification, parents' occupation, etc. Almost every study has included these attributes to predict students' academic outcomes (Tsai et al., 2011; Nandeshwar et al., 2011; Márquez-Vera et al., 2012; Osmanbegorvic & Suljic, 2012; Kabakchieva, 2013; Pal & Pal, 2013; Márquez-Vera et al., 2013; Gray et al., 2014; Mishra et al., 2014; Sharabiani et al., 2014; Thammasiri et al., 2014; Alfiani & Wulandari, 2015; Amrieh et al., 2016; Mueen et al., 2016; Pandey & Taruna, 2016; Costa et al., 2017; Estrera et al., 2017; Adejo & Connolly, 2018; Ahmad & Shahzadi, 2018; Ashraf et al., 2018; Hashim et al., 2018; Hussain et al., 2018; Kiu, 2018; Yaakub et al., 2018; Bindhia & Suvanam, 2019; Fernandes et al., 2019; Kamal & Ahuja, 2019; Masud et al., 2019; Tomasevic et al., 2020; Verma & Yadav, 2020; Delen et al., 2020; Ha et al., 2020; Karthikeyan et al., 2020; Musso et al., 2020; Ashraf et al., 2020; Gil et al., 2021; Alnassar et al., 2021; Desiani et al., 2021; Nahar et al., 2021; Rahman et al., 2021; Yuliansyah et al., 2021; Begum & Padmannavar, 2022; Kumar et al., 2022; Ramaswami et al., 2022; Teoh et al., 2022; Verma et al., 2022; Roslan & Chen, 2023). The collection of background data can be achieved from the student as well as the institution database easily and correctly (Saa et al., 2019) reported that gender is one of the background attributes that may affect students' performance as the existence of gender differentiation may affect students' degrees of motivation. In the findings of Parajuli & Thapa (Parajuli & Thapa, 2017). and Ioannis & Maria (Ioannis & Maria, 2019) female students outperformed male students. In today's scenario, all the courses in higher education are conducted in the English medium. Students having medium other than English in past academics, face difficulties in understanding concepts and meanings. So, the medium of the previous study of students plays a significant role in their academic performance (Verma & Yadav, 2020; Verma et al., 2022). Bhardwaj & Pal (Bhardwaj & Pal, 2011) revealed that the residential location or surroundings in which students live may affect their academic achievements because students living in urban areas have better facilities for learning than those who live in rural regions. Moreover, parents' education affects the academic performance of the students; well-qualified parents may help their children by giving them the right counseling and support. Muller (Muller, 2018), Idris & Ahmad (Idris et al., 2020)., Verma & Yadav (Verma & Yadav, 2020)., and Roslan & Chen (Roslan & Chen, 2023). discovered that there is a significant relationship between parents' qualifications and the academic performance of their children. In several situations, the occupation of parents may also affect the performance of students. Generally, highly busy parents do not spend their time with their children and do not provide sufficient educational support to their children directly. The living status of the parents may also affect the student's academic achievement. Sometimes, students do not perform well when their parents are living apart. Adeyemo & Kuyoro (Adeyemo & Kuyoro, 2013). found that parents' conditions, including their marital status, occupation, and education, also have a high impact on students' academic performance. Ayinde et al. (Avindi et al., 2013). have explored associations between the number of children in a family and children's academic achievement.

Social Activity Attributes

This category considered all the attributes related to the participation of the students in social activities such as, extra-curricular activities, number of friends, use of the internet, etc. Ribeiro et al. (Ribeiro et al., 2024). found the positive impact of extracurricular activities on the students' academic performance and their employability. Various studies have been carried out by using social activities attributes to identify students' academic. Hardy & Bryson (Hardy & Bryson, 2016). examined that students having strong social network with their peers and teaching staff, achieved better in their academics. It is observed that by participating in extracurricular activities, students' self-confidence and self-expression skills are developed. With the help of these skills, they may perform well in their studies. Mat et al. (Mat et al., 2013)include extracurricular activities in their dataset for predicting students' academic performance. Kiu (Kiu, 2018). investigated the role of students' background, social activities, and coursework results in predicting their academic achievements in mathematics. He found that students' backgrounds and social activities play a significant role in prediction. Rahman et al. (Rahman et al., 2020). also discovered a positive correlation between extra-curricular activities and academic performance.

Psychological Attributes

It includes all the attributes related to students' psychology such as, a reason to choose a program, homesickness, motivation, interest in the study, etc. From past data, it was observed that very limited studies were performed using psychological factors. It may be due to the reason that the accurate collection of psychological data is not easy as students don't want to share their psychological information with anyone and may possibly give incorrect or misleading information. A student may pursue higher education with self-motivation or motivation from others like parents or family friends and self-motivated students may perform well in their studies. Gray et al. (Gray et al., 2014). and Musso et al. (Musso & Cascallar, 2020). included motivation as an input attribute to determine the academic performance of students. Homesickness may affect students' academic performance as highly homesick students may not perform well in their academics (Sun & Hagedorn, 2016). Adaptability is directly related to homesickness, students have to go from outside their homes to learn and they have to adjust to a new environment. Adjustable students feel comfortable and may focus on their studies, hence may achieve better results. Mental stress is also an important factor that may influence the academic performance of students. Sembiring et al. (Sembiring et al., 2011). concluded that the mental condition and academic performance of students are highly correlated. Sinval et al. (Sinval et al., 2025). concluded that the depression, anxiety, and stress have a strong relation with poor educational performance. Moreover, students who have a great interest in the study may have high academic achievement (Sembiring et al., 2011; Ahmad & Shahzadi, 2018). Health is also an important factor that may influence the academic outcome of a student (Kiu, 2018). A healthy student may focus on his/her studies and may achieve better results. Introvert or extrovert personalities may also affect students' performance. Every personality has its own characteristics, and introverted students may focus more on their studies at the higher education level. Gray et al. (Gray et al., 2014). used personality as one of the input attributes to determine the academic achievement of students. It is observed that a confident student outperforms a less confident student and has a better oral presentation (Al-Hebaish, 2012). Further, curiosity can develop learning habits as students eager to learn new things may cultivate the tendency of study, which can improve their academic performance (Maksum & Khory, 2020). Khan & Rehman (Khan & Rehman, 2017). concluded that punctuality is the key to success and plays a very significant role in students' academic performance. It may be due to the fact that punctual students work on a pre-planned schedule and achieve their goals, such as successfully completing their course.

Behavioural Activities

Online activities, video interactions, raising hands, being interactive, student engagement, etc. can also play an important role in predicting the academic performance of students. This category of attributes is mainly important in an online learning environment in which students' performance is analyzed with their engagement. Mueen et al. (Mueen et al., 2016)., Costa et al. (Costa et al., 2017)., Oeda & Hashimoto (Oeda & Hashimoto, 2017., Ahmad & Shahzadi (Ahman & Shahzadi, 2018)., Bindhia & Suvanam (Bindhiya & Suvanam 2019)., Kamal & Ahuja (Kamal & Ahuja 2019)., Hasan et al. (Hasan et al., 2020)., Karthikeyan et al. (Karthikeyan et al., 2020)., Tomasevic et al. (Tomasevic et al., 2020), Nkomo & Nat (Nkomo & Nat, 2021)., Nahar et al. (Nahar et al., 2021)., and Ramaswami et al. (Ramaswami et al., 2022). have predicted the student's performance in higher educational institutions with the help of data related to behavioral features. Amrieh et al. (Amrieh et al., 2016). accomplished a 22.1% improvement in results when they used behavioral features in an online learning environment. Teoh et al. (Teoh et al., 2022). predicted student academic performance by using background, video watch engagement, and their learning style. Aouifi et al. (Aouifi et al., 2021). observed a strong connection between learning performances and the viewing of video sequences.

In previous work, various authors have taken different parameters to predict student academic performance which is related to background data, student engagement data, student academic data, etc. So, it will be legitimate to predict important attribute categories that can be used to classify good and poor performers correctly. The distribution of all the five categories of attributes is presented in Figure 5. From this figure; it is observed that academic attributes are the most widely used attribute category for predicting students' academic achievement, after that background attributes were used in past literature. Social activity and psychological attributes are in third and fourth place respectively while behavioral attributes were the least used attribute category in past literature. So, in the future, there is a need to elaborate on the effect of Social activity, psychological and behavioral attributes on students' academic performance and it may be a focus area of research for upcoming researchers.

Figure 5. Distribution of Attributes

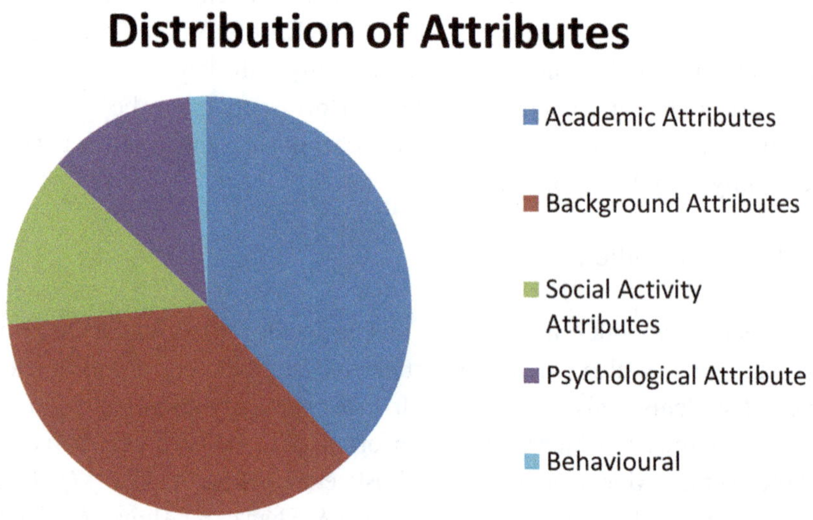

3.3 RQ3: What are the State of The Art Techniques Available to Predict Students' Academic Performance?

In this section, various data mining techniques used by authors for the prediction of students' academic performance are described as follows:

3.3.1 Association Rule Mining

Association rule mining is a data mining technique used to find out rules that will predict the occurrence of an item based on the occurrences of other items. Association rule algorithms such as the Apriori algorithm were used by previous authors that produce essential rules to predict students' academic outcomes (Buldu & Üçgün, 2010; Hashim, Hamoud, & Awadh, 2018; Hussain, Dahan, Ba-Alwi, & Ribata, 2018). By association rule mining, correlations, causal relationships, frequent patterns, and associations among the set of items are discovered in transaction databases, relational databases, or other data repositories. Association rules

(AR) are often used in several areas, such as telecommunication networks, market and risk management, inventory control, cross-marketing, classification, catalog design, etc. There are two measures in the Association rule A→B: (i) Support and (ii) Confidence, where Support is defined as the percentage of records that contain both A and B out of the total number of records, and confidence is also known as conditional probability implies the percentage of records that contain both A & B to the records that contain A. The rules that fulfill the minimum threshold value of support and confidence are called interesting and useful rules. Badr et al. (Badr et al., 2016). predicted student performance in a programming course based on the performance in the English and Mathematics subjects with the help of Classification based on Association rule algorithm. The input parameters were Graduation year, student identification number, course code, and grade point average. They concluded that the Mathematics subject grade did not impact the programming course.

3.3.2 Clustering Techniques

After classification techniques, clustering techniques are popular to determine high and low performers and could provide necessary suggestions and take beneficial action for students' gain. Clustering is an unsupervised data mining technique applied to the unlabeled dataset and is a useful method to analyze high-dimensional data and large-scale databases. In this technique, similar types of instances are grouped together based on similar characteristics, and clusters are formed based on the principles of high intra-class similarity and low inter-class similarity. Thus the objects in one cluster are very different from those in other clusters. Tsai et al. (Tsai et al., 2011). have made predictions to assess students' computer literacy before admission into higher education by using k-means, self-organizing maps (SOM), and two-step clustering (i.e. BIRCH). After finding the best clustering result they used the Decision Tree algorithm to extract useful rules from each of the identified clusters. From the study, they have concluded that these rules can be used by universities to identify, warn, or counsel students who have a higher probability of failing the test. Alfiani & Wulandari (Alfiani & Wulandari, 2015). used a dataset of 300 students consisting of demographic and attendance information. They mapped students into their actual condition by using a k-mean clustering algorithm and found three clusters namely smart, standard, and low performer. Oeda & Hashimoto (Oeda & Hashimoto, 2017., Asif et al. (Asif et al,. 2017), Bindhia & Suvanam (Bindhiya & Suvanam 2019)., and Vasuki & Revathy (Vasuki & Revathy, 2022). also used clustering techniques to analyze students' progress. Nkomo & Nat (Nkomo & Nat, 2021). applied k-mean clustering in an online learning platform for making groups of similar characteristics. Feng et al. (Feng et al., 2022). determined the number of clusters while using the k-mean algorithm.

3.3.3 Regression

The regression technique is a numeric value prediction data mining technique used to find out the relationship between dependent variables and one or more independent variables, as well as to determine how such relationships can be used to predict students' academic achievements. EDM researchers have applied several regression techniques like linear regression, multiple linear regression, etc. for predicting students' academic performance and identifying factors that could predict the students' performance (Adekitan & Salau, 2019; Masud et al., 2019; Huang & Fang, 2013; Tomasevic et al., 2020; Hussain et al., 2021). To validate the performance of students, Adekitan & Salau (Adekitan & Salau, 2019) applied linear as well as pure-quadratic-based regression models and achieved R^2 values were 0.955 and 0.957 respectively. Huang & Fang (Huang & Fang, 2013). predicted student scores with the help of SVM and students' dataset having 323 undergraduate students with eight cognitive factors.

3.3.4 Genetic Algorithm

Genetic Algorithms plan to integrate concepts of natural evolution and can be used for classification as well as optimization problems. Hamsa et al. (Hamsa et al., 2016). have proposed a model to predict students' academic performance by using a Decision Tree and fuzzy genetic algorithm. To conduct their study they have chosen internal marks, sessional marks, and admission scores as input parameters. Márquez-Vera et al. (Márquez-Vera et al., 2021) proposed a genetic programming algorithm and different white-box data mining algorithms to predict student failure on a dataset of 670 high school students from Mexico, they also selected the best attribute to reduce the high dimensionality of data and applied rebalancing to overcome the problem of imbalanced data.

3.3.5 Classification Technique

Classification is a supervised, most popular, and most widely used (because of the availability of labeled data) technique for predicting students' academic performance. It classifies objects into predefined classes by using a labeled training dataset. Classification algorithms are implemented to build a classifier that performs binary or multiclass classification of students by using the previous history of students. The accuracy of a classifier depends on various factors like input attributes, database size, model validation technique, types of classification, etc. For building classification models, various supervised machine learning algorithms may be implemented that are discussed as follows:

Decision Tree (DT)

DT is one of the most popular and simple algorithms for predicting students' academic performance. Its structure is like a flowchart or a tree where each internal node tests an attribute, each branch corresponds to an outcome or an attribute value, and each leaf node assigns a classification. Decision Trees may be converted into rules for the classification of a dataset. The Iterative Dichotomizer (ID3), C4.5, Classification and Regression Tree (CART), and Chi Square Automatic Interaction Detection (CHAID) algorithms come under the Decision Tree algorithm. From previous research, it is found that Decision Tree is the most common classification algorithm because of its understandability and interpretability (Kabakchieva, 2013; Pal & Pal, 2013; Adeyemo & Kuyoro, 2013; Márquez-Vera et al., 2013; Mishra et al., 2014). The work of previous authors who preferred the Decision Tree algorithm or found it to be most suitable to predict students' academic performance is summarised in Table 1.

Table 1. Results of Decision Tree

Author	Attributes	Dataset Size	Model validation	Classification Type	Performance result
Nasiri & Minaei (Nasiri & Minaei, 2012).	Academic	Records are not specified and 13 attributes	Not specified	Multiclass (5 classes)	C5.0 achieved 88.5% accuracy
Pal & Pal (Pal, A. & Pal, S., 2013).	Academic and background attributes	200 records with 15 attributes	Hold-out	Multiclass (4 classes)	ID3 gave classification accuracy of 78%
Kabakchieva (Kabakchieva, 2013).	Academic and background attributes	10330 records with 20 parameters	Hold-out	Multiclass (5 classes)	C4.5 obtained 66.59% accuracy
Adeyemo & Kuyoro (Adeyemo & Kuyoro, 2013).	Academic and background attributes	240 records and 18 attributes	10-fold cross-validation	Multiclass (7 classes)	C4.5 algorithm achieved 71% accuracy
Estrera et al. (Estrera et al., 2017).	Academic, psychological, and background attributes	150 records and 9 attributes	10-fold cross-validation	Multiclass (4 classes)	Accuracy of 90.67% by using ID3
Kiu (Kiu, 2018).	Background, social activity, and academic	395 records with 33 attributes	Not specified	2-level and 5-level	C4.5 achieved 92.4% accuracy with 2-level and 79.1% accuracy with 5-level
Bindhia & Suvanam (Bindhiya & Suvanam 2019).	Academic, behavioral, and extra features	16 features and records are not specified	Hold-out	Binary	75.47% accuracy with ID3

continued on following page

Table 1. Continued

Author	Attributes	Dataset Size	Model validation	Classification Type	Performance result
Gray & Perkins(Gray & Perkins, 2019).	Numeric codes for the student's program, the school, and the year of study	4970 records and 32 features	LOOCV	Multiclass (5 classes)	86.20% accuracy with C4.5
Yuliansyah et al. (Yuliansyah et al., 2021).	Academic and background attributes	740 records and 4 features namely department, GPA, English score, and age	Hold-out	Binary	C4.5 algorithm achieved 90% accuracy
Roslan & Chen (Roslan & Chen, 2023).	Academic, demographics, and psychological attributes	159 students' data and 29 factors	Hold-out (80:20)	Multiclass (3 classes)	ID3 achieved 87% accuracy
Kumar et al. (Kumar et al., 2022).	Academic, background, social, and school-related characteristics	649 records with 33 attributes	10-fold cross-validation	Multiclass (5 classes)	DT had 96.76% accuracy

Bayesian Classifiers (Naïve Bayes, Bayesian Network)

These are other popular supervised classification algorithms to predict students' academic performance which can determine class membership probabilities and are based on the Bayes Theorem. To simplify the computation, Naïve Bayes (NB) assumes independence among attributes, so it is called 'Naïve'. It is easy to use and requires a small amount of training data to estimate the parameters necessary for classification. Various authors concluded that Naïve Bayes perform better than other classifiers with small data sets (Mueen, Zafar, & Manzoor, 2016; Ha et al., 2020; Ashraf, Zaman, & Ahmed, 2020; Verma, Yadav, & Kholiya, 2022; Roslan & Chen, 2023). The Bayesian Network (Bayes Net) is a probabilistic graphical model that uses a directed acyclic graph and a set of conditional probability tables (Delen et al., 2020). Its directed acyclic graph represents a set of variables and their conditional dependencies. Table 2 and Table 3 compile the brief work of previous researchers who showed the Naïve Bayes and the Bayesian Network as the best-performing classifiers, respectively.

Table 2. Results of Naive Bayes

Author	Attributes	Dataset Size	Model validation	Classification	Performance result
Osmanbegorvic & Suljic (Osmanbegorvic & Suljic, 2012).	Academic, background, and social activities	257 records and 12 features	3-fold cross-validation	Binary	76.65% accuracy
Mueen et al. (Mueen et al., 2016).	Academic record and forum participation	60 records with 38 attributes	10-fold cross-validation	Binary	Overall prediction accuracy of 86%
Asif et al. (Asif et al,. 2017)	Academic attributes	210 records with 8 variables	The model established for one cohort generalizes to the following cohort	Multiclass (5 classes)	With 83.65% accuracy
Ashraf et al. (Ashraf et al., 2020).	Background, academic attributes	23 attributes and 1793 records	10-fold cross-validation	Multiclass (3 classes)	95.50% accuracy
Ha et al. (Ha et al., 2020).	Academic and background attributes	525 records & 9 attributes	Holdout method with 70: 30 split ratio	Multiclass (5 classes)	Accuracy of 86.19%

Table 3. Results of Bayesian Network

Author	Attributes	Dataset Size	Model validation	Classification	Performance result
Sharabiani et al. (Sharabiani et al., 2014).	Academic and background attributes	300 records with 10 attributes	Train-test split (70:30)	Multiclass (4 classes)	Accuracy of 60%
Delen et al. (Delen et al., 2020)	Demographic attributes, pre-college records, college performance characteristics, scholarship data, and financial support-related attributes	Data consisting of 35,021 records and 33 factors	10-fold cross-validation	Binary	87% accuracy

Artificial Neural Network (ANN)

It is a computational algorithm that bears a resemblance to a biological neural network. It is a network of processing units (neurons) with connections (synapses) between them, and each connection has a weight present with it. During the learning phase, the network learns by adjusting weights so as to be able to predict the correct class labels of the input records. The main benefit of using this classification algorithm is that it can probe all possible relations between input variables and can also detect a complex nonlinear relationship between dependent and independent variables. The neural network algorithm has been acknowledged as one of the best prediction techniques by some previous authors (Ahmad & Shahzadi, 2018; Tomasevic, Gvozdenovic, & Vranes, 2020; Ha et al., 2020; Musso, Hernandez, & Cascallar, 2020; Agarwal & Agarwal, 2023). It is less used than any other classification algorithm in the field of EDM because of its complexity and less interpretability. In Table 4, the results of the researchers who found ANN to be a preferable algorithm are compiled.

Table 4. Results of Artificial Neural Network

Author	Attributes	Dataset Size	Model Validation	Classification	Performance result
Ahmad & Shahzadi (Ahman & Shahzadi, 2018).	Background, academic, and behavioral characteristics	300 records and 8 attributes	Hold-out	Binary	MLP achieved Accuracy of 85%
Ha et al. (Ha et al., 2020).	Academic, background	525 records and 9 attributes	Holdout method with 70: 30 split ratio	Multiclass (5 classes)	Accuracy of 86.19%
Agarwal & Agarwal (Agarwal & Agarwal, 2023)	Background, Educational and Psychological attributes	689 records and 19 attributes	Holdout method with 70: 30 split ratio	Multiclass (3 classes)	Achieved 96% accuracy

Support Vector Machine (SVM)

SVM is an algorithm that is used for classifying linear as well as nonlinear data. This algorithm transforms training objects into a higher dimension and then uses a decision boundary to classify objects into different classes. Huang & Fang (Huang & Fang, 2013)., Pande (Pande, 2023). and Lacerda et al. (Lacerda et al., 2023). proved that SVM also produced good results with 90%, 84.37%, 93% accuracy respectively, while predicting the academic performance of students. Costa et al. (Costa et al., 2017). also compared four EDM algorithms (Decision Tree, SVM, Neural Network, and Naïve Bayes) applied to two different datasets of 262 distance and 161 on-campus students to predict failure in an introductory programming course at an early stage in Brazil Public University. Their results indicated that all the algorithms are able to predict students' failure at an early stage while SVM was the highest-performing algorithm. Table 5 summarizes the work of Gil et al. (Gil et al., 2021). and Gray et al. (Gray et al., 2014). who preferred SVM for predicting academic outcomes of students.

Table 5. Results of Support Vector Machine

Author	Attributes	Dataset Size	Model Validation	Classification	Performance result
Gil et al. (Gil et al., 2021).	Socio-demographic, social origin, previous education, special statutes, and educational path dimensions	9652 records and 68 features	10-fold cross-validation	Binary	AUC around 0.94 (94%)
Gray et al. (Gray et al., 2014).	Academic, background, and psychological attributes	914 records and 4 attributes	10-fold cross-validation	Binary	SVM achieved the accuracy of 73.3%

Logistic Regression (LR)

It is a simple supervised classification algorithm used to predict the probability of a target variable being in a discrete set of classes. Poturić et al. (Poturić et al., 2023). achieved 94.74% accuracy with LR for predicting academic success of students with the help of 93 students' dataset consisting quizzes and project activities. The results achieved by the authors who find out Logistic Regression as the best algorithm for predicting students' academic performance are presented in Table 6.

Table 6. Results of Logistic Regression

Author	Attributes	Dataset size	Model validation	Classification	Performance result
Adekitan & Salau (Adekitan & Salau, 2019)	Academic attributes	Dataset of 1841 students and 4 factors	Hold-out (70:30) method with stratified sampling	Multiclass (4 class label)	Accuracy of 89.15%
Rahman et al. (Rahman et al., 2020)	Academic and extracurricular based attributes	850 records and 13 factors	Train-test split	Binary	The highest accuracy obtained was 99.5294%
Verma et al. (Verma et al., 2022).	Academic, background, social activity. and psychological	550 records with 27 attributes	Hold-out method	Binary	92.72% accuracy

k-Nearest Neighbour (k-NN)

k-NN is a supervised lazy learner algorithm that assigns the class of a test sample to the majority of the classes of its k nearest training samples. The value of k in k-NN should be higher than zero. If k=1, then k-NN is called Nearest-Neighbor. With the help of k-NN, Tanner &Toivonen (Tanner & Toivonen, 2010). conducted a study in which they classified students who are taking an online course into two categories, "students at the risk of dropout" and "students who have the capability to succeed in the exam". They also predicted the actual scores of the final test. Gray & Perkins (Gray & Perkins, 2019). predicted students' performance as early as week 3 with the help of 1-NN, C4.5 Trees with Leave One Out Cross-Validation (LOOCV). The work of the previous researchers who used k-NN and found it to be most suitable algorithm for academic performance prediction is briefed in Table 7.

Table 7. Results of k-Nearest Neighbour

Author	Attributes	Dataset Size	Model Validation	Classification	Performance result
Aouifi et al. (Aouifi et al., 2021).	Video courses' details such as learners' ID, click sequences, and grade	The number of learners was (N=66) and 3 factors	3-fold cross-validation	Binary	Average accuracy of 65.07%.
Alnassar et al. (Alnassar et al., 2021).	Background, academic, and behavioral attributes	32,593 records with 7 attributes	Train-test split	Binary	99.93% accuracy with past academic data

The different classifiers used by the previous authors as the best-performing classifiers are represented in Figure 6. From this figure, it is observed that the Decision Tree is widely used by the previous authors and gives the best results for predicting students' educational achievement as most authors observed that it is better than other classifiers. After that, Bayesian classifiers especially Naive Bayes were used by the previous authors and achieved the best results. The highest achieved accuracy by a classifier is plotted in Figure 7. It may be observed that although the maximum achieved accuracy with k-NN is 99.93% with past academic attributes (Alnassar et al., 2021)., but it can't be declared as best performing algorithm as in some other studies it underperformed in comparison to other classifiers (Tomasevic, Gvozdenovic, & Vranes, 2020; Yagci, 2022). Thus, there is no unanimous results regarding the best-performing algorithm and for different dataset different algorithm may give the best prediction result. This raises the need of a scalable model which may outperform for various datasets.

Figure 6. Distribution of Classifiers

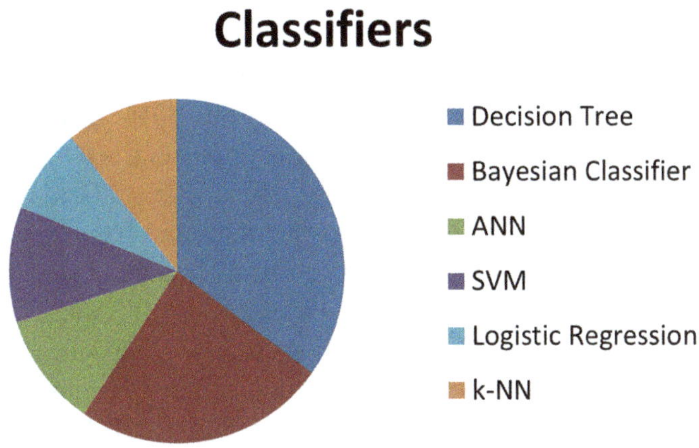

Figure 7. Comparison of classification algorithms

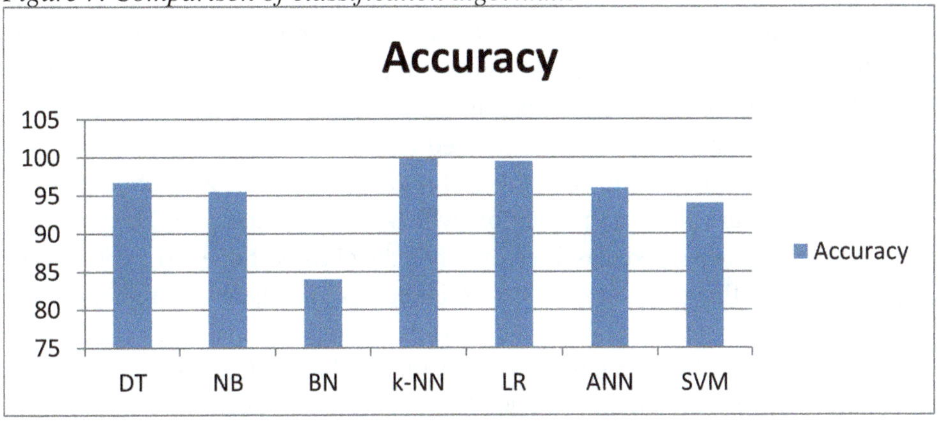

3.4 RQ4: Which techniques may improve the performance results of the predicting methods?

An ensemble model construction is a very popular technique to improve classification accuracy. Besides this, Data preprocessing steps such as feature selection, balancing the dataset, etc. may also improve the performance results (Costa et al., 2017; Ashraf et al., 2020). These techniques are discussed as follows:

3.4.1 Ensemble Model

Ensemble-based classifiers use multiple classifiers to obtain better predictions than any of the ingredient classifiers alone and usually give more accurate results than other algorithms (Amrieh et al., 2016; Ashraf et al., 2018; Injadat et al., 2020; Verma et al., 2022; Wunnasri et al., 2023). Bagging, Boosting, and Random Forests are the main examples of ensemble classifiers. Besides this, Stacking is also the most popular heterogeneous ensemble technique that may achieve better results than other ensemble methods (Adejo & Connolly, 2018). An ensemble model is an integration of traditional classifiers in which every classifier's vote counts and a class label prediction is returned by the ensemble based on the collection of votes. Ensemble models are more reliable than single classifiers because they reduce the variance of predictions. Amrieh et al. (Amrieh et al., 2016). gained a 25.8% enhancement in accuracy with the ensemble model. There are various types of ensemble models, such as:

- Bagging (Bootstrap Aggregation): For merging of k models M_1, M_2 ...M_k, training sets D_1, D_2..... D_k are formed by sampling with replacement from an original dataset D, where each training set is a bootstrap sample in which there is a possibility to include some of the instances of *D* more than once, while others may not be included in the training sets. In the training phase, each classification model, M_i is learned for each training set, D_i. In the testing phase, to predict the class label of a testing instance, *X*, each individual classifier returns its class prediction, which counts as one vote. The votes are counted by the bagged classifier, and then the class with the most votes is assigned to *X*. Bagging can also be used in regression, where the prediction of a test sample's continuous value output is the average value of each continuous value predicted by each individual classifier. Nahar et al. (Nahar et al., 2021). and Verma et al. (Verma et al., 2022). found Bagging as the best-performing technique in finding students' academic outcome.
- Boosting: Boosting is another ensemble model in which weights are allocated to every training instance of dataset D. A sequence of *k* classifiers is iteratively learned. After a classifier M_i is learned, the weights are updated to allow the subsequent classifier, M_{i+1}, to "pay more attention" to the training instances that were misclassified by M_i. In this way, the next model tries to correct the error that was made by the prior models. The main boosting model merges the voting results of every classification model and the weight of every classification model's vote depends on the function of its accuracy. AdaBoost is the first successfully implemented boosting technique and uses a Decision Tree (also called a decision stump) as a base model. Gradient

Boosting is also another example of a boosting technique that is an extension of AdaBoost and uses an arbitrary loss function to make boosting more flexible. Asselman et al. (Asselman et al., 2021). demonstrated that XGBoosting also obtained the best results. Ramaswami et al. (Ramaswami et al., 2022). applied k-fold cross-validation with the CatBoost algorithm that achieved an average accuracy of 75% in predicting students' academic outcomes.

- Random Forest: As the forest is a collection of trees, in the same way, the Random Forest is a collection of Decision Trees. It is an extension of the Bagging Classifier in which Decision Trees are the individual classifiers that are trained by the bootstrap samples and Random Forest also samples the attributes of the dataset. Hussain et al. (Hussain et al., 2018). evaluated the performance of J48, PART, Random Forest, and Bayesian Network, and concluded that Random Forest has achieved the highest accuracy.

The work of the authors, who achieved maximum accuracy by using Ensemble classifiers for students' academic achievement prediction, is briefly mentioned in Table 8.

Table 8. Results of Ensemble Classifiers

Author	Attributes	Database Size	Model-validation	Classification	Performance result
Mishra et al. (Mishra et al., 2014)	Academic, demographic, and psychological	250 records with 25 attributes	10-fold cross-validation	Multiclass (4 classes)	Random Tree obtained 94.418% accuracy
Adejo & Connolly (Adejo & Connolly, 2018).	Demographic, personality, psychological, employment, and institutional factors	141 records. With 14 attributes	70:30 with 10-fold cross-validation	Binary	Hybrid model with Stacking achieved 81.67% accuracy.
Hussain et al. (Hussain et al., 2018).	Socio-economic, demographic as well as academic information	300 students with 24 attributes	Bootstrap	Multiclass (5 classes)	Random Forest with 99% accuracy
Kamal & Ahuja (Kamal & Ahuja 2019).	Background, academic, social, and behavioral factors	499 records with 51 factors	10-fold cross-validation	Binary	Ensemble of Decision Tree, Naive Bayes, and SVM achieved 98.5% accuracy.
Hasan et al. (Hasan et al., 2020).	Student academics, behavioral activities	772 records with 19 attributes	10-fold cross-validation	Binary	Random Forest outperformed with an accuracy of 88.3%
Karthikeyan et al. (Karthikeyan et al., 2020).	Academic & background	232 records with 14 attributes	Cross-validation	Binary	HEDM model, a combination of J48 & Naive Bayes, produced an accuracy of 98.6%

continued on following page

Table 8. Continued

Author	Attributes	Database Size	Model-validation	Classification	Performance result
Nahar et al. (Nahar et al., 2021).	Student academics, behavioural activities	Dataset1 contains 80 records & 14 factors, dataset2 contains 80 records and 12 factors	Train-test split	Multiclass (3 classes)	For dataset1, bagging algorithm outperformed with 67.9% accuracy, while for dataset2, boosting algorithm achieved highest accuracy of 75.4%.
Yagci (Yagci, 2022).	Academic (midterm exam grade)	1854 records with 4 factors	10-fold cross-validation	Multiclass (4 classes)	Random Forest outperformed with 74.6% accuracy
Kumar et al. (Kumar et al., 2022).	Academic, background, social, and school-related characteristics	649 records with 33 attributes	10-fold cross-validation	Multiclass (5 classes)	Stacking Ensemble Learning Model including Logistic Regression and Simple Logistic achieved maximum accuracy of 99.53%.
Begum & Padmannavar (Begum & Padmannavar, 2022).	Demographic, social, and academic characteristics	Two datasets containing 33 attributes namely Mathematics (395 records) and Portuguese (649 records and)	10-fold cross-validation	Multiclass	Boosting has achieved maximum accuracy of 98.11 and 96.82% for Mathematics and Portuguese dataset respectively.

3.4.2 Improving Classification Accuracy of Class-Imbalanced Data

The dataset is called class-imbalanced if the main class of interest (Positive class) is rare. There are various approaches for improving the classification accuracy of class-imbalanced data. These approaches include (i) Oversampling, (ii) Undersampling, and (iii) Hybrid Sampling.

Oversampling and Undersampling change the distribution of tuples in the training set so that the rare (positive) class is well represented. In oversampling, the training set acquires the same number of positive and negative tuples by resampling or increasing the positive tuples. Undersampling works by decreasing the number of negative tuples and to make the number of positive and negative tuples the same, it arbitrarily eliminates tuples from the majority (negative) class. This sampling method is applied when the amount of collected data is sufficient. Hybrid Sampling is the combination of the oversampling and the undersampling techniques. On a pedagogical dataset taken from the University of Kashmir, Ashraf et al. (Ashraf et al., 2018). predicted student performance using an ensemble classifier named Stacking C and also applied filtering procedures for improvements in results. They achieved 95.65% accuracy with undersampling (spread sub-sampling), and 96.11% accuracy with oversampling (SMOTE).

Oversampling raises the weight of the minority class by replicating or creating new minority-class samples and it is worth noting that the oversampling approach is generally applied more frequently than other approaches. On the basis of how tuples are added, there are various types of oversampling techniques such as Random Oversampling, SMOTE (Synthetic Minority Oversampling Technique), Borderline SMOTE, SVM SMOTE, and ADASYN (Adaptive Synthetic).

- Random Over-Sampling: This method increases the size of the dataset by the repetition of the original samples of the minority class until the samples of the minority class become equal to the samples of the majority class. The random over-sampling does not generate new instances; hence it does not make the variety of instances. The drawback of this technique is that the classification models may be overfitted because of the repetitions of the minority samples, and hence they do not learn efficiently. Ghorbani & Ghousi (Ghorbani & Ghousi, 2020). implemented various resampling technique including Random Over-Sampling to get balanced dataset and achieved better results than unbalanced dataset.
- SMOTE: This method is a statistical technique that increases the number of minority samples in the dataset by generating new instances. This algorithm takes samples of the feature space for each target class and its nearest neighbors and then creates new samples that combine features of the target case with those of its neighbors (Márquez-Vera et al., 2012, 2013; Mueen et al., 2016). It may be noted that the new instances in this method are not copies of existing minority samples. Desiani et al. (Desiani et al., 2021). observed that the accuracy of the classifiers was significantly improved after applying SMOTE. Verma et al. (Verma et al., 2022). also pointed out that the accuracy of logistic regression was improved from 92.72% to 94.54% and the accuracy of the ensemble model is improved from 93.63% to 95.45% while using SMOTE.
- Borderline SMOTE: Generally, samples and the neighboring ones are more likely to be misclassified than the ones far from the borderline. To resolve this problem Borderline SMOTE method uses the number of majority neighbors of each minority sample to divide the minority samples into three groups, namely: safe, dangerous, and noise. It should be noted that the danger group is used to generate new instances. Ghorbani & Ghousi (Ghorbani & Ghousi, 2020). applied Borderline SMOTE to balance the dataset.
- SVM-SMOTE: This method generates the new minority class samples along with directions from existing minority class instances towards their nearest neighbors. The SVM-SMOTE focuses on creating new minority class samples near borderlines using the SVM model to help set boundaries between

classes. Ghorbani & Ghousi (Ghorbani & Ghousi, 2020). achieved highest accuracy with Random Forest while SVM-SMOTE was used as a resampling technique.

- ADASYN: This method generates a different number of samples depending on an estimate of the local distribution of the class to be oversampled. Depending on the level of difficulty in learning, different weights are assigned to minority class samples. The main advantage of ADASYN is that instead of generating the same copy of minority data, it generates synthetic samples for "harder to learn" samples. For balancing a dataset, (Verma et al., 2022). applied ADASYN oversampling technique.

Findings from previous work in which the accuracy of classifiers was enhanced by using data balancing techniques are depicted in Table 9.

Table 9. Results of Classifiers with Balanced Dataset

Author	Attributes	Dataset Size	Model Validation	Classification	Performance result
Thammasiri et al. (Thammasiri et al., 2014).	Background, academic attributes	34 variables, and 21,654 examples	10-fold cross-validation	Binary	SVM obtained the best results with 90.24% accuracy when combined with SMOTE
Pandey & Taruna (Pandey & Taruna, 2016).	Academic information and demographic information	960 records with 18 attributes	Train-test split	Binary	Integrated classifier consisting of Decision Tree, k-Nearest Neighbour, and Aggregating One-Dependence Estimators achieved an accuracy of 98.86% with SMOTE
Ashraf et al. (Ashraf et al., 2018).	Background, academic attributes	23 attributes and 1793 records	10-fold cross-validation	Multiclass (3 classes)	Ensemble StackingC classifier achieved 96.11% accuracy with SMOTE
Ashraf et al. (Ashraf et al., 2020).	Background, academic attributes	23 attributes and 1793 records	10-fold cross-validation	Multiclass (3 classes)	Naïve Bayes achieved 97.15% accuracy with SMOTE
Ghorbani & Ghousi (Ghorbani & Ghousi, 2020).	Background, academic attributes	In the Iran dataset, 650 students with 19 different attributes, in the Portugal dataset, 394 students with 19 different attributes.	hold-out and 5-fold cross-validation	Multiclass (4 classes)	Random Forest obtained the highest accuracy of 73% for the Iran dataset and 81.27% for the Portugal dataset with 5-fold cross-validation and SVM-SMOTE

continued on following page

Table 9. Continued

Author	Attributes	Dataset Size	Model Validation	Classification	Performance result
Teoh et al. (Teoh et al., 2022).	Background and b behavioral attributes in online learning	110 records with 22 features	Train-test split (70:30)	Binary	AdaBoost has achieved a maximum accuracy of more than 90% with SMOTE
Desiani et al. (Desiani et al., 2021).	students' educational background records	2243 records and 11 attributes	Not specified	Multiclass (5 classes)	k-NN achieved 83.71% accuracy with SMOTE.
Verma et al. (Verma et al., 2022).	past academic, background, psychological, and social factors	550 records with 27 attributes	Hold-out method	Binary	Bagging Classifier achieved 95.45% accuracy with SMOTE

3.4.3 Feature Selection Techniques

In feature selection techniques, only relevant and influential features that contribute to the output/target variable are selected automatically, so that machine learning algorithms may train faster and their accuracy may be improved (Xiang et al., 2024. The features that are used to train the model play a significant role in the performance of the model. Irrelevant attributes have a negative impact on the model performance, especially for linear models like Linear Regression and Logistic Regression. Delen et al. (Delen et al., 2020) eliminated unnecessary features without losing the performance of the model. Márquez-Vera et al. (Márquez-Vera et aL., 2013). reduced the 77 features to the top 15 influential features with various feature selection algorithms. Hasan et al. (Hasan et al., 2020). applied genetic search and principle component analysis to reduce the features in an online learning environment. Overfitting can be reduced by feature selection; i.e., there is less chance of making decisions based on noise. There are basically two types of feature selection techniques: (i) Supervised and (ii) Unsupervised. Supervised feature selection techniques consider target attributes, such as methods that remove irrelevant attributes, while unsupervised feature selection ignores the target attribute, such as methods that select relevant attributes by using correlation. Kumar et al. (Kumar et al., 2022). implemented a correlation-based feature selection technique for finding influential attributes and then applied various machine learning algorithms. They found that after using the feature selection technique the accuracy of Logistic Regression increased from 87.51 to 97.84%. Further, there are three classifications of supervised feature selection methods:

- Filter-based: It uses statistical methods and feature importance methods to choose the most influential attributes. The selection of these attributes is independent of any learning algorithm. There are five filter-based feature selection methods: gain-ratio, information gain, symmetrical uncertainty, ReliefF Algorithm, and Chi-Square (Pandey & Taruna, 2016; Hashim et al., 2018; Verma et al., 2022; Teoh et al., 2022). For finding influential attributes, Amrieh et al. (Amrieh et al., 2016). and Costa et al. (Costa et al., 2017). used a filter-based method named information gain-based feature selection. Hashim et al. applied the ReliefF algorithm for feature selection (Hashim et al., 2018). . Recently, Verma et al. (Verma, 2025). proposed an optimized chi-square-based feature selection method that dynamically selects the optimal features for each classification algorithm to ensure their high efficacy.
- Wrapper-based: It builds various models with different subsets of input attributes and selects those features that result in the best-performing model. This technique may be computationally expensive. Ex: Recursive Feature Elimination (RFE), Sequential Forward Selection (SFS). Gray & Perkins (Gray & Perkins, 2019). applied SFS to determine important attributes.
- Intrinsic Methods: Some machine learning algorithms select the influential attributes automatically. Ex. Decision Tree, Random Forest, etc. The classification of different feature selection techniques is shown in Figure 8.

Figure 8. Classification of Feature Selection Technique

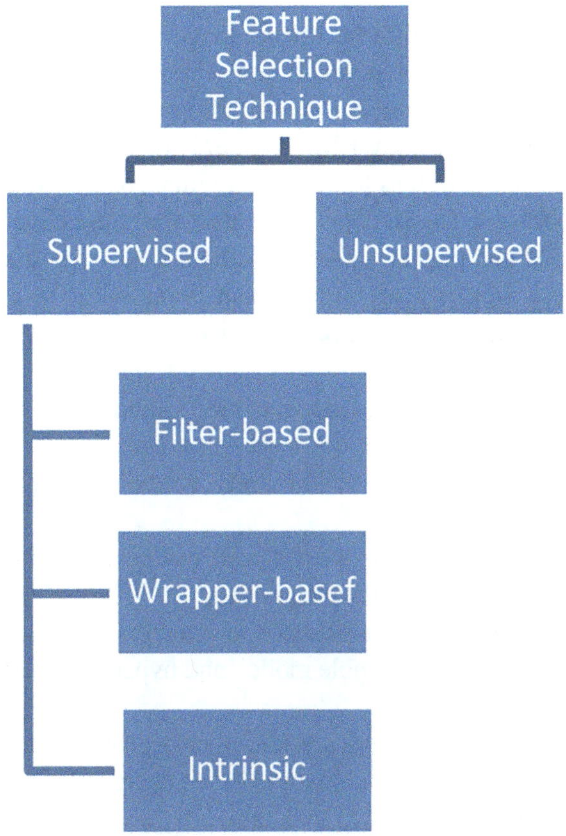

3.4.4 Fine-tuning of Parameters

During the classification, the hyperparameter tuning of EDM techniques may be performed to achieve the best results. There are various methods for setting the parameters optimally such as GridSearchCV, RandomizedSearchCV, Bayesian Optimization, etc. To boost the performance of the classifier, Costa et al. (Costa et al., 2017). and Verma et al. (Verma et al., 2022). applied GridSearchCV algorithm to set the optimal parameters and they also analyzed the impact of data pre-processing and fine-tuning on all of the data mining algorithms' performance. They have found that the performance of some of the data mining techniques was improved after applying pre-processing and fine-tuning. Wunnasri et al. (Wunnasri et al., 2023). used the Bayesian optimization technique to fine-tune the optimal parameter set.

4. CONCLUSION

The information given by different data mining techniques could facilitate educational institutions to predict students' future achievements with higher accuracy, make better decisions and planning in directing them, and may be useful in assigning resources and staff more efficiently. EDM will be also helpful for teachers to improve their teaching practices as well as for learners to improve their learning process by receiving timely feedback. This in-depth study of previous researchers' work revealed that the use of EDM can play an important role in improving students' academic achievements. This article focuses on four main strategies which are necessary for the prediction of students' academic performance:

(i) Steps for predicting low performers: data preparation, data preprocessing, data mining, and result evaluation.
(ii) Attribute category: academic, background, social activity, psychological, and behavioural.
(iii) Different data mining techniques: classification, association rule mining, clustering, regression, and genetic algorithm.
(iv) Techniques for improving classification accuracy: feature selection, data resampling, construction of ensemble model, and hyper-parameter tuning

Most of the authors suggested past academic performance as the most influential attribute category for the prediction of students' academic performance. This may be due to the fact that the past academic parameters gave a glimpse of the learning capability and a basic understanding or knowledge required for higher study. Also, other parameters like background, behaviour, etc. may affect the past academic parameters inherently. Thus, it is justifiable to give admission or grade on the basis of past performance in COVID-19 pandemic-like situations. Also, it is the duty of every instructor to examine the past and current performance of every student and try to support weaker students timely so that the overall institution can proceed to give quality education. Also, background attributes were proven to be influential factors in the student's academic performance (Fernandes et al., 2019; Tsai et al., 2011; Nandeshwar et al., 2011; Alfiani & Wulandari, 2015; Pandey & Taruna, 2016; Yaakub et al., 2018; Roslan & Chen, 2023). Social Activities Attributes and Psychological Attributes are least used by the researchers because the collection of these types of parameters requires careful planning and surveys among the students. Besides this it may be very challenging to get accurate data as most of the students may be unwilling to share their social (Ex: in a romantic relationship) and psychological information (Ex: Homesickness or not, Interest in the study, etc.).

As per the meta-analysis of the previous works, classification is the most commonly used data mining technique for predicting students' academic performance as the classification technique can predict student academic performance efficiently. Furthermore, among classification algorithms, the Decision Tree is widely used and most authors observed that it is better than other classifiers. But, there is no unanimous result exists, i.e. we can't say the Decision Tree algorithm always performs better and the suitability of the classifiers varies with the nature of the dataset. After classification, clustering is popular to identify classes of students according to their similarities. Statistics, Association Rule Mining, and Regression techniques are also widely used by previous authors. Genetic algorithm is less used by the authors; it may be due to the fact that it is computationally expensive (time-consuming). By selecting the right data mining technique on the students' dataset, better accuracy in the prediction may be achieved so that any educational organization may concentrate on and promote the students' outcome. It may also be concluded that prediction accuracy can be enhanced by using feature selection, data balancing, hyperparameter tuning, and the construction of an ensemble model. Finally, the present review emphasize on the need of a scalable model that gives the highest accuracy in predicting students' academic performance for all kind of datasets.

DECLARATIONS

Data Availability

Not applicable.

Conflict of interest

No conflict of interest to declare.

REFERENCES

Adejo, O. W., & Connolly, T. (2018). Predicting student academic performance using multi-model heterogeneous ensemble approach. *Journal of Applied Research in Higher Education, 10*(1), 61–75.

Adekitan, A. I., & Salau, O. (2019). The impact of engineering students' performance in the first three years on their graduation result using educational data mining. *Heliyon, 5*(2), 1–20.

Adeyemo, A. B., & Kuyoro, S. O. (2013). Investigating the Effect of Students Socio-Economic/ Family Background on Students Academic Performance in Teritary Institutions Using Decision Tree Algorithms. *Acta Phytotaxonomica et Geobotanica, 4*(2), 61–78.

Agarwal, M., & Agarwal, B. B. (2023). Methodological Implementation of Data Miming Classifiers and ANN for Prediction of Accomplishment of Student Education. *International Journal of Performability Engineering, 19*(9), 587–597.

Ahmad, Z., & Shahzadi, E. (2018). Prediction of students' academic performance using artificial neural netwok. *Bulletin of Education and Research, 40*(3), 157–164.

Al-Hebaish, S. M. (2012). The correlation between general self-confidence and academic achievement in the oral presentation course. *Theory and Practice in Language Studies, 2*(1).

Albreiki, B., Zaki, N., & Alashwal, H. A. (2021). A Systematic Literature Review of Student' Performance Prediction Using Machine Learning Techniques. *Education Sciences, 11*(552), 1–27.

Alfiani, A. P., & Wulandari, F. A. (2015). Mapping Student's Performance Based on Data Mining Approach. *Agriculture and Agricultural Science Procedia, 3*, 173–177.

Algarni, A. (2016). Data mining in Education. *International Journal of Advanced Computer Science and Applications, 7*(6), 456–461.

Alnassar, F., Blackwell, T., Homayounvala, E., & Yee-king, M. (2021). How Well a Student Performed? A Machine Learning Approach to Classify Students' Performance on Virtual Learning Environment, 2nd *IEEE International Conference on Intelligent Engineering and Management (ICIEM),* 1-6.

Amrieh, E. A., Hamtini, T., & Aljarah, I. (2016). Mining Educational Data to predict Students' academic performance using Ensemble methods. *International Journal of Database Theory and Application, 9*(8), 119–136.

Aouifi, H. E., Hajji, M. E., Es-Saady, Y., & Douzi, H. (2021). Predicting learner's performance through video sequences viewing behavior analysis using educational data-mining. *Education and Information Technologies*, *26*(5), 5799–5814.

Ashraf, M., Zaman, M., & Ahmed, M. (2018). Using Ensemble StackingC Method and Base Classifiers to Ameliorate Prediction Accuracy of Pedagogical Data. *Procedia Computer Science*, *132*, 1021–1040.

Ashraf, M., Zaman, M., & Ahmed, M. (2020). An Intelligent Prediction System for Educational Data Mining Based on Ensemble and Filtering approaches. *Procedia Computer Science*, *167*, 1471–1483.

Asif, R., Merceron, A., Ali, S. A., & Haider, N. G. (2017). Analyzing undergraduate students' performance using educational data mining. *Computers & Education*, *113*, 177–194.

Asselman, A., Khaldi, M., & Aammou, S. (2021). Enhancing the prediction of student performance based on the machine learning XGBoost algorithm. *Interactive Learning Environments*, *31*(3), 1–20.

Ayindi, A. Q., Odeniyi, O. A., & Sarumi, O. A. (2013). Mining Parent Socio-Economic Factors to predict Students' Academic Performance in Osum State College of Technology, Esa Oke. *International Journal of Engineering Research & Technology (Ahmedabad)*, *2*(12), 1677–1683.

Badr, G., Algobail, A., Almutairi, H., & Almutery, M. (2016). Predicting Student's Performance in University Courses: A Case Study and Tool in KSU Mathematics Department, Symposium on Data Mining Applications. *Procedia Computer Science*, *82*, 80–89.

Bakhshinategh, B., Zaiane, O. R., ElAtia, S., & Ipperciel, D. (2018). Educational data mining applications and tasks: A survey of the last 10 years. *Education and Information Technologies*, *23*, 537–553.

Begum, S., & Padmannavar, S. S. (2022). Genetically optimized ensemble classifiers for multiclass student performance prediction. *Int. J. Intell. Eng. Syst*, *15*(2), 316–328.

Bhardwaj, B. K., & Pal, S. (2011). Data Mining: A prediction for performance improvement using classification. *International Journal of Computer Science and Information Security*, *9*(4), 136–140.

Bindhia, K. F., & Suvanam, S. B. (2019). Predicting Academic Performance of Students Using a Hybrid Data Mining Approach. *Journal of Medical Systems*, *43*(162), 1–15.

Buldu, A., & Üçgün, K. (2010). Data mining application on students' data. *Procedia: Social and Behavioral Sciences*, *2*, 5251–5259.

Costa, E. B., Fonseca, B., Santana, M. A., Araujo, F. F. D., & Rego, J. (2017). Evaluating the effectiveness of educational data mining techniques for early prediction of students' academic failure in introductory programming courses. *Computers in Human Behavior*, *73*, 247–256.

Delen, D., Topuz, K., & Eryarsoy, E. (2020). Development of a Bayesian Belief Network-based DSS for predicting and understanding freshmen student attrition. *European Journal of Operational Research*, *281*(3), 575–587.

Desiani, A., Yahdin, S., & Kartikasari, A., & Irmeilyana. (2021). Handling the imbalanced data with missing value elimination SMOTE in the classification of the relevance education background with graduates employment. *International Journal of Artificial Intelligence*, *10*(2), 346–354.

Estrera, P. J. M., Natan, P. E., Rivera, B. G. T., & Colarte, F. B. (2017). Student Performance Analysis for Academic Ranking Using Decision Tree Approach in University of Science and Technology of Southern Philippines Senior High School. *International Journal of Engineering and Techniques*, *3*(5), 147–154.

Feng, G., Fan, M., & Chen, Y. (2022). Analysis and prediction of students' academic performance based on educational data mining. *IEEE Access : Practical Innovations, Open Solutions*, *10*, 19558–19571.

Fernandes, E., Holanda, M., Victorino, M., Borges, V., Carvalho, R., & Van Erven, G. (2019). Educational data mining: Predictive analysis of academic performance of public school students in the capital of Brazil. *Journal of Business Research*, *94*, 335–343.

Ghorbani, R., & Ghousi, R. (2020). Comparing Different Resampling Methods in Predicting Student's Performance Using Machine Learning Techniques. *IEEE Access : Practical Innovations, Open Solutions*, *8*, 7899–67911.

Gil, P. D., da Cruz Martins, S., Moro, S., & Costa, J. M. (2021). A data-driven approach to predict first-year students' academic success in higher education institutions. *Education and Information Technologies*, *26*(2), 2165–2190.

Gray, C. C., & Perkins, D. (2019). Utilizing early engagement and machine learning to predict student outcomes. *Computers & Education*, *131*, 22–32.

Gray, G., McGuinness, C., & Owende, P. (2014). An application of classification models to predict learner progression in tertiary education, *IEEE International Advance Computing Conference (IACC)*, 549–554.

Ha, D. T., Giap, C. N., Loan, P. T. T., & Huong, N. T. L. (2020). An Empirical Study for Student Academic Performance Prediction Using Machine Learning Techniques. *International Journal of Computer Science and Information Security*, *18*(3), 21–28.

Hamsa, H., Indiradevi, S., & Kizhakkethottam, J. J. (2016). Student academic performance prediction model using decision tree and fuzzy genetic algorithm. *Procedia Technology*, *25*, 326–332.

. Hardy, C., & Bryson, C. (2016). The salience of social relationships and networks in enabling student engagement and success. *Student Engagement in Higher Education Journal, 1*(1).

Hasan, R., Palaniappan, S., Mahmood, S., Abbas, A., Sarker, K. U., & Sattar, M. U. (2020). Predicting Student Performance in Higher Educational Institutions using Video Learning Analytics and Data Mining Techniques. *Applied Sciences (Basel, Switzerland)*, *10*(3894), 1–20.

Hashim, A. S., Hamoud, A. K., & Awadh, W. A. (2018). Analyzing students' answers using association rule mining based on feature selection. *Journal of Southwest Jiaotong University*, *53*(5).

Huang, S., & Fang, N. (2013). Predicting student academic performance in an engineering dynamics course: A comparison of four types of predictive mathematical models. *Computers & Education*, *61*, 133–145.

Hussain, S., Dahan, N. A., Ba-Alwi, F. M., & Ribata, N. (2018). Educational Data Mining and Analysis of Students' Academic Performance Using WEKA. *Indonesian Journal of Electrical Engineering and Computer Science*, *9*(2), 447–459.

Hussain, S., Gaftandzhieva, S., Maniruzzaman, M., Doneva, R., & Muhsin, Z. F. (2021). Regression analysis of student academic performance using deep learning. *Education and Information Technologies*, *26*, 783–798.

Idris, M., Hussain, S., & Ahmad, N. (2020). Relationship between parents' education and their children's academic achievement [JASS]. *Journal of Arts & Social Sciences*, *7*(2), 82–92.

Injadat, M., Moubayed, A., Nassif, A. B., & Shami, A. (2020). Systematic ensemble model selection approach for educational data mining. *Knowledge-Based Systems*, *200*, 1–16.

Ioannis, B., & Maria, K. (2019). Gender and student course preferences and course performance in computer science departments: A case study. *Education and Information Technologies*, *24*(2), 1269–1291.

. Issah, I., Appiah, O., Appiahene, P., & Inusah, F. (2023). A systematic review of the literature on machine learning application of determining the attributes influencing academic performance, *Decision Analytics Journal*, 100204.

Kabakchieva, D. (2013). Predicting Student Performance by Using Data Mining Methods for Classification. *Cybernetics and Information Technologies*, *13*(1), 61–72.

Kamal, P., & Ahuja, S. (2019). An ensemble-based model for prediction of academic performance of students in undergrad professional course, *Journal of Engineering. Defence Technology*, *17*(4), 769–781.

Karthikeyan, V. G., Thangaraj, P., & Karthik, S. (2020). Towards developing hybrid educational data mining model (HEDM) for efficient and accurate student performance evaluation. *Soft Computing*, *24*(24), 18477–18487.

Khan, A., & Ghosh, S. K. (2020). Student performance analysis and prediction in classroom learning: A review of educational data mining studies. *Education and Information Technologies*, *26*(1), 205–240.

Khan, S. K. S. M. H., & Rehman, K. (2017). Comparison Between the Academic Performance of Public and Private Schools' Male Students on the Basis of Regularity, Discipline, Punctuality and Environment at Secondary School Level in District Lakki Marwat. *Humanity & Social Sciences Journal*, *7*(7), 53–57.

Kiu, C.-C. (2018). Data Mining Analysis on Student's Academic Performance through Exploration of Student's Background and Social Activities, *IEEE Conference*, 1-5.

Kumar, M., Quteishat, A., & Qtaishat, A. (2022). Performance Comparison of the Optimized Ensemble Model with Existing Classifier Models. *International Journal of Modern Education & Computer Science*, *14*(3).

Lacerda, B., Marcondes, F. S., Lima, H., Durães, D., & Novais, P. (2023). Prediction of Students' Grades Based on Non-academic Data. *International Conference in Methodologies and intelligent Systems for Technology Enhanced Learning,* 87-95.

. Maksum, A., & Khory, F. D. (2020). Effect of learning climate, thinking pattern, and curiosity on academic performance in higher education. *Problems of Education in the 21st Century, 78*(1), 102.

Márquez-Vera, C., Cano, A., Romero, C., & Ventura, S. (2012). Predicting student failure at school using genetic programming and different data mining approaches with high dimensional and imbalanced data. *Applied Intelligence*, *38*(3), 315–330.

Márquez-Vera, C., Morales, C. R., & Soto, S. V. (2013). Predicting school failure and dropout by using data mining techniques. *IEEE Journal of Latin-American Learning Technologies*, *8*(1), 7–14.

Masud, S., Mufarrih, S. H., Qureshi, N. Q., Khan, F., Khan, S., & Khan, M. N. (2019). Academic performance in adolescent students: The role of parenting styles and socio-demographic factors–a cross sectional study from eshawar, Pakistan. *Frontiers in Psychology*, *10*, 2497.

Mat, U. B., Buniyamin, N., Arsad, P. M., & Kassim, R. (2013). An overview of using academic analytics to predict and improve students' achievement: A proposed proactive intelligent intervention, in Engineering Education, *2013 IEEE 5th Conference*, 126-130.

Mishra, T., Kumar, D., & Gupta, S. (2014). Mining students' data for prediction performance, *Advanced Computing & Communication Technologies (ACCT), Fourth International Conference IEEE*, 255-262.

Mohamad, S. K., & Tasir, Z. (2013). Educational data mining: A review. *Procedia: Social and Behavioral Sciences*, *97*, 320–324.

Mueen, A., Zafar, B., & Manzoor, U. (2016). Modeling and Predicting Students' Academic Performance Using Data Mining Techniques, *I. J. Modern Education and Computer Science*, *11*, 36–42.

Muller, C. (2018). Parent involvement and academic achievement: An analysis of family resources available to the child, Parents, their children, and schools, *Routledge*, 77–114.

Musso, M. F., Hernandez, C. F. R., & Cascallar, E. C. (2020). Predicting key educational outcomes in academic trajectories: A machine learning approach. *Higher Education*, *80*(5), 875–894.

Nahar, K., Shova, B. I., Ria, T., Rashid, H. B., & Islam, A. S. (2021). Mining educational data to predict students performance: A comparative study of data mining techniques. *Education and Information Technologies*, *26*(5), 6051–6067.

Nandeshwar, A., Menzies, T., & Nelson, A. (2011). Learning patterns of university student retention. *Expert Systems with Applications*, *38*(12), 14984–14996.

Nasiri, M., & Minaei, B. (2012). Predicting GPA and academic dismissal in LMS using educational data mining: A case mining, In E-Learning and E-Teaching (ICELET), *Third International Conference, IEEE*, 53-58.

Nkomo, L. M., & Nat, M. (2021). Student Engagement Patterns in a Blended Learning Environment: An Educational Data Mining Approach. *TechTrends*, *65*(5), 808–817.

Oeda, S., & Hashimoto, G. (2017). Log-data clustering analysis for dropout prediction in beginner programming classes. *Procedia Computer Science*, *112*, 614–621.

Osmanbegorvic, E., & Suljic, M. (2012). Data Mining Approach for Predicting Student Performance, Economic Review-. *Journal of Economics and Business*, *10*(1), 3–12.

Pal, A. K., & Pal, S. (2013). Analysis and Mining of Educational Data for Predicting the Performance of Students. *International Journal of Electronics Communication and Computer Engineering*, *4*(5), 1560–1565.

Pande, S. M. (2023). Machine Learning Models for Student Performance Prediction. *International Conference on Innovative Data Communication Technologies and Application,* 27-32.

Pandey, M., & Taruna, S. (2016). Towards the integration of multiple classifier pertaining to the Student's performance prediction. *Perspectives in Science*, *8*, 364–366.

Parajuli, M., & Thapa, A. (2017). Gender differences in the academic performance of students. *Journal of Development and Social Engineering*, *3*(1), 39–47.

Poturić, V. Č., Čandrlić, S., & Dražić, I. (2023). Predicting Students' Academic Success Based on Various Course Activities: A Case Study. In *The Learning Ideas Conference,* 665-675.

Rahman, S. R., Islam, M. A., Akash, P. P., Parvin, M., Moon, N. N., & Nur, F. N. (2021). Effects of co-curricular activities on student's academic performance by machine learning. *Current Research in Behavioral Sciences*, *2*(100057), 1–7.

Ramaswami, G., Susnjak, T., & Mathrani, A. (2022). On developing generic models for predicting student outcomes in educational data mining. *Big Data and Cognitive Computing*, *6*(1), 1–16.

Ribeiro, N., Malafaia, C., Neves, T., & Menezes, I. (2024). The impact of extra-curricular activities on university students' academic success and employability. *European Journal of Higher Education*, *14*(3), 389–409.

Rodrigues, M. W., Zárate, L. E., & Isotani, S. (2018). Educational Data Mining: A review of evaluation process in the e-learning. *Telematics and Informatics*, •••, 1–25.

Roslan, M. H. B., & Chen, C. J. (2023). Predicting students' performance in English and Mathematics using data mining techniques. *Education and Information Technologies*, *28*(2), 1427–1453.

Saa, A. A., Al-Emran, M., & Shaalan, K. (2019). Factors affecting students' performance in higher education: A systematic review of predictive data mining techniques, *Technology. Knowledge and Learning*, *24*(4), 567–598.

Sarker, S., Paul, M. K., Thasin, S. T. H., & Hasan, M. A. M. (2024). Analyzing students' academic performance using educational data mining. *Computers and Education: Artificial Intelligence*, *7*, 100263.

Sembiring, S., Zarlis, M., Hartama, D., Ramliana, S., & Wani, E. (2011). Prediction of student academic performance by an application of data mining techniques, *International Conference on Management and Artificial Intelligence IPEDR*, 6, 110-114.

Shahiri, A. M., Husain, W., & Rashid, N. A. (2015). A Review on Predicting Student's Performance using Data Mining Techniques. *Procedia Computer Science*, *72*, 414–422.

Sharabiani, A., Karim, F., Sharabiani, A., Atanasov, M., & Darabi, H. (2014). An Enhanced Bayesian Network Model for Prediction of Students' Academic Performance in Engineering Programs, *IEEE Global Engineering Education Conference*, 832-837.

Sinval, J., Oliveira, P., Novais, F., Almeida, C. M., & Telles-Correia, D. (2025). Exploring the impact of depression, anxiety, stress, academic engagement, and dropout intention on medical students' academic performance: A prospective study. *Journal of Affective Disorders*, *368*, 665–673.

Sun, J., & Hagedorn, L. S. (2016). Homesickness at college: Its impact on academic performance and retention. *Journal of College Student Development*, *57*(8), 943–957.

Tanner, T., & Toivonen, H. (2010). Predicting and preventing student failure – using the k-nearest neighbour method to predict student performance in an online course environment. *International Journal of Learning Technology*, *5*(4), 356–377.

Teoh, C. W., Ho, S. B., Dollmat, K. S., & Tan, C. H. (2022). Ensemble-Learning techniques for predicting student performance on video-based learning. *International Journal of Information and Education Technology (IJIET)*, *12*(8), 741–745.

Thammasiri, D., Delen, D., Meesad, P., & Kasap, N. (2014). A critical assessment of imbalanced class distribution problem: The case of predicting freshmen student attrition. *Expert Systems with Applications*, *41*(2), 321–330.

Tomasevic, N., Gvozdenovic, N., & Vranes, S. (2020). An overview and comparison of supervised data mining techniques for student exam performance prediction. *Computers & Education*, *143*, 1–18.

Tsai, C. F., Tsai, C. T., Hung, C. S., & Hwang, P. S. (2011). Data mining techniques for identifying students at risk of failing a computer proficiency test required for graduation. *Australasian Journal of Educational Technology*, *27*(3), 481–498.

Vasuki, M., & Revathy, S. (2022). Analyzing performance of placement students record using different clustering algorithm [IJCSE]. *Indian Journal of Computer Science and Engineering*, *13*(2), 410–419.

Verma, S., Kholiya, K., & Bala, K. (2025). Leveraging Ensemble Model and Optimized Feature Selection to Boost Prediction Accuracy in Educational Data Mining. *SN Computer Science*, *6*(5), 487.

Verma, S., & Yadav, R. K. (2020). Effect of different attributes on the academic performance of engineering students, *IEEE International Conference on Advent Trends in Multidisciplinary Research and Innovation (ICATMRI),* 1-4.

Verma, S., Yadav, R. K., & Kholiya, K. (2022). A Scalable Machine Learning-based Ensemble Approach to Enhance the Prediction Accuracy for Identifying Students at-Risk. *International Journal of Advanced Computer Science and Applications*, *13*(8), 185–192.

Widyahastutia, F., & Tjhin, V. U. (2018). Performance Prediction in Online Discussion Forum: State-of-the-art and comparative analysis. *Procedia Computer Science*, *135*, 302–314.

Wong, S. L. (2019). Relationship between interest and mathematics performance in a technology-enhanced learning context in Malaysia. *Research and Practice in Technology Enhanced Learning*, *14*(1), 1–13.

Wunnasri, W., Musikawan, P., & So-In, C. (2023). A Two Phase Ensemble-Based Method for Predicting Learners' Grade in MOOCs. *Applied Sciences (Basel, Switzerland)*, *13*(1492), 1–15.

Xiang, F., Zhao, Y., Zhang, M., Zuo, Y., Zou, X., & Tao, F. (2024). Ensemble learning-based stability improvement method for feature selection towards performance prediction. *Journal of Manufacturing Systems*, *74*, 55–67.

Xiao, W., Ji, P., & Hu, J. (2021). A survey on educational data mining methods used for predicting students' performance. *Engineering Reports*, *4*(5), 1–23.

Yaakub, T. N. T., Ahmad, W. R. W., Husaini, Y., & Burham, N. (2018). Influence Factors in Academic Performance among Electronics Engineering Student: Geographic Background, Mathematics Grade and Psycographic Characteristics, *IEEE 10th International Conference on Engineering Education Kuala Lumpur, Malaysia*, 30-33.

Yagci, M. (2022). Educational data mining: Prediction of students' academic performance using machine learning algorithms. *Smart Learning Environments*, *9*(1), 1–19.

Yuliansyah, H., Imaniati, R. A. P., Anggit Wirasto, A., & Wibowo, M. (2021). Predicting Students Graduate on Time Using C4.5 Algorithm. *Journal of Information Systems Engineering and Business Intelligence*, *7*(1), 67–73.

Chapter 8 Quantitative Modeling of Dietary Behavior, Obesity, and Perceived Stress Among University Students: A Cross-Sectional Study Using Logistic and Multinomial Decision Models

Mohamed Rektouti
https://orcid.org/0009-0004-5560-4925
Faculty of Sciences, Ibn Tofail University, Kenitra, Morocco

Said Bouchefra
Instituts Supérieurs des Professions Infirmières et Techniques de Santé, Morocco

Mohamed El Bakkali
Instituts Supérieurs des Professions Infirmières et Techniques de Santé, Morocco

Brahim El Ouardi
https://orcid.org/0009-0000-4559-6141
Faculty of Sciences, Ibn Tofail University, Kenitra, Morocco

Mohamed Belomaria
Instituts Supérieurs des Professions Infirmières et Techniques de Santé, Morocco

Yassine Mouniane
https://orcid.org/0000-0002-6037-5301
Faculty of Sciences, Ibn Tofail University, Kenitra, Morocco

DOI: 10.4018/979-8-3373-6746-0.ch008

Wafaa Benkayba
https://orcid.org/0000-0001-5196-1678
Laboratory of Chemistry Analytical and Bromatology, Faculty of Medicine and Pharmacy, University Mohammed V, Rabat, Morocco

Soad Khal Layoun
https://orcid.org/0009-0004-0783-5841
Faculty of Sciences, Ibn Tofail University, Kenitra, Morocco

Abdellatif Bour
Faculty of Sciences, Ibn Tofail University, Kenitra, Morocco

ABSTRACT

This study applies quantitative decision-making and numerical modeling to analyze interactions between dietary behavior, obesity, and perceived stress among 1,036 students at Ibn Tofail University, Morocco. Using binary and multinomial logistic regression models, key determinants were identified for adopting a balanced diet, developing obesity, and experiencing high stress. Age, home-prepared meals, and health awareness increased the likelihood of a balanced diet, whereas sugary drinks and long engagement time reduced it. Regular physical activity and better nutritional knowledge lowered obesity risk, while prolonged daily engagement increased it. Obesity strongly predicted high perceived stress, supported by an E-value of 6.075, indicating robust associations. These findings highlight the relevance of numerical analysis in understanding student health dynamics and support evidence-based decision-making for targeted university health interventions.

INTRODUCTION

The eating habits, nutritional status, and psychological well-being of university students are topics of growing concern due to their impact on immediate and long-term health. This transitional period, often characterized by changes in lifestyle, increased academic responsibilities, and stress, puts students at increased risk for dietary imbalances and mental health issues (Almutairi et al., 2018; Al-Qahtani, 2022; José et al., 2025). The university environment, while offering educational and social opportunities, can exacerbate these risks due to time constraints, academic stress, and limited availability of balanced food options (Alzamil et al., 2019; El Ansari et al., 2012). Adopting a balanced diet is a fundamental pillar of public health, particularly among university students. This population often faces significant transitions in their eating habits due to increased independence, time constraints, and social influences (Abraham et al., 2018; Sogari et al., 2018). A balanced diet rich in fruits, vegetables, lean proteins, and whole grains not only protects against obesity and metabolic diseases, but is also associated with better mental health.

Systematic studies have shown that a high-quality diet is linked to reduced symptoms of depression and anxiety, highlighting the importance of eating habits for students' psychological well-being and academic performance (O'Neil et al., 2014). However, in university settings, the prevalence of unhealthy eating habits remains high, partly due to increased consumption of ultra-processed foods and fast food (Navarro-Prado et al., 2024).

Obesity, defined as a body mass index (BMI) greater than or equal to 30 kg/m^2, is a common consequence of dietary imbalances and is a growing concern in academic settings. This condition is associated with increased risks of chronic diseases, including type 2 diabetes, cardiovascular disorders, and depression (World Health Organization, 2024). Among students, several factors contribute to obesity, including unbalanced dietary choices characterized by increased consumption of ultra-processed foods and fast foods that are high in calories but low in essential nutrients (Dicken & Batterham, 2024). Ultra-processed foods (UPFs) are generally high in added sugars, unhealthy fats, sodium, and chemical additives such as colorings, preservatives, flavorings, and flavor enhancers, while being low in essential nutrients and fiber. This diet is linked not only to excessive calorie intake but also to unfavorable metabolic outcomes. Research shows that adolescents consuming a higher proportion of UPF tend to have a high body mass index (BMI), as well as high levels of inflammatory markers and components of metabolic syndrome (de Miranda et al., 2021; Magalhães et al., 2021). In addition, UPFs often contribute to poor eating habits, reinforcing a cycle of unhealthy eating that increases the risk of obesity and associated comorbidities. The effect of ultra-processed foods on adolescent obesity can also be observed in the context of behavioral and psychological factors. Finally, academic pressures and perceived stress can promote impulsive eating behaviors, exacerbating nutritional imbalances and increasing the likelihood of weight gain (Al-Naimi et al., 2025).

Finally, perceived stress is a pervasive issue among university students, due to academic, social, and financial pressures. Stress has a direct impact on food choices, often promoting the consumption of foods high in fat and sugar, which contributes to obesity and nutritional imbalances (Tomiyama, 2019). In turn, studies have shown that obesity, through its physiological and social implications, can also exacerbate stress levels (Puhl et al., 2020). This bidirectional dynamic highlights the urgency of studying the complex relationships between balanced diet, obesity, and stress among university students in order to better understand the determinants of their health. Applications using artificial intelligence-based machine learning algorithms are being implemented with a high degree of accuracy in areas such as disease prediction and big data analysis related to health (Derici and Doğan, 2025). In this study, conducted among students at Ibn Tofail University in Kenitra, Morocco, we explored the distribution of BMI, perceived stress, and consumption of a balanced diet. Using a multivariate logistic model, we analyzed the factors associated with adopting a balanced diet, as well as the determinants of obesity and its link to stress. The objective of this research is to provide a comprehensive view of the interactions between the nutritional and psychological dimensions of student health in order to guide appropriate interventions in this context.

METHODOLOGY

Study Design and Participants

This cross-sectional study was conducted between January 1, 2021, and January 1, 2024, at Ibn Tofaïl University, Kenitra, Morocco, a public institution welcoming students from various neighboring regions. Participants, aged 18 to 30, were recruited from different faculties, including scientific, economic, legal, literary, and technical disciplines, to ensure a representative diversity of academic profiles. A total of 1,036 undergraduate students enrolled in bachelor's degree programs were included. Inclusion criteria included: being enrolled at Ibn Tofaïl University during the study period, being an undergraduate student, and providing informed consent. Graduate students (Master's, Doctorate) were excluded due to their often part-time enrollment status. Data were collected using a self-administered questionnaire distributed directly in the lecture halls and classrooms of the main faculties, such as the Faculty of Sciences, the Faculty of Arts, the Faculty of Legal, Economic, and Social Sciences, and the National School of Applied Sciences. Participants completed the questionnaire anonymously without time constraints, and no financial or material incentives were offered. The questionnaires were collected immediately after completion. Before participating, the objectives of the study were explained, and informed consent was obtained. Participation was entirely voluntary, respecting anonymity and guaranteeing the confidentiality of responses.

Bias

To minimize bias in this study conducted among students at Ibn Tofail University, Kenitra, Morocco, several measures were taken. Selection bias was reduced by ensuring that a representative sample of the target student population was recruited. Classification bias was controlled by pre-checking the data collected from participants, including self-reported information, to ensure accuracy and consistency. Finally, confounding factors were taken into account using multivariate logistic analyses, thereby yielding robust and reliable conclusions.

Inclusion and Exclusion Criteria

The inclusion and exclusion criteria for this study conducted among students at Ibn Tofail University, Kenitra, Morocco, were defined to ensure sample homogeneity and guarantee the validity of the results. Students enrolled at the university during the current academic year, aged 18 or over, who gave their informed consent to participate were included. Participants had to be available to complete the questionnaires and provide complete data on the variables studied, including their eating habits, physical activity, and level of awareness of health and nutrition. However, students with diagnosed medical conditions that could influence their eating habits or stress levels, such as chronic metabolic disorders, psychiatric disorders, or medically prescribed diets, were excluded. Similarly, participants who provided incomplete or inconsistent responses to the questionnaires or who refused to participate in any of the required measurements were excluded from the final analysis.

Data Collection and Measurements

Data were collected using a semi-structured self-administered questionnaire, which was pre-tested on 10 undergraduate students at Ibn Tofail University to ensure its clarity and relevance. This questionnaire was used to collect information on several key variables: sociodemographic characteristics (age, gender, socioeconomic status), dietary practices (frequency of consumption of specific food groups, regular meals, consumption of sugary drinks), levels of physical activity (frequency and duration of physical activity per week), stress levels (assessed using a validated scale adapted to the student context), and anthropometric measurements (weight and height). Anthropometric measurements included self-reported values for weight (in kilograms, kg) and height (in centimeters, cm), supplemented by direct measurements where possible. Weight was measured using electronic scales accurate to 100 g, and height was measured using a measuring tape. Body mass index (BMI), expressed in kilograms per square meter (kg/m^2), was calculated using the standard formula: weight (kg) divided by height squared (m). Participants were then classified according to World Health Organization (WHO) categories: underweight (BMI < 18.5 kg/m^2), normal weight (BMI between 18.5 and 24.9 kg/m^2), overweight (BMI between 25 and 29.9 kg/m^2), and obese (BMI ≥ 30 kg/m^2).

Stress levels were assessed using a validated scale adapted to the student context, allowing participants to be classified into categories: low, moderate, or high. This variable was included to analyze its potential interactions with eating behaviors, physical activity, and anthropometric indicators.

All data were collected after official approval from Ibn Tofail University and with the informed consent of participants. This methodology ensures the reliability and validity of the data, while complying with international scientific standards.

Ethical Considerations

This study was conducted in accordance with the ethical principles established by the Declaration of Helsinki. Official authorization was obtained from Ibn Tofail University after rigorous evaluation of the research protocol by the relevant authorities. Informed consent was obtained from each participant prior to their inclusion in the study. The objectives of the research, the nature of the information to be collected, and any risks or discomfort associated with participation were explained clearly and in detail. Participants were informed of their right to withdraw from the study at any time, without having to provide justification and without this having any negative impact on them. The confidentiality and anonymity of the data collected were strictly respected, in accordance with national and international regulations on the protection of personal data. The information collected was stored securely and used exclusively for scientific research purposes, thus ensuring optimal protection of participants' rights and privacy.

Statistical Analysis

All statistical analyses were performed using R software (version 4.2.2) (available at https://cran.r-project.org), Python (version 3.9) (available at https://www.python.org), and Stata (StataCorp LLC, 4905 Lakeway Drive, College Station, TX, USA). Data were organized and tabulated using Microsoft Excel prior to analysis in Stata. Descriptive statistics were expressed as frequencies and percentages for categorical variables. Comparisons between qualitative variables were performed using the Chi-square test, while Fisher's exact test was used for dichotomous variables or when sample sizes were small ($n < 5$). A p-value ≤ 0.05 was considered statistically significant.

To explore associations between independent variables (e.g., sociodemographic and behavioral factors) and the outcomes studied (balanced diet, obesity, stress levels), binary and multinomial logistic regression models were used. Variables were selected for inclusion in the multivariate models based on their clinical relevance, data from the literature, and the results of univariate analyses, with a significance threshold set at $p \leq 0.20$ for the initial selection of variables. The multivariate approach followed a three-step process: (1) variable selection, (2) assessment of potential interactions, and (3) control of confounding factors. Model adequacy was assessed using the Hosmer-Lemeshow test to ensure their goodness of fit, and results were reported as adjusted odds ratios (AORs) with their 95% confidence intervals (CI). A p-value ≤ 0.05 was used to interpret statistical significance in the multivariate models. The adjusted combined effects of the explanatory variables were estimated using the lincom command in multinomial logistic regression models. This method allowed us to calculate specific linear contrasts, expressed in terms of adjusted odds ratios, in order to assess the joint impact of factors such as age, meal preparation, knowledge about healthy eating, and consumption of sugary drinks. These analyses facilitated the interpretation of complex interactions between the variables studied and highlighted the combined influences of behavioral and contextual factors.

Finally, graphs of the results were generated using Python's matplotlib and seaborn libraries and R software to clearly visualize the associations and combined effects in the adjusted models. This integrated approach ensures methodological rigor and reproducibility of results, in accordance with scientific standards.

RESULTS

The figure presents an analytical summary of three key dimensions relating to participant characteristics (Figure 1): perceived stress levels, body mass index (BMI) distribution, and consumption of a balanced diet. Data on perceived stress levels indicate that the majority of participants report low stress levels (47.9%), followed by those with moderate stress levels (34.7%), while a minority report high stress levels (17.5%). This distribution reflects a relatively balanced overall stress management within the population, although a significant proportion experience significant stress. In terms of BMI, the majority of participants fall into the normal weight category (74.0%), indicating a generally healthy weight status. However, smaller proportions are classified as underweight (6.2%), overweight (9.9%), or obese (9.8%), highlighting the two extremes of the BMI spectrum and the importance of targeted strategies. With regard to eating habits, a slight majority of participants (52.6%) adhere to a balanced diet, while a significant proportion (47.4%) do not.

Figure 1. Profiles of Stress, BMI, and Dietary Habits Among Participants.

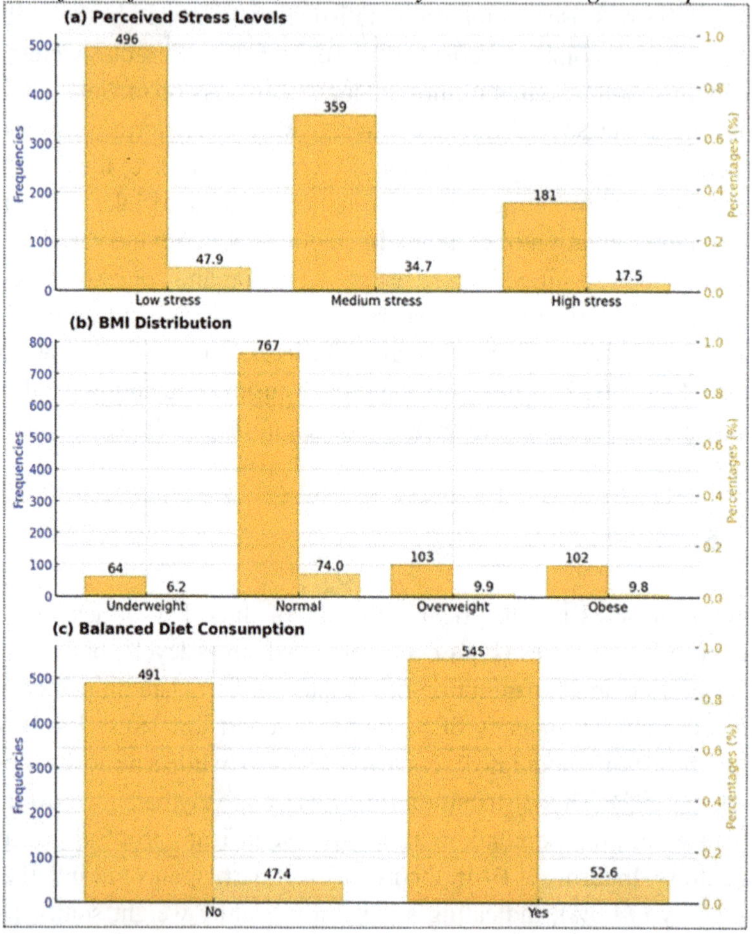

Table 1 presents the results of the multivariate analysis identifying significant factors influencing the consumption of a balanced diet. With regard to age, individuals aged 24 to 26 (adjusted OR = 1.54; 95% CI = 1.02-2.33; p<0.05) and those aged 27 and over (adjusted OR = 3.41; 95% CI = 2.23-5.22; p<0.001) were significantly more likely to follow a balanced diet than those aged 18-20. Students who prepared their own meals were significantly more likely to eat a balanced diet than those who ate purchased meals (adjusted OR = 3.78; 95% CI = 2.79-5.12; p<0.001) compared to meals prepared by themselves. Knowledge of healthy eating is also a key factor (adjusted OR = 2.10; 95% CI = 1.56-2.84; p<0.001). Moderate awareness of health and nutrition is associated with an increased likelihood of a balanced diet (adjusted OR = 1.49; 95% CI = 1.01-2.20; p<0.05), while a high level of awareness increases this likelihood even more significantly (adjusted OR =

3.33; 95% CI = 2.33-4.77; p<0.001). Individuals who spend more than 8 hours/day on activities are 37% less likely to adopt a balanced diet compared to the reference group (less than 4 hours/day) (adjusted OR = 0.63 (CI = [0.44; 0.90], p < 0.05). Finally, consumption of sugary drinks in excess of 5 liters per week is associated with a significant decrease in the likelihood of adopting a balanced diet (adjusted OR = 0.14; 95% CI = 0.09-0.21; p<0.001).

Table 1. Univariate and multivariate analysis of factors associated with balanced diet consumption among students at Ibn Tofail University, Kenitra, Morocco.

Consumption of a balanced diet	Univariate analysis		Multivariate analysis	
	Unadjusted		Adjusted	
	Odds Ratio	95% IC	Odds Ratio	95% IC
Age (Years)				
18-20	1		1	
21-23	0.95	0.67-1.36	0.72	0.47-1.09
24-26	1.49*	1.06-2.12	1.54*	1.02-2.33
27 and above	2.02***	1.44-2.83	3.41***	2.23-5.22
Sex				
Males	1		1	
Females	1.09	0.84-1.40	0.99	0.73-1.34
Socioeconomic Status				
Low	1		1	
Moderate	0.87	0.65-1.16	0.84	0.59-1.20
High	0.78	0.57-1.05	0.74	0.51-1.06
Meal Preparation				
Purchased meals	1		1	
Self-prepared	3.30***	2.55-4.26	3.78***	2.79-5.12
Knowledge of Healthy Eating				
No	1		1	
Yes	2.01***	1.56-2.59	2.10***	1.56-2.84
Physical Activity				
Irregular	1		1	
Regular	1.30*	1.02-1.66	1.22	0.91-1.65
Consumption of Traditional Moroccan Dishes				
No	1		1	
Yes	1.42**	1.11-1.82	1.27	0.95-1.71

continued on following page

Table 1. Continued

Consumption of a balanced diet	Univariate analysis		Multivariate analysis	
	Unadjusted		Adjusted	
	Odds Ratio	95% IC	Odds Ratio	95% IC
Time spent engaged (Hours)				
Less than 4 hours/day	1		1	
4-8 hours/day	0.69*	0.51-0.94	0.70	0.48-1.01
More than 8 hours/day	0.55**	0.41-0.74	0.63*	0.44-0.90
Sugary drinks consumption (liters per week)				
0-2 liters/week	1		1	
3-5 liters/week	0.87	0.65-1.16	0.81	0.58-1.12
More than 5 liters/week	0.18***	0.12-0.26	0.14***	0.09-0.21
Level of awareness about health and nutrition				
Low awareness	1		1	
Moderate awareness	1.69**	1.21-2.36	1.49*	1.01-2.20
High awareness	3.90***	2.86-5.33	3.33***	2.33-4.77

*: $p < 0.05$ (statistically significant association at the 5% level).
**: $p < 0.01$ (statistically significant association at the 1% level).
***: $p < 0.001$ (statistically significant association at the 0.1% level).

Table 2 presents the results of the multivariate analysis of factors associated with obesity among students. With regard to age, a significant reduction in the probability of obesity was observed in the 24-26 age group (adjusted OR = 0.43; 95% CI = 0.22-0.80; p<0.01) and 27 years and older (adjusted OR = 0.55; 95% CI = 0.30-0.98; p<0.05) compared to the 18-20 age group. With regard to gender, women are significantly less likely to be obese than men (adjusted OR = 0.52; 95% CI = 0.33-0.84; p<0.01). The method of meal preparation was also a determining factor, with purchased meals being associated with a lower probability of obesity (adjusted OR = 0.61; 95% CI = 0.39-0.95; p<0.05). Knowledge of healthy eating significantly reduces the likelihood of obesity (adjusted OR = 0.59; 95% CI = 0.37-0.92; p<0.05). Regular physical activity is strongly associated with a reduced risk of obesity (adjusted OR = 0.50; 95% CI = 0.32-0.78; p<0.01). Furthermore, working more than 8 hours per day is associated with a significant increase in the likelihood of obesity (adjusted OR = 1.80; 95% CI = 1.07-3.03; p<0.05). Consumption of sugary drinks in excess of 5 liters per week was associated with a significant decrease in the risk of obesity (adjusted OR = 0.44; 95% CI = 0.24-0.82; p<0.05). Finally, a high level of health and nutrition awareness significantly reduces the likelihood of obesity (adjusted OR = 0.54; 95% CI = 0.32-0.91; p<0.05).

Table 2. Univariate and multivariate analysis of factors associated with obesity among students at Ibn Tofail University, Kenitra, Morocco.

Obesity	Univariate analysis		Multivariate analysis	
	Unadjusted		Adjusted	
	Odds Ratio	95% IC	Odds Ratio	95% IC
Age (Years)				
18-20	1		1	
21-23	0.87	0.51-1.49	0.77	0.44-1.36
24-26	0.47	0.26-0.87	0.43**	0.22-0.80
27 and above	0.53	0.30-0.92	0.55*	0.30-0.98
Sex				
Males	1		1	
Females	0.55	0.35-0.86	0.52**	0.33-0.84
Socioeconomic Status				
Low	1		1	
Moderate	1.01	0.61-1.65	0.86	0.51-1.44
High	1.10	0.67-1.81	1.06	0.63-1.78
Meal Preparation				
Purchased meals	1		1	
Self-prepared	0.61	0.40-0.94	0.61*	0.39-0.95
Knowledge of Healthy Eating				
No	1		1	
Yes	0.60	0.38-0.92	0.59*	0.37-0.92
Physical Activity				
Irregular	1		1	
Regular	0.54	0.35-0.84	0.50**	0.32-0.78
Consumption of Traditional Moroccan Dishes				
No	1		1	
Yes	0.79	0.53-1.20	0.79	0.52-1.22
Time spent engaged (Hours)				
Less than 4 hours/day	1		1	
4-8 hours/day	1.33	0.76-2.31	1.30	0.73-2.29
More than 8 hours/day	1.97	1.19-3.25	1.80*	1.07-3.03
Level of awareness about health and nutrition				
Low awareness	1		1	

continued on following page

Table 2. Continued

Obesity	Univariate analysis		Multivariate analysis	
	Unadjusted		Adjusted	
	Odds Ratio	**95% IC**	**Odds Ratio**	**95% IC**
Moderate awareness	0.69	0.41-1.15	0.80	047-1.37
High awareness	0.48	0.29-0.78	0.54*	0.32-0.91

*: p < 0.05 (statistically significant association at the 5% level).
**: p < 0.01 (statistically significant association at the 1% level).

Combined effects of behavioral and contextual factors on the adoption of a balanced diet

Figures 2 and 3 analyze the combined effects of several behavioral and contextual factors on the likelihood of adopting a balanced diet, expressed in terms of adjusted odds ratios (OR) with their 95% confidence intervals. The results highlight how favorable factors, such as regular meals, knowledge about healthy eating, and awareness of health and nutrition, interact with unfavorable behaviors, such as high time commitment and excessive consumption of sugary drinks, to modulate this probability. In Figure 2, the CE1 model, which includes regular meals, knowledge about healthy eating, and average awareness of health and nutrition, shows a significant effect with an OR of 11.97 (CI = [6.68; 21.45]; p<0.01). However, when unfavorable behaviors are added, the positive effect gradually decreases. For example, adding high consumption of sugary drinks (>5 L/week) in CE2 reduces this effect to an OR of 7.59 (CI = [3.81; 15.10]; p<0.01). Similarly, the introduction of high screen time (>8 hours/day) in CE3 further reduces the effect to an OR of 1.70 (CI = [0.86; 3.35]; p>0.05), which is not significant. Finally, when these two unfavorable behaviors are combined (CE4), the association becomes even weaker with an OR of 1.07 (CI = [0.50; 2.32]; p>0.05), revealing the cumulative impact of unfavorable behaviors on the probability of adopting a balanced diet. Figure 3 expands on this analysis by examining the combined effects in the context of high health and nutrition awareness. The CE5 model, which includes regular meals, knowledge about healthy eating, and high awareness, shows a very marked effect with an OR of 26.64 (CI = [14.77; 48.06]; p<0.01). However, as in Figure 2, the addition of unfavorable behaviors attenuates this effect. For example, in CE6, the introduction of high consumption of sugary drinks reduces the OR to 3.77 (CI = [1.88; 7.56]; p<0.01). Similarly, the addition of high time spent in CE7 lowers the OR to 16.89 (CI = [8.47; 33.67]; p<0.01). Finally, the CE8 model, which simultaneously combines high time spent and sugary drink consumption, shows an OR of 2.39 (CI = [1.10; 5.22]; p<0.05), indicating a significant reduction in the initial effect.

Figure 2. Combined effects – Adjusted Odds Ratios with 95% Confidence Intervals (CE1 to CE4).

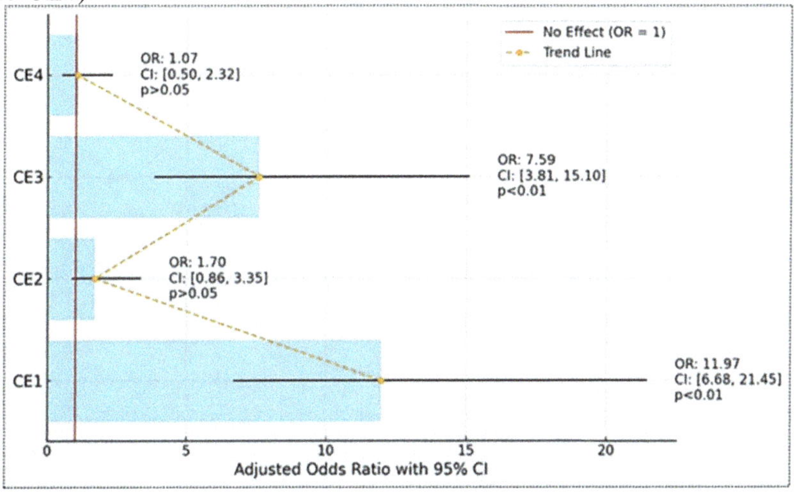

CE1: Combined effect of regular meals, knowledge about healthy eating, and average awareness of health and nutrition; CE2: Combined effect of regular meals, knowledge about healthy eating, average awareness, and consumption of sugary drinks (>5 L, high); CE3: Combined effect of regular meals, knowledge about healthy eating, average awareness, and time spent (>8 hours); CE4: Combined effect of regular meals, knowledge about healthy eating, average awareness, time spent (>8 hours), and consumption of sugary drinks (>5 L, high). CI: confidence interval

Figure 3. Combined effects – Adjusted Odds Ratios with 95% Confidence Intervals (CE5 to CE8).

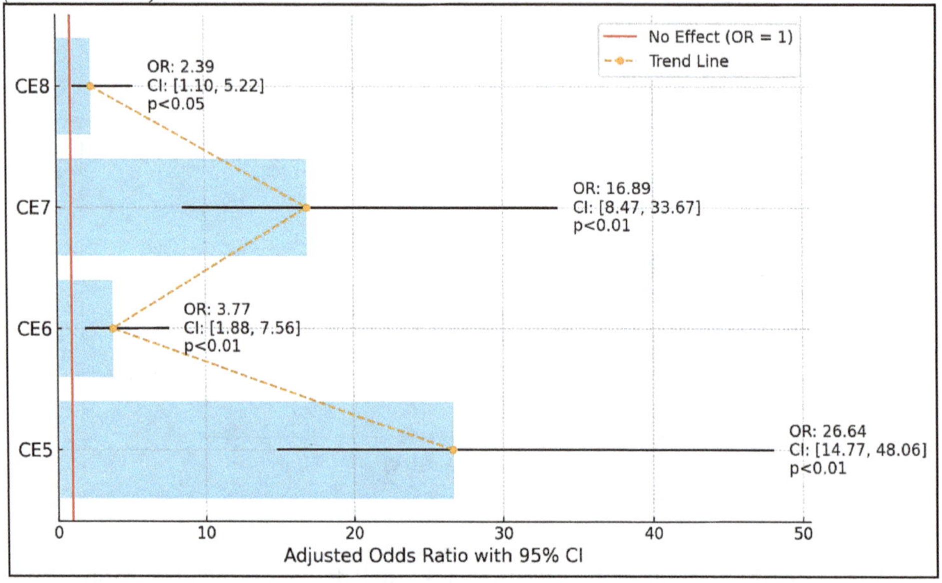

CE5: Combined effect of regular meals, knowledge about healthy eating, and high awareness of health and nutrition; CE6: Combined effect of regular meals, knowledge about healthy eating, high awareness, and consumption of sugary drinks (>5 L, high); CE7: Combined effect of regular meals, knowledge about healthy eating, high awareness, and time spent (>8 hours); CE8: Combined effect of regular meals, knowledge about healthy eating, high awareness, time spent (>8 hours), and consumption of sugary drinks (>5 L, high). CI: confidence interval

Table 3 presents the results of the multinomial logistic model, analyzing the factors associated with stress levels (medium and high) based on body mass index (BMI), adjusted for several confounding factors (age, gender, socioeconomic status, physical activity, consumption of sugary drinks, time spent, and level of awareness of health and nutrition). With regard to average stress levels, underweight individuals have an adjusted odds ratio (AOR) of 0.98 (CI = [0.56; 0.72]), indicating no significant association compared to those with a normal BMI. Overweight individuals have an AOR of 0.72 (CI = [0.44; 1.19]), which is also not significant. Obese individuals, on the other hand, have an AOR of 1.15 (CI = [0.7; 1.96]), which does not allow us to conclude that there is a significant association with average stress. The associated E-values, whether for point estimates or confidence intervals, remain low, suggesting limited robustness of the results to unmeasured factors. With regard to high stress, the results show more marked differences. Underweight individuals have an OR of

0.55 (CI = [0.22; 1.36]), indicating a non-significant trend toward a reduced risk of high stress compared to those with a normal BMI. Overweight individuals have an OR of 1.07 (CI = [0.66; 1.92]), which remains non-significant. However, obese individuals have an OR of 3.30 (CI = [1.95; 5.59]), revealing a significant association between obesity and an increased risk of high stress. This association is reinforced by high E-values, including a point estimate of 6.075, indicating that the results are unlikely to be biased by unmeasured confounding factors.

Table 3. Multinomial logistic model of factors associated with stress levels, adjusted for confounding factors.

BMI (kg/m²)	Stress moyen$^{\Omega}$			Stress élevé$^{\Omega}$		
	RRR with 95%CI	E-value (Point Estimate)	E-value (CI)	RRR with 95%CI	E-value (Point Estimate)	E-value (CI)
Normal	1			1		
Underweight	0.98 [0.56-0.72]	1.135	1.000	0.55 [0.22-1.36]	3.009	1.000
Overweight	0.72 [0.44-1.19]	2.082	1.000	1.07 [0.66-1.92]	1.362	1.000
Obese	1.15 [0.7-1.96]	1.565	1.000	3.30 [1.95-5.59]	6.075	3.324

$^{\Omega}$ The multinomial logistic model was adjusted for the following confounding factors: age, gender, socioeconomic status, physical activity, consumption of sugary drinks, time spent, and level of awareness of health and nutrition. RRR: relative risk ratio; CI: confidence interval.

DISCUSSION

The descriptive results of this study provide crucial insight into stress levels, BMI distribution, and adherence to a balanced diet among university students. These data highlight notable disparities in these key dimensions, which warrant further analysis in the context of students' lifestyles and pressures. Perceived stress levels show that 47.9% of participants report low stress levels, while a significant proportion report moderate (34.7%) or high (17.5%) stress levels. This prevalence of high stress is consistent with existing literature, which reports that university students are often exposed to multiple stressors, such as academic workload, financial uncertainty, and the transition to independence (Bayram & Bilgel, 2008). These results are also consistent with a study conducted by Beiter et al. (2015), which found that nearly 20% of university students in the United States experienced high levels of stress

(Beiter et al., 2015). Stress, as a psychological factor, can directly or indirectly influence eating behaviors and physical health, particularly through compulsive eating or alterations in appetite regulation (Torres & Nowson, 2007).

With regard to BMI, the results show that 74.0% of participants are of normal weight, reflecting a generally healthy weight status. However, nearly 10% of participants are overweight (9.9%) and 9.8% are obese, reflecting a worrying prevalence of weight imbalances in this young population. These results are consistent with the nutritional transition observed in several developing countries, where modern eating habits rich in sugars and fats coexist with increased sedentary lifestyles (Popkin et al., 2020). Furthermore, although the proportion of underweight individuals is relatively low (6.2%), it also warrants consideration, as it may reflect nutritional imbalances or restrictive eating behaviors, often influenced by social pressures and beauty ideals (Rodgers & DuBois, 2016). With regard to eating habits, a slight majority of participants (52.6%) reported adhering to a balanced diet, while 47.4% did not. This low adherence could be linked to a lack of awareness of the importance of nutrition or to financial and social constraints limiting access to healthy foods, as highlighted in previous studies (Collado-Soler et al., 2023; Darmon & Drewnowski, 2008). Unbalanced diets can increase the risk of overweight, obesity, or underweight, and their prevalence in this university population highlights the urgent need for educational programs to raise awareness about nutrition.

These descriptive results highlight important public health issues. They suggest that, although the majority of students have relatively balanced indicators (normal weight and low stress), a significant proportion of the population remains at risk due to high stress levels, weight imbalances, or inadequate diet. These dimensions are interconnected and can amplify their negative effects on students' overall health, justifying targeted interventions to improve their physical and mental well-being. The results of the multivariate analysis highlight several significant determinants of balanced diet adoption among students. These factors include age, dietary practices (method of meal preparation), knowledge and awareness of healthy eating, and specific behaviors such as consumption of sugary drinks.

In terms of age, participants aged 24 to 26 and those aged 27 and over are significantly more likely to adopt a balanced diet. This result is consistent with previous studies showing that older young adults, having gained more independence and experience, tend to make more responsible and thoughtful food choices (Larson et al., 2011). In contrast, younger students (aged 18-20), who often face major transitions, such as moving into independent living or increased academic demands, may have less structured eating habits. Preparing meals oneself is strongly associated with an increased likelihood of adopting a balanced diet. This result confirms previous observations that food autonomy promotes better nutritional quality, as opposed to meals prepared by others, which are often associated with less regularity and lower quality meals (Harrington et al., 2023). In addition, a pilot program focused on meal prep revealed a significant increase in the consumption of home-cooked meals, highlighting the role of practical strategies in overcoming time constraints and improving nutritional quality (Mendez et al., 2021). These findings underscore that mastering and practicing home cooking are essential levers for promoting healthy eating behaviors among students.

Healthy eating knowledge and increased awareness of health and nutrition are emerging as key factors in adopting a balanced diet. These findings are consistent with the conclusions of Spronk et al. (2014), who demonstrated that educational interventions that reinforce nutritional knowledge can significantly improve food choices among young adults (Spronk et al., 2014). Moderate awareness (adjusted OR = 1.49) also has a positive impact, although it is less pronounced than that of high awareness, highlighting the importance of deeper levels of understanding and motivation in encouraging balanced food choices. On the other hand, excessive consumption of sugary drinks (>5 liters/week) is strongly associated with a decreased likelihood of adopting a balanced diet. This result is consistent with the literature, which shows that sugary drinks, due to their low nutritional value and high added sugar content, are indicators of unbalanced diets and excessive calorie intake (Vartanian et al., 2007). These behaviors may also be influenced by food marketing and preferences for high-energy-density products, which often target young adults (Smith et al., 2019). Furthermore, individuals who spend more than 8 hours/day on activities are 37% less likely to have a balanced diet than those who spend less than 4 hours/day. This association could be related to time constraints and academic stress, which push these students towards quick, often low-nutrient food choices, thus compromising the adoption of a balanced diet (Mikolajczyk et al., 2009).

These results highlight that adopting a balanced diet depends on a complex mix of individual and behavioral factors, as well as the food environment in which students live. This underscores the need for combined strategies to promote healthy eating behaviors, such as nutrition education, improving access to balanced food options, and reducing consumption of unhealthy products such as sugary drinks. In the CE1 model, regular meals, knowledge about healthy eating, and moderate awareness of health and nutrition significantly increase the likelihood of adopting a balanced diet. These results confirm the positive impact of structured eating practices and moderate awareness, in line with previous studies that have shown that regular meals and basic nutrition education contribute to improving eating behaviors (Spronk et al., 2014). However, this positive effect is significantly attenuated by unfavorable behaviors. For example, adding excessive consumption of sugary drinks (>5 L/week) to the CE2 model reduces the effect to an OR of 7.59. Similarly, introducing high screen time (>8 hours/day) into the CE3 model further lowers the positive effect to an OR of 1.70, rendering the association non-significant. This highlights the role of negative behaviors, which can not only counteract positive efforts but also have a cumulative impact on health outcomes, as reported by Malik et al. (2019) in the context of excessive sugar consumption (Malik et al., 2010).

When both unfavorable behaviors (high time spent and excessive consumption of sugary drinks) are combined in the CE4 model, the effect becomes even weaker. This result highlights the cumulative negative interaction of unfavorable behaviors and reveals their ability to neutralize the positive effects of favorable practices. These observations underscore the importance of integrated interventions that simultaneously target favorable and unfavorable behaviors to improve nutritional outcomes. The CE5 model, which includes regular meals, knowledge about healthy eating, and high awareness, shows an even more pronounced effect. This result confirms that high awareness is a powerful lever for promoting healthy eating behaviors, corroborating the findings of Deliens et al. (2016), who demonstrated that nutrition education and awareness strengthen intrinsic motivation to adopt a balanced diet (Deliens et al., 2014). However, as in the previous model, the addition of unfavorable behaviors reduces the effect. For example, introducing sugary drink consumption in CE6 lowers the OR to 3.77, while adding high time commitment in CE7 lowers the effect to 16.89. Finally, when these two unfavorable behaviors are combined in the CE8 model, the effect is reduced to an OR of 2.39, although it remains significant. These results show that, even in the presence of high awareness, unfavorable behaviors have a negative, albeit smaller, impact on the adoption of a balanced diet. This highlights the importance of educational strategies that incorporate consistent messages about reducing harmful behaviors.

Overall, these results highlight that students' eating behaviors are influenced by a dynamic balance between favorable and unfavorable factors. While factors such as high awareness and regular meals have a protective effect, they can be compromised by negative habits. These observations are consistent with previous studies suggesting that combined interventions, which simultaneously address favorable and unfavorable behaviors, are necessary to optimize nutritional outcomes in student populations (Deliens et al., 2014; MJ, 2013; Sprake et al., 2018). The results of the multivariate analysis highlight several factors significantly associated with obesity among students, underscoring the impact of sociodemographic characteristics, eating behaviors, and lifestyle habits on the risk of weight imbalances.

Age appears to be a protective factor against obesity, with a significant reduction in the likelihood of obesity among participants aged 24 to 26 and 27 and older, compared to those aged 18 to 20. These results suggest that young undergraduate students, who often face major changes in their lifestyle, may be more vulnerable to weight gain. As they get older, students may develop better eating habits and healthier behaviors, consistent with observations from previous studies, which report that maturity is associated with more balanced weight management (Vilaro et al., 2018;

Zhuchkova et al., 2024). With regard to gender, women are significantly less likely to be obese than men. This observation is consistent with studies suggesting that women are more conscious of their physical appearance and more likely to adopt restrictive or balanced behaviors to avoid weight gain (Anfal et al., 2024; Shafiee et al., 2024; Vilaro et al., 2018). However, this difference in risk could also reflect biological variations, such as differences in fat distribution or metabolic regulation between the sexes. The method of meal preparation emerges as a key factor, with participants who prepare their own meals having a reduced likelihood of obesity. This finding is consistent with previous research showing that home-cooked meals are often associated with better nutritional quality and increased portion control (Wolfson & Bleich, 2015). This highlights the importance of promoting home meal preparation as a key strategy for improving the quality of food choices, particularly in university settings where ready-to-eat options are often high in calories and low in essential nutrients.

Familiarity with healthy eating and a high level of awareness about health and nutrition significantly reduce the likelihood of obesity. These findings are consistent with the work of Spronk et al. (2014), which highlights the effectiveness of educational interventions on nutrition in improving eating behaviors and preventing obesity (Spronk et al., 2014). Increased awareness can encourage individuals to make healthier food choices and reduce their consumption of ultra-processed foods. Regular physical activity stands out as a major protective factor against obesity. This finding is consistent with a wealth of literature demonstrating that regular exercise plays an essential role in preventing obesity by contributing to energy expenditure and regulating metabolism (Warburton et al., 2006). On the other hand, working more than 8 hours a day is associated with a significant increase in the likelihood of obesity. This factor could be linked to a sedentary lifestyle, particularly among university students who spend long hours studying or using electronic devices. Overall, these findings reveal several modifiable levers for reducing the risk of obesity among students, including promoting physical activity, nutrition education, and strategies to limit sedentary time. They also highlight the importance of specific food contexts, such as university environments, in obesity prevention.

Analysis of the results shows that obesity is significantly associated with an increased risk of high stress among university students, who are more than three times more likely to report high stress levels than their peers with a normal BMI. This relationship can be explained by psychosocial mechanisms, including social stigma and weight discrimination, which amplify psychological distress while limiting social interactions (Brewis et al., 2018). In academic settings, obese students may feel increased pressure to conform to social norms regarding weight, further exacerbating their stress levels (Puhl et al., 2020). In addition, they often face explicit and implicit discrimination, reinforcing feelings of marginalization and stress (Puhl

& Heuer, 2010). These negative social experiences can also intensify activation of the hypothalamic-pituitary-adrenal (HPA) axis, exacerbating physiological responses to stress, a mechanism that is particularly pronounced in obese individuals (Tomiyama et al., 2011). These findings underscore the urgent need to reduce weight-related stigma in academic settings and to implement appropriate supports to improve the mental health of obese students.

Limitations and Strengths of the Study

This study has certain limitations that should be highlighted. The data are based primarily on participant self-reports, which may introduce biases related to self-reporting, particularly for dietary habits and physical activity. In addition, the cross-sectional nature of the study does not allow for the establishment of causal relationships between the variables studied. The absence of certain data, such as participants' income or precise measures of physical activity, also limits the in-depth analysis of certain health determinants. Finally, although the sample is representative of Ibn Tofail University, the results may vary in other Moroccan university contexts. However, the study has several strengths. It provides a detailed and contextualized analysis of eating behaviors and stress in a student population that is often under-studied in Morocco. The use of a multivariate logistic model made it possible to identify key factors and their combined effects, thus providing concrete avenues for guiding targeted interventions.

CONCLUSION

This study revealed significant links between eating behaviors, nutritional status, and perceived stress among university students at Ibn Tofail University in Morocco. It shows that practices such as preparing meals independently and increased awareness of nutrition play a key role in preventing obesity. On the other hand, unfavorable behaviors, such as excessive consumption of sugary drinks or a sedentary lifestyle, increase the risk of weight imbalances. Regular physical activity, strongly associated with a reduced risk of obesity, stands out as an essential protective factor. Furthermore, the study highlights a relationship between obesity and high stress levels, underscoring the psychosocial and psychological impact of this condition and the need for strategies to improve students' mental and physical health. These findings underscore the importance of combined interventions targeting eating habits, promoting physical activity, and reducing harmful behaviors. To further explore these observations, longitudinal studies incorporating complementary variables would be necessary to better understand the complex relationships between these dimensions and guide appropriate actions to improve the well-being of university students.

Publisher's Note

Correction Notice: A correction to this manuscript has been published

Reason: Figures needed to be replaced

The publisher was notified after publication that the images were incorrect and some were missing. We have replaced the figures as the author(s) requested.

Correction Date: January 30th, 2026

REFERENCES

Abraham, S., Noriega, B. R., & Shin, J. Y. (2018). College students eating habits and knowledge of nutritional requirements. *Journal of Nutrition and Human Health*, *2*(1), 13–17. DOI: 10.35841/nutrition-human-health.2.1.13-17

Al-Naimi, S., Sooud, M., Alyahyaee, M., Al-Saedi, Z., & Feroz, H. (2025). Relationship Between Academic Stress and Eating Habits Among University of Sharjah Medical Students. *Clinical Psychology Review*, *119*, 103890. DOI: 10.70706/j.cpr.2024.103890

Al-Qahtani, A. M. (2022). Lifestyle habits among Najran University students, Najran, Saudi Arabia. *Frontiers in Public Health*, *10*, 938062. DOI: 10.3389/fpubh.2022.938062

Almutairi, K. M., Alonazi, W. B., Vinluan, J. M., Almigbal, T. H., Batais, M. A., Alodhayani, A. A., Alsadhan, N., Tumala, R. B., Moussa, M., & Aboshaiqah, A. E. (2018). Health promoting lifestyle of university students in Saudi Arabia : A cross-sectional assessment. *BMC Public Health*, *18*, 1–10. DOI: 10.1186/s12889-018-5999-z

Alzamil, H. A., Alhakbany, M. A., Alfadda, N. A., Almusallam, S. M., & Al-Hazzaa, H. M. (2019). A profile of physical activity, sedentary behaviors, sleep, and dietary habits of Saudi college female students. *Journal of Family & Community Medicine*, *26*(1), 1–8. DOI: 10.4103/jfcm.JFCM_58_18

Anfal, A.-D., Batarseh, N., Al-Bashabsheh, Z., Batarseh, D., & Karablieh, A. (2024). Stress levels and eating behavior among university students in Jordan : A cross-sectional study. *Journal of Education and Health Promotion*, *13*(1), 285. https://doi.org/DOI. DOI: 10.4103/jehp.jehp_168_24

Bayram, N., & Bilgel, N. (2008). The prevalence and socio-demographic correlations of depression, anxiety and stress among a group of university students. *Social Psychiatry and Psychiatric Epidemiology*, *43*, 667–672. DOI: 10.1007/s00127-008-0345-x

Beiter, R., Nash, R., McCrady, M., Rhoades, D., Linscomb, M., Clarahan, M., & Sammut, S. (2015). The prevalence and correlates of depression, anxiety, and stress in a sample of college students. *Journal of Affective Disorders*, *173*, 90–96. DOI: 10.1016/j.jad.2014.10.054

Brewis, A., SturtzSreetharan, C., & Wutich, A. (2018). Obesity stigma as a globalizing health challenge. *Globalization and Health*, *14*, 1–6. DOI: 10.1186/s12992-018-0337-x

Collado-Soler, R., Alférez-Pastor, M., Torres, F. L., Trigueros, R., Aguilar-Parra, J. M., & Navarro, N. (2023). A systematic review of healthy nutrition intervention programs in kindergarten and primary education. *Nutrients*, *15*(3), 541. DOI: 10.3390/nu15030541

Darmon, N., & Drewnowski, A. (2008). Does social class predict diet quality? *The American Journal of Clinical Nutrition*, *87*(5), 1107–1117. DOI: 10.1093/ajcn/87.5.1107

de Miranda, R. C., Rauber, F., & Levy, R. B. (2021). Impact of ultra-processed food consumption on metabolic health. *Current Opinion in Lipidology*, *32*(1), 24–37. https://doi.org/DOI. DOI: 10.1097/MOL.0000000000000728

Deliens, T., Clarys, P., De Bourdeaudhuij, I., & Deforche, B. (2014). Determinants of eating behaviour in university students : A qualitative study using focus group discussions. *BMC Public Health*, *14*, 1–12. DOI: 10.1186/1471-2458-14-53

Derici, S., & Dogan, N. O. (2025). A decision model based on artificial intelligence for disease prediction and patient treatment process planning. *Asia Pacific Journal of Health Management*, *20*(2), 1–12.

Dicken, S. J., & Batterham, R. L. (2024). Ultra-processed Food and Obesity : What Is the Evidence? *Current Nutrition Reports*, *13*(1), 23–38. DOI: 10.1007/s13668-024-00517-z

El Ansari, W., Stock, C., & Mikolajczyk, R. T. (2012). Relationships between food consumption and living arrangements among university students in four European countries-a cross-sectional study. *Nutrition Journal*, *11*, 1–7. DOI: 10.1186/1475-2891-11-28

Harrington, E., Parsons, K., Neptune, L., Winn, C., Watras, E., Raymond, K., Zhou, W., Barr, M., & McNamara, J. (2023). Cooking Self-efficacy Predicts Diet Quality in College Students. *Journal of the Academy of Nutrition and Dietetics*, *123*(9), A59.

José, M. E. R., Ramos, I. E. C., de Sousa, T. M., & Canella, D. S. (2025). Food consumption associated with depression, anxiety and stress in students entering a public university. *Journal of Nutritional Science*, *14*, e3. https://doi.org/doi:10.1017/jns.2024.90

Larson, N., Neumark-Sztainer, D., Laska, M. N., & Story, M. (2011). Young adults and eating away from home : Associations with dietary intake patterns and weight status differ by choice of restaurant. *Journal of the American Dietetic Association*, *111*(11), 1696–1703. DOI: 10.1016/j.jada.2011.08.007

Magalhães, V., Severo, M., Correia, D., Torres, D., de Miranda, R. C., Rauber, F., Levy, R., Rodrigues, S., & Lopes, C. (2021). Associated factors to the consumption of ultra-processed foods and its relation with dietary sources in Portugal. *Journal of Nutritional Science*, *10*, e89. https://doi.org/doi:10.1017/jns.2021.61

Malik, V. S., Popkin, B. M., Bray, G. A., Després, J.-P., & Hu, F. B. (2010). Sugar-sweetened beverages, obesity, type 2 diabetes mellitus, and cardiovascular disease risk. *Circulation*, *121*(11), 1356–1364. DOI: 10.1161/CIRCULATIONAHA.109.8761

Mendez, S., Kubota, J., Widaman, A. M., & Gieng, J. (2021). Advance quantity meal preparation pilot program improves home-cooked meal consumption, cooking attitudes, and self-efficacy. *Journal of Nutrition Education and Behavior*, *53*(7), 608–613. DOI: 10.1016/j.jneb.2020.12.014

Mikolajczyk, R. T., El Ansari, W., & Maxwell, A. E. (2009). Food consumption frequency and perceived stress and depressive symptoms among students in three European countries. *Nutrition Journal*, *8*, 1–8. DOI: 10.1186/1475-2891-8-31

MJ, G. M. (2013). Food habits and nutritional assessment in a university population. *Nutrición Hospitalaria*, *28*(2), 438–446. DOI: 10.3305/nh.2013.28.2.6303

Navarro-Prado, S., Schmidt-RioValle, J., Fernández-Aparicio, Á., Montero-Alonso, M. Á., Perona, J. S., & González-Jiménez, E. (2024). Assessing the Impact of Religion and College Life on Consumption Patterns of Ultra-Processed Foods by Young Adults : A Cross-Sectional Study. *Nutrients*, *16*(11). Advance online publication. DOI: 10.3390/nu16111619

O'neil, A., Quirk, S. E., Housden, S., Brennan, S. L., Williams, L. J., Pasco, J. A., Berk, M., & Jacka, F. N. (2014). Relationship between diet and mental health in children and adolescents : A systematic review. *American Journal of Public Health*, *104*(10), e31–e42. DOI: 10.2105/AJPH.2014.302110

Obesity and overweight. (s. d.). Consulté 18 janvier 2025, à l'adresse https://www.who.int/news-room/fact-sheets/detail/obesity-and-overweight

Popkin, B. M., Corvalan, C., & Grummer-Strawn, L. M. (2020). Dynamics of the double burden of malnutrition and the changing nutrition reality. *Lancet*, *395*(10217), 65–74. DOI: 10.1016/S0140-6736(19)32497-3

Puhl, R. M., & Heuer, C. A. (2010). Obesity stigma : Important considerations for public health. *American Journal of Public Health*, *100*(6), 1019–1028.

Puhl, R. M., Himmelstein, M. S., & Pearl, R. L. (2020). Weight stigma as a psychosocial contributor to obesity. *The American Psychologist*, *75*(2), 274. DOI: 10.1037/amp0000538

Rodgers, R. F., & DuBois, R. H. (2016). Cognitive biases to appearance-related stimuli in body dissatisfaction : A systematic review. *Clinical Psychology Review*, *46*, 1–11. DOI: 10.1016/j.cpr.2016.04.006

Shafiee, A., Nakhaee, Z., Bahri, R. A., Amini, M. J., Salehi, A., Jafarabady, K., Seighali, N., Rashidian, P., Fathi, H., & Esmaeilpur Abianeh, F. (2024). Global prevalence of obesity and overweight among medical students : A systematic review and meta-analysis. *BMC Public Health*, *24*(1), 1673. DOI: 10.1186/s12889-024-19184-4

Smith, R., Kelly, B., Yeatman, H., & Boyland, E. (2019). Food marketing influences children's attitudes, preferences and consumption : A systematic critical review. *Nutrients*, *11*(4), 875. DOI: 10.3390/nu11040875

Sogari, G., Velez-Argumedo, C., Gómez, M. I., & Mora, C. (2018). College students and eating habits : A study using an ecological model for healthy behavior. *Nutrients*, *10*(12), 1823. DOI: 10.3390/nu10121823

Sprake, E., Russell, J., Cecil, J., Cooper, R., Grabowski, P., Pourshahidi, L. K., & Barker, M. (2018). Dietary patterns of university students in the UK: A cross-sectional study. *Nutrition Journal*, *17*, 1–17. DOI: 10.1186/s12937-018-0398-y

Spronk, I., Kullen, C., Burdon, C., & O'Connor, H. (2014). Relationship between nutrition knowledge and dietary intake. *British Journal of Nutrition*, *111*(10), 1713–1726. https://doi.org/doi:10.1017/S0007114514000087

Tomiyama, A. J. (2019). Stress and obesity. *Annual Review of Psychology*, *70*(1), 703–718. DOI: 10.1146/annurev-psych-010418-102936

Tomiyama, A. J., Dallman, M. F., & Epel, E. S. (2011). Comfort food is comforting to those most stressed : Evidence of the chronic stress response network in high stress women. *Psychoneuroendocrinology*, *36*(10), 1513–1519. DOI: 10.1016/j.psyneuen.2011.04.005

Torres, S. J., & Nowson, C. A. (2007). Relationship between stress, eating behavior, and obesity. *Nutrition (Burbank, Los Angeles County, Calif.)*, *23*(11-12), 887–894. DOI: 10.1016/j.nut.2007.08.008

Vartanian, L. R., Schwartz, M. B., & Brownell, K. D. (2007). Effects of soft drink consumption on nutrition and health : A systematic review and meta-analysis. *American Journal of Public Health*, *97*(4), 667–675.

Vilaro, M. J., Colby, S. E., Riggsbee, K., Zhou, W., Byrd-Bredbenner, C., Olfert, M. D., Barnett, T. E., Horacek, T., Sowers, M., & Mathews, A. E. (2018). Food choice priorities change over time and predict dietary intake at the end of the first year of college among students in the US. *Nutrients*, *10*(9), 1296. DOI: 10.3390/nu10091296

Warburton, D. E., Nicol, C. W., & Bredin, S. S. (2006). Health benefits of physical activity : The evidence. *Canadian Medical Association Journal*, *174*(6), 801–809. https://doi.org/DOI. DOI: 10.1503/cmaj.051351

Wolfson, J. A., & Bleich, S. N. (2015). Is cooking at home associated with better diet quality or weight-loss intention? *Public Health Nutrition*, *18*(8), 1397–1406. DOI: 10.1017/S1368980014001943

Zhuchkova, S. M., Busalaeva, E. I., Markova, T. N., & Semenova, V. A. (2024). Obesity in medical students : What has changed in 18 years. *Therapy*, *10*(3), 146–152. DOI: 10.18565/therapy.2024.3.146-152

Chapter 9
Risk Factors for Health and Their Psychosocial Dimensions:
Evidence From OECD Countries

Fatih Çemrek
https://orcid.org/0000-0002-6528-7159
Eskişehir Osmangazi Üniversity, Turkey

Ebrucan İslamoğlu
https://orcid.org/0009-0009-2974-9350
Nevşehir Hacı Bektaş Veli University, Turkey

Okan Tiring
https://orcid.org/0000-0001-9253-8403
Marmara University, Turkey

ABSTRACT

This study examines the impact of tobacco-related risk factors on public health within OECD countries by focusing on smoking and vaping prevalence across different demographic groups. Eight variables were analyzed, including daily smoking rates by sex and age groups (15–24 years and 15 years or over), regular vaping product usage among males, and overall tobacco consumption. Data were derived from OECD health statistics and covered both total and gender-specific populations. Multivariate analysis was conducted to explore the relationship between these variables and key health outcomes, highlighting the psychosocial dimensions associated with tobacco use, such as age, gender disparities, and youth vulnerability. Results reveal that smoking prevalence remains significant among young adults, particularly males, while vaping products are emerging as a critical factor influencing tobacco

DOI: 10.4018/979-8-3373-6746-0.ch009

consumption patterns. The findings emphasize the need for integrated public health policies targeting both traditional and alternative tobacco products, with special focus on preventive strategies for youth and gender-sensitive interventions to mitigate long-term health risks.

1. INTRODUCTION

Tobacco use and electronic cigarettes (e-cigarettes) are among the leading causes of preventable diseases and premature deaths worldwide. According to the World Health Organization (WHO, 2023), tobacco exposure results in approximately 8 million deaths annually, of which 1.3 million are attributable to secondhand smoke. In 2023, tobacco exposure was reported to cause more than 7 million deaths globally, ranking as the leading cause of death among men and the seventh among women (Global Burden of Disease, 2023; Guardian, 2025).

In the context of OECD countries, the daily smoking rate among individuals aged 15 and above has been recorded at an average of 15.9% (OECD, 2023a). The prevalence of e-cigarette use is estimated at 23% for lifetime use and 11% for current use, with increasing trends observed particularly among younger populations (BioMed Central, 2022). Studies conducted in Europe and the United States indicate that between 2011 and 2018, e-cigarette use among adolescents increased by more than 19% (Barrington-Trimis et al., 2019). These findings highlight the rapid spread of e-cigarette use, especially among youth, and underscore its potential long-term health risks.

Psychosocial factors play a critical role in influencing tobacco and e-cigarette use. Stress levels, social support mechanisms, work–life balance, and gender differences can serve as significant determinants of individuals' inclination toward tobacco products (WHO, 2023; OECD, 2023b; CDC, 2023). These psychosocial dimensions particularly heighten the risk of addiction and vulnerability among younger populations. In OECD countries, tobacco and e-cigarette use among youth exhibits considerable variation. While the average daily smoking rate among individuals aged 15 and above is 15.9%, this figure differs markedly across countries; for instance, it exceeds 25% in France and Turkey, whereas it remains below 10% in countries such as Iceland, Norway, and Canada (OECD, 2023a).

E-cigarette use among individuals aged 15–24 averages 6.1% and exhibits significant increases in certain countries; in Estonia and New Zealand, this rate exceeds 20%. The rise in e-cigarette use indicates that, despite declines in conventional smoking in some countries, nicotine exposure persists among youth, creating new risks for addiction (OECD, 2023b). Gender differences in e-cigarette use are also notable across OECD countries. Among young people aged 15–24, e-cigarette use is 7.1% in males and 5.1% in females, suggesting that young men are more likely to use e-cigarettes than young women. In the general population, e-cigarette use among individuals aged 15 and above is 3.9% for men and 2.5% for women, indicating that men across all age groups engage in higher e-cigarette use compared to women (OECD, 2023b). In this context, the present study aims to analyze the risk factors associated with tobacco and e-cigarette use in OECD countries and to evaluate the psychosocial dimensions of such use from a public health perspective. The study seeks to contribute scientifically to the development of policies addressing both traditional and alternative tobacco products (WHO, 2023; OECD, 2023a; CDC, 2023; BioMed Central, 2022; Barrington-Trimis et al., 2019).

Tobacco-related risk factors and psychosocial dimensions constitute a significant area of research within public health. Studies conducted in OECD countries indicate that socioeconomic indicators exert a decisive influence on health outcomes. Cilhoroz (2024) found that per capita gross domestic product (GDP) and literacy rates were associated with reductions in deaths related to mental and behavioral disorders, whereas urbanization, alcohol consumption, and obesity were linked to increases in such deaths. Similarly, Akdemirel (2024) demonstrated that alcohol consumption, tobacco use, and sulfur oxide emissions negatively affected life expectancy, while total health expenditures had a positive impact on health outcomes. International literature corroborates these findings; Pikhart and Pikhartova (2015a) reported that psychosocial factors, including high job demands, low autonomy, interpersonal conflicts, and limited social support, were independently associated with chronic diseases. These findings suggest that the relationship between socioeconomic characteristics and health outcomes may be partially mediated by psychosocial factors. In another study encompassing OECD countries, Yigit et al. (2021) identified significant associations among noncommunicable disease risk factors, health indicators, and income distribution, emphasizing the importance of comprehensive policies addressing obesity, tobacco, and alcohol consumption.

The data utilized in this study were obtained from multiple sources. OECD data, including the *Health at a Glance* reports and the OECD data platform, provide statistics on cigarette and e-cigarette (vaping) use disaggregated by age, gender, and country (OECD, 2023a; OECD, 2023b). Information on international health policies was derived from the World Health Organization (WHO), MPOWER reports, and the Global Burden of Disease (IHME GBD) studies, which collectively cover global health impacts, tobacco-related mortality, and policy implementations (WHO, 2023; Global Burden of Disease, 2023). Academic studies and meta-analyses, through scientific articles and systematic reviews, offer insights into the prevalence, trend analyses, and issues related to youth and addiction concerning tobacco and e-cigarette use (BioMed Central, 2022; Barrington-Trimis et al., 2019). Additionally, youth- and addiction-focused data, sourced from the CDC and relevant literature, provide detailed information on vaping behaviors, associated health effects, and cognitive risk factors among young populations (CDC, 2023).

This study aims to analyze tobacco-related risk factors and their psychosocial dimensions in OECD countries, assessing their impact on public health. The analysis considers demographic variables such as daily cigarette smoking rates, e-cigarette use, and age- and gender-related differences, alongside psychosocial factors including stress, social support, and the vulnerability of the youth population. Through the findings, the study seeks to provide scientific insights to inform the development of public health policies targeting both conventional and alternative tobacco products (WHO, 2023; OECD, 2023a; CDC, 2023; BioMed Central, 2022; Barrington-Trimis et al., 2019).

2. LITERATURE REVIEW

From a tobacco use perspective, the health impacts associated with daily cigarette smoking are well-established. Cigarette consumption constitutes a major risk factor for cardiovascular diseases, respiratory disorders, and various forms of cancer (WHO, 2020). In addition, electronic cigarette (e-cigarette) use has emerged as an increasingly significant public health concern. Recent studies indicate that while e-cigarettes generally contain lower levels of toxic substances compared to conventional cigarettes, they maintain nicotine dependence and thereby elevate the risk of addiction, particularly among the youth population (Glantz & Bareham, 2018).

Demographic differences also play a significant role in tobacco use. Various studies have demonstrated that age and gender substantially influence both cigarette and electronic cigarette consumption rates. In particular, young males are more prone to using cigarettes and e-cigarettes, whereas females tend to exhibit higher cessation rates as they age (Palipudi et al., 2016). These findings underscore the necessity for targeted public health policies that account for age- and gender-specific patterns of tobacco use.

Finally, psychosocial dimensions influencing tobacco use are strongly emphasized in the literature. Elevated stress levels, work–life balance challenges, and insufficient social support contribute to increased cigarette and electronic cigarette consumption (Pikhart & Pikhartova, 2015a; Umberson & Montez, 2010). Therefore, tobacco control policies should address not only individual addictive behaviors but also the broader socioeconomic and psychosocial conditions that shape these behaviors.

2.1. Socioeconomic Determinants and Health

Research examining health-related risk factors and psychosocial dimensions in OECD countries indicates significant associations between socioeconomic factors and health outcomes. Aras (2024) found that per capita GDP and literacy rates were negatively associated with mortality due to mental and behavioral disorders, whereas urban population share, alcohol consumption, and obesity exerted positive effects on such mortality. Similarly, Akdemirel (2024) demonstrated that alcohol consumption, tobacco use, and sulfur oxide emissions negatively impacted life expectancy, while total health expenditures had a positive effect on health outcomes.

International evidence supports these findings. Pikhart and Pikhartova (2015b) identified psychosocial factors—such as high job demands, low autonomy, interpersonal conflicts, and low social support—as independently associated with chronic diseases. These factors partially mediate the relationship between socioeconomic characteristics and health outcomes, contributing to the understanding of the social gradient in health. Similarly, Yıgıt et al. (2021) confirmed significant associations between non-communicable disease risk factors, health indicators, and income distribution across OECD countries, emphasizing the necessity of comprehensive policies addressing obesity, tobacco use, and alcohol consumption to improve population health.

Şahman and Gün (2024) analyzed health-related social determinants in OECD countries using the MOORA method and found that Norway, Sweden, Ireland, Switzerland, and Finland ranked the highest, whereas Turkey, Chile, Greece, Slovakia, and Poland ranked the lowest. This finding underscores the significant impact of social determinants on health outcomes.

2.2. Health Status and Expenditures

Ersöz (2008) compared the health status and expenditures of Turkey with those of OECD countries, noting that Turkey exhibited similar health variables to South Korea, Mexico, Poland, and Slovakia, while showing substantial differences from Germany, Norway, and Austria. Sayılı et al. (2017) reported that Turkey's health indicators and expenditures remained below the OECD average, particularly in terms of life expectancy, infant and maternal mortality rates, and overall health spending. Kırcı Çevik (2021) demonstrated that Turkey's health indicators consistently clustered with those of other developing OECD countries from 2002 to 2015. Furthermore, Tatlı et al. (2021) confirmed that education, per capita income, and health expenditures have positive effects on life expectancy across OECD countries.

Beceren, Balcı İzgi, and Fırat (2021) examined the determinants of health expenditures in OECD countries, finding income elasticity to be –0.07 (negative) and the aging effect to be 0.45 (positive). These findings indicate that an aging population is a key driver of increasing health expenditures.

2.3. Health Risk Factors

Cilhoroz (2022) compared OECD countries in terms of lifestyle determinants of health, including smoking, alcohol consumption, diet, and physical activity, reporting that Austria demonstrated the best performance in healthy lifestyle indicators, whereas Costa Rica exhibited the weakest performance. Sağin and Güler (2020) found that in OECD countries, increases in obesity were influenced by calorie intake, urbanization, unemployment, and alcohol consumption, while smoking had a negative impact on obesity. Koç et al. (2020) demonstrated that COVID-19 cases and mortality were associated with health expenditures, obesity, elevated blood sugar levels, smoking prevalence, and the stringency of policy measures.

2.4. Psychosocial Dimensions

Tekin Epik and Öztürk (2020) identified the psychosocial risks encountered by healthcare professionals—including stress, burnout, mobbing, workplace violence, long working hours, and substance abuse—and reported that these factors negatively impact both physical and mental health. Demiral et al. (2023) found that high work pace, lack of recognition, and insufficient social support adversely affect the mental health of employees in the textile sector, with women and white-collar workers being at higher risk. Filiz et al. (2022) reported that 31.6% of healthcare workers experienced burnout, 28.5% experienced mobbing, and 56% experienced stress and violence, highlighting that risk factors vary across occupational groups.

Gümüşdaş et al. (2014) found that the psychosocial health levels of high-risk pregnant women were lower compared to low-risk pregnant women, with stress, anxiety, and the need for psychological support being particularly pronounced. Additionally, Gizir (2016) examined the concept of psychological resilience, emphasizing the importance of risk and protective factors for children and adolescents, and highlighted the need for programs aimed at enhancing resilience.

2.5. Health System Efficiency

Tokatlıoğlu and Altın (2020) assessed the efficiency of OECD health sectors using Data Envelopment Analysis, identifying Estonia, Iceland, Israel, Luxembourg, and Poland as efficient, while ranking Turkey 34th, indicating inefficiency. Güneş and Ercan (2021) classified OECD countries based on health indicators and found a strong relationship between health expenditures and the perceived control of corruption. Alkaya and Alkaş (2021) clustered and classified OECD countries according to health indicators, reporting that Turkey lagged behind many developed countries across numerous indicators, whereas Japan and Korea stood out with favorable health outcomes.

3. METHOD

The aim of this study is to rank OECD member countries based on statistical data on cigarette and electronic cigarette consumption using Factor Analysis, and to reveal the similarities and differences among them graphically through Multidimensional Scaling (MDS) analysis. Accordingly, data on cigarette and electronic cigarette consumption for 37 OECD countries were obtained from the official OECD website, and the dataset corresponds to the year 2022.

3.1. Variables Used and Their Definitions

The variables examined in the study are presented below.

$\mathbf{X_1}$: Share of population who are daily smokers (Total, From 15 to 24 years (%)
$\mathbf{X_2}$: Share of population who are regular vaping product users (Male, 15 years or over) (%)
$\mathbf{X_3}$: Share of population who are regular vaping product users (Male- From 15 to 24 years) (%)
$\mathbf{X_4}$: Share of population who are daily smokers (Total -15 years or over) (%)
$\mathbf{X_5}$: Share of population who are daily smokers (Female-From 15 to 24 years) (%)
$\mathbf{X_6}$: Share of population who are daily smokers (Female -15 years or over) (%)
$\mathbf{X_7}$: Tobacco consumption (Total-15 years over) (%)
$\mathbf{X_8}$: Tobacco consumption (Total-Age Group: Total) (%)

3.2. Methods Used

3.2.1. Factor Analysis

The foundations of Factor Analysis, which is frequently used in many fields, particularly in the social sciences, were established by Charles Spearman. In 1904, Spearman developed factor analysis to investigate the relationship between students' performance in various subjects and their levels of intelligence (Sharma, 1996, p. 90).

The aim of factor analysis is to identify a smaller number of new, uncorrelated variables—called factors—from a set of p interrelated variables. This analysis seeks to achieve dimension reduction and to eliminate dependency structures among the variables (Tatlıdil, 2002, p. 167).

The raw data matrix (Xpxn) is standardized to obtain the standardized data matrix (Zpxn).

The linear model representing the relationship between the standardized variables (Zj) and the common factors is expressed as follows:

$$z_j = a_{j1}f_1 + a_{j2}f_2 + \dots + a_{jm}f_m + b_j u_j; j = 1,2,\dots,p \tag{1}$$

a_{jm}: The loading or weight of the variable on the *m*th factor,
f_m: m-factor,
u_j: residual factor,
b_j: shows the coefficient related to the residual factor,

It is also possible to write model (1) in matrix form as follows:

$$Z= AF + BU \tag{2}$$

After performing factor analysis, with factor rotation;
Dimensionality reduction,
Ensuring orthogonality or independence,
Providing conceptual meaningfulness.

In factor analysis, the minimum number of factors that best explain or account for the correlations among variables are identified; the most reasonable factor solution is obtained through factor rotation; factor loadings are estimated, and interpretations are made for the common factor or factors. If necessary, factor scores are also estimated (Anderson, 1974; Johnson & Wichern, 1992; Sharma, 1996, p. 99).

3.2.2. Multidimensional Scaling (MDS)

In Multidimensional Scaling (MDS), the aim is to represent the positions of n objects (individuals or observations) in a multidimensional space by using the distance values determined according to p variables, thus revealing the relational structure among these objects (Özdamar, 1999, p. 479).

The general purpose of multidimensional scaling (mds) analysis is to represent the structure of objects (using distance values) as closely as possible to the original form with fewer dimensions. MDS is one of the Q-type analysis techniques, such as cluster and discriminant analysis; additionally, due to its dimensionality reduction feature, it is considered among R-type analysis techniques (Tatlıdil, 2002, p. 353).

This analysis provides a graphical representation that represents the distances among objects with minimal error and aims to determine these distances through a suitable function (Oğuzlar, 1995).

Multidimensional Scaling (MDS) is a multivariate statistical analysis method used for analyzing behavioral data such as personal preferences, attitudes, tendencies, beliefs, and expectations. The geometric meaning of MDS is to represent the relevant individuals or stimuli as points in a space where each variable constitutes a dimension (Kurtuluş, 2006, p. 357–358). It is included among classification and grouping analyses, such as cluster analysis and discriminant analysis (Hair et al., 1998; Tatlıdil, 2002; İpekçi Çetin, 2003).

Multidimensional Scaling (MDS) is divided into two types: metric and non-metric scaling techniques. In the application of the metric scaling technique, an approach similar to principal components is used. When a distance matrix S among n objects is given, these objects are represented as n points in an (n-1)-dimensional E space. Here, the S_{ij} values,

$$Sij < Sik + Skj \text{ for all i, j, and k} \tag{3}$$

meets the condition.

The geometric model of any n objects that satisfies this condition can be constructed. A problem encountered here is representing the model with a reduced number of dimensions. Therefore, when performing scaling, in order to reduce the number of dimensions, new d_{ij} values are calculated in the new space to represent these original S_{ij} values.

The steps followed in the application of the non-metric scaling technique are:

All elements of the D similarity matrix (excluding the diagonal elements) are ranked.

$$d_{r_1 s_1} < \ldots\ldots < d_{r_m s_m};\ m = \frac{1}{2}n(n-1) \tag{4}$$

The stress value, which is an expression of the difference between the true configuration in the multidimensional space and the configuration estimated in the reduced-dimensional space, is calculated. This process continues until the minimum stress value is obtained (Kalaycı, 2006, p. 380-381; Jobson, 1992, p. 568-580).

In determining whether the number of dimensions used in the graphical arrangement obtained as a result of MDS (Multidimensional Scaling) analysis is appropriate, the stress value emerges as an important criterion. The stress measure, which indicates the goodness of fit or adequacy of representation, is widely used in multidimensional scaling analysis.

For the stress coefficient, which serves as a measure of the fit between the obtained configuration and the actual configuration, the tolerance ratios developed by Kruskal are used.

stress $\geq$ 0.20, it indicates a poor fit;
stress = 0.10, it indicates a moderate fit;
stress = 0.05, it indicates a good fit;
stress = 0.00, it indicates a perfect fit.
it is considered to be so (Tatlıdil, 2002, p. 363).

3.2.3. Findings

Table 1 presents the descriptive statistics of the data obtained for the countries regarding the variables used in the study.

Table 1. Descriptive Statistics of the Variables Used

Variable	Min.	Max.	Mean	Standart Deviation
x1	2,00	4462,71	129,4187	721,97643
x2	1,00	13,30	4,4513	2,77670
x3	0,60	1876582,55	49390,9129	304420,6950
x4	6,23	28,00	15,7429	5,19121
x5	1,00	190,12	14,6300	29,69699
x6	2,33	22,30	12,2403	4,52702
x7	18,90	2256,00	1018,1578	380,95828
x8	7,10	22405,50	601,9895	3632,59161

3.2.3.1. Results Obtained from the Factor Analysis

KMO and Bartlett's Test		
Kaiser-Meyer-Olkin Measure of Sampling Adequacy		**0,643**
Bartlett's Test of Sphericity	Approx. Chi-Square	1103,823
	df	28
	Sig.	<0,01

As a result of the factor analysis, two eigenvalues greater than one were obtained. The total variance explained by these two factors was found to be 87.055%. This means that by reducing the 12 original variables to two factors, there is approximately a 13% loss of variance.

Table 2. Eigenvalues and Variance Explained for the Obtained Factors

Factor	Eigenvalue	Variance Explained (%)
1	4.470	55.878
2	1.512	18.906
Total	5.982	74.784

As shown in Table 2, two factors were obtained as a result of the analysis. The eigenvalue of the first factor (4.470) explains 55.878% of the variance, indicating that a large proportion of the variables are represented by this factor. The eigenvalue of the second factor (1.512) explains 18.906% of the variance. Together, the two factors account for 74.784% of the total variance. This ratio is quite high in factor analysis and demonstrates that the model represents the data structure well. Generally, a total explained variance above 60% is considered sufficient for an acceptable factor solution. Therefore, these results indicate that the factor structure is strong and that the variables are meaningfully grouped under two factors.

The factors in which the examined variables are grouped can be determined by referring to the component matrix presented in Table 3.

Table 3. Component Matrix

Variable	Component 1	Component 2
x3	0.978	0.186
x8	0.978	0.186
x1	0.977	0.192
x5	0.946	0.300
x7	-0.485	0.132
x4	-0.463	0.822
x6	-0.503	0.750
x2	0.055	-0.246

Extraction Method: Principal Component Analysis

a. Two Components Were Extracted

As seen in Table 3, the variables x1, x3, x5, and x8 have high loadings on the first component, indicating that they are primarily represented by Factor 1. On the other hand, the variables x4 and x6 show strong loadings on the second component, suggesting their association with Factor 2. The relatively low or negative loadings of x2 and x7 imply that these variables contribute less distinctly to either factor. This distribution supports the two-factor structure obtained in the analysis. From Table 3, it is clearly observed which factor each variable belongs to. Accordingly, the factors and the variables included in each factor are presented in the table below.

Table 4. Variables Included in Each Factor

Faktor	Variable
1	X_1, X_3, X_5, X_7, X_8
2	X_2, X_4, X_6

According to Table 4, the variables X_1 (Share of population who are daily smokers — Total, aged 15–24 years (%)), X_3 (Share of population who are regular vaping product users — Male, aged 15–24 years (%)), X_5 (Share of population who are daily smokers — Female, aged 15–24 years (%)), X_7 (Tobacco consumption — Total, aged 15 years and over (%)), and X_8 (Tobacco consumption — Total, all age groups (%)) are included in Factor 1.

The variables included in Factor 2 are X_2 (Share of population who are regular vaping product users — Male, aged 15 years and over (%)), X_4 (Share of population who are daily smokers — Total, aged 15 years and over (%)), and X_6 (Share of population who are daily smokers — Female, aged 15 years and over (%)).

The factor score coefficient matrix was obtained as shown in Table 5.

Table 5. Factor Score Coefficient Matrix

Variable	Component 1	Component 2
X1	0.247	0.054
X2	-0.038	-0.158
X3	0.246	0.049
X4	0.069	0.549

continued on following page

Table 5. Continued

Variable	Component 1	Component 2
X5	0.263	0.124
X6	0.046	0.506
X7	-0.276	0.116
X8	0.246	0.050

Extraction Method: Principal Component Analysis

Rotation Method: Varimax with Kaiser Normalization

Component Scores

As observed in Table 5, the variables with the highest weights according to the factor score coefficients remain unchanged.

Table 6. Factor 1 Scores for Countries

Country	F1 Score	Country	F1 Score	Country	F1 Score
Australia	-0.15892	Hungary	0.11585	Norway	-0.33465
Austria	0.04086	Iceland	-0.33556	Poland	-0.15228
Belgium	-0.20943	Ireland	-0.17053	Portugal	-0.19283
Canada	-0.35434	Israel	-0.09922	Slovak Republic	0.00686
Chile	-0.05797	Italy	-0.05887	Slovenia	-0.06041
Costa Rica	-0.44826	Japan	-0.15978	Spain	-0.05689
Czechia	-0.43693	Korea	-0.38616	Sweden	-0.33107
Denmark	-0.25394	Latvia	-0.05641	Switzerland	-0.12416
Estonia	-0.08059	Lithuania	-0.15366	Turkey	-0.00597
Finland	-0.14150	Luxembourg	-0.17993	United Kingdom	0.00030
France	0.19588	Mexico	-0.36401	United States	-0.30766
Germany	-0.30937	Netherlands	-0.14305	****	****
Greece	0.07010	New Zealand	-0.23578	*****	*****

*Note: The missing values (*** / ****) should be filled in with the corresponding countries and their F1 scores.*

As seen in Table 6, based on the Factor 1 scores, the countries that are the most distant from each other are France and Costa Rica.

Table 7. Factor 2 Scores for Countries

Country	F2 Score	Country	F2 Score	Country	F2 Score
Australia	-0.83377	Hungary	2.36336	Norway	-1.17047
Austria	1.31958	Iceland	-1.34844	Poland	0.41965
Belgium	-0.15832	Ireland	-0.43652	Portugal	-0.42518
Canada	-1.10796	Israel	-0.04101	Slovak Republic	1.07140
Chile	0.93834	Italy	0.64181	Slovenia	0.66261
Costa Rica	-1.75430	Japan	-0.44525	Spain	0.86801
Czechia	0.54622	Korea	-0.68773	Sweden	-0.50031
Denmark	-0.65486	Latvia	0.69685	Switzerland	0.36969
Estonia	-0.96077	Lithuania	0.11576	Turkey	1.64816
Finland	-0.49941	Luxembourg	0.02313	United Kingdom	-0.11805
France	1.65648	Mexico	-1.90423	United States	-0.89326
Germany	-0.06727	Netherlands	0.17731	****	
Greece	1.75020	New Zealand	-0.51414	*****	

Table 7 presents the ranking of countries based on their Factor 2 scores. When examining the scores for the second factor, it can be observed that Hungary and the United States are the most distant countries from each other. According to this ranking, Hungary has the highest score, while the United States has the lowest score.

3.2.3.2. Results Obtained from the Multidimensional Scaling Analysis

Multidimensional Scaling (MDS) analysis was conducted for the eight variables examined to analyze statistical data on cigarette and e-cigarette consumption among OECD member countries. The MDS was performed in both two and three dimensions, yielding a stress value of 0.0181 for the two-dimensional solution and 0.00026 for the three-dimensional solution. These results indicate that the three-dimensional solution provides a near-perfect fit. Therefore, the three-dimensional MDS results are presented, and the interpretations are based on this solution.

Table 8. Coordinate Values of Each Country on the Dimensions in 3-Dimensional MDS

Country	Dimension			Country	Dimension		
	1	2	3		1	2	3
Australia	0,1681	0,77	-0,6129	Korea	-0,2666	0,9789	-0,7919
Austria	-0,4894	-1,0495	-0,5105	Latvia	-0,3732	-0,6248	0,0776
Belgium	-0,1945	0,1579	0,1935	Lithuania	-0,2502	-0,101	-0,5768
Canada	0,0136	0,9151	0,0776	Luxembourg	-0,2209	0,0942	2,0654
Chile	-0,4688	-0,7811	-0,5671	Mexico	0,3325	1,5138	-0,6323
Costa Rica	0,1626	1,4465	0,1632	Netherlands	-0,2539	-0,1495	-0,586
Czechia	**-0,903**	-0,3884	2,3001	New Zealand	-0,0784	0,4476	0,2887
Denmark	-0,0408	0,5945	0,5219	Norway	0,0667	0,9585	-0,2914
Estonia	**0,3714**	1,0178	1,9552	Poland	-0,3666	-0,3141	-0,1803
Finland	0,0397	0,4129	-0,5228	Portugal	-0,0663	0,3258	-0,6541
France	-0,4204	-1,3612	0,8884	Slovak Republic	-0,4278	-0,8432	-0,5216
Germany	-0,4558	0,1847	-0,1713	Slovenia	-0,3472	-0,5333	-0,3937
Greece	-0,6347	-1,4137	-0,1852	Spain	-0,4221	-0,6518	0,1891
Hungary	-0,8343	-1,8529	-0,0055	Sweden	-0,2427	0,4865	**-1,0088**
Iceland	0,1414	1,0919	-0,2311	Switzerland	-0,3134	-0,3008	-0,3565
Ireland	-0,0207	0,4122	0,3303	Türkiye	-0,7101	-1,3846	-0,0925
Israel	-0,1022	0,021	-0,436	United Kingdom	0,0939	0,1616	0,8201
Italy	-0,331	-0,4886	-0,1344	United States	-0,2666	0,9789	-0,7919
Japan	-0,012	0,3985	-0,4711				

Figure 1. 3-Dimensional MDS Plot

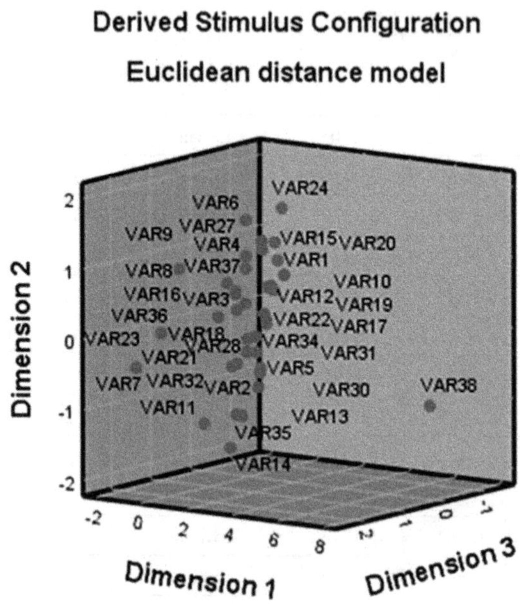

If the countries are ranked according to the first dimension, Table 9 is obtained. According to Table 9, the countries that are the most distant from each other on the first dimension are Estonia and Czechia. The variables that contribute to this difference for these two countries are presented in Table 10, and the corresponding plot is shown in Figure 2.

Table 9. Ranked Coordinate Values of Countries According to the First Dimension

Country	Coordinat Value	Country	Coordinat Value	Country	Coordinat Value
Estonia	0,3714	Portugal	-0,0663	Latvia	-0,3732
Mexico	0,3325	New Zealand	-0,0784	France	-0,4204
Australia	0,1681	Israel	-0,1022	Spain	-0,4221
Costa Rica	0,1626	Belgium	-0,1945	Slovak Republic	-0,4278
Iceland	0,1414	Luxembourg	-0,2209	Germany	-0,4558

continued on following page

Table 9. Continued

Country	Coordinat Value	Country	Coordinat Value	Country	Coordinat Value
United Kingdom	0,0939	Sweden	-0,2427	Chile	-0,4688
Norway	0,0667	Lithuania	-0,2502	Austria	-0,4894
Finland	0,0397	Netherlands	-0,2539	Greece	-0,6347
Canada	0,0136	Korea	-0,2666	Türkiye	-0,7101
Japan	-0,012	Switzerland	-0,3134	Hungary	-0,8343
United States	-0,0139	Italy	-0,331	Czechia	-0,903
Ireland	-0,0207	Slovenia	-0,3472		
Denmark	-0,0408	Poland	-0,3666		

Table 10. Data for Estonia and Czechia, the Most Dissimilar Countries on the First Dimension

Country	X1	X2	X3	X4	X5	X6	X7	X8
Estonia	4,9	11,2	16,5	13,2	10,7	12,6	18,9	12
Czechia	14,8	10,9	14	18,1	9,9	15,2	2256	12,7

Figure 2. Comparison Plot for Estonia and Czechia

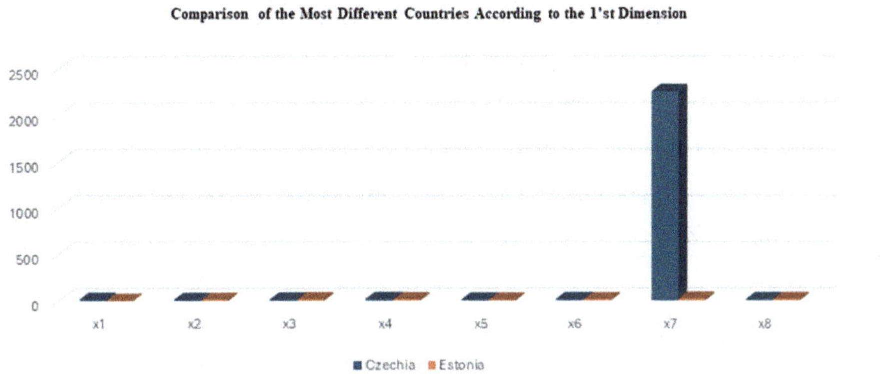

According to Figure 2, in the comparison of Estonia and Czechia, the countries that are the most distant on the first dimension, the variable that contributes the most to the difference between these two countries is X7 (Tobacco consumption — Total, aged 15 years and over (%)). If the countries are ranked according to the second dimension, Table 11 is obtained. According to Table 11, the countries that are the most distant from each other on the second dimension are Mexico and Hungary. The variables that contribute to this difference for these two countries are presented in Table 12, and the corresponding plot is shown in Figure 3.

Table 11. Ranked Coordinate Values of Countries According to the Second Dimension

Country	Coordinate Value	Country	Coordinate Value	Country	Coordinate Value
Mexico	1.5138	Ireland	0.4122	Italy	-0.1344
Costa Rica	1.4465	Japan	0.3985	Slovenia	-0.3937
Iceland	1.0919	Portugal	0.3258	Latvia	0.0776
Estonia	1.0178	Germany	0.1847	Spain	0.1891
Korea	0.9789	United Kingdom	0.1616	Chile	-0.5671
Norway	0.9585	Belgium	0.1579	Slovak Republic	-0.5216
Canada	0.9151	Luxembourg	0.0942	Austria	-0.5105
Australia	0.7700	Israel	0.0210	France	0.8884
United States	0.7405	Lithuania	-0.1010	Turkey	-0.0925
Denmark	0.5945	Netherlands	-0.1495	Greece	-0.1852
Sweden	0.4865	Switzerland	-0.3008	Hungary	-0.0055
New Zealand	0.4476	Poland	-0.3141		
Finland	0.4129	Czechia	-0.3884		

Table 12. Data for Mexico and Hungary, the Most Dissimilar Countries on the Second Dimension

Country	X1	X2	X3	X4	X5	X6	X7	X8
Mexico	4.9	11.2	16.5	13.2	10.7	12.6	18.9	12.0
Hungary	14.8	10.9	14.0	18.1	9.9	15.2	2256	12.7

Figure 3. Comparison Plot for Mexico and Hungary

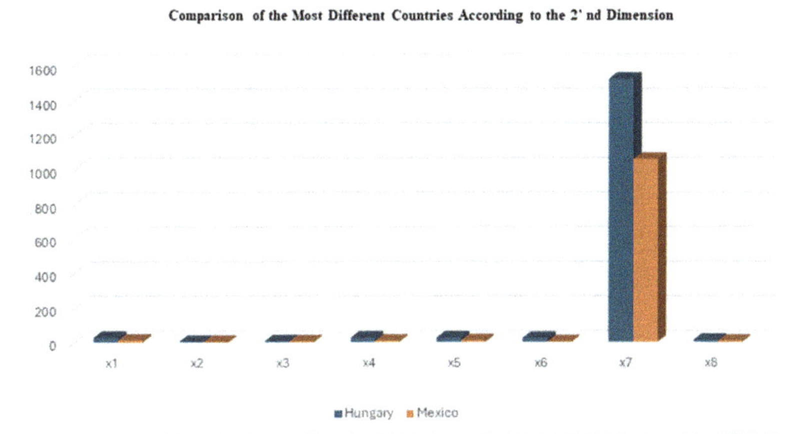

According to Figure 3, in the comparison of Mexico and Hungary, the countries that are the most distant on the second dimension, the variable that contributes the most to the difference between these two countries is X_7 (Tobacco consumption — Total, aged 15 years and over (%)). If the countries are ranked according to the third dimension, Table 13 is obtained. According to Table 13, the countries that are the most distant from each other on the third dimension are Czechia and Sweden. The variables that contribute to this difference for these two countries are presented in Table 14, and the corresponding plot is shown in Figure 4.

Table 13. Ranked Coordinate Values of Countries According to the Third Dimension

Country	Coordinate Value	Country	Coordinate Value	Country	Coordinate Value
Czechia	**2.3001**	United States	0.0688	Austria	-0.5105
Luxembourg	2.0654	Hungary	-0.0055	Slovak Republic	-0.5216
Estonia	1.9552	Turkey	-0.0925	Finland	-0.5228
France	0.8884	Italy	-0.1344	Chile	-0.5671
United Kingdom	0.8201	Germany	-0.1713	Lithuania	-0.5768
Denmark	0.5219	Poland	-0.1803	Netherlands	-0.5860
Ireland	0.3303	Greece	-0.1852	Australia	-0.6129
New Zealand	0.2887	Iceland	-0.2311	Mexico	-0.6323

continued on following page

Table 13. Continued

Country	Coordinate Value	Country	Coordinate Value	Country	Coordinate Value
Belgium	0.1935	Norway	-0.2914	Portugal	-0.6541
Spain	0.1891	Switzerland	-0.3565	Korea	-0.7919
Costa Rica	0.1632	Slovenia	-0.3937	**Sweden**	**-1.0088**
Canada	0.0776	Israel	-0.4360		
Latvia	0.0776	Japan	-0.4711		

Table 14. Data for the Most Dissimilar Countries in Dimension 3: Czechia and Sweden

Country	X1	X2	X3	X4	X5	X6	X7	X8
Czechia	14.8	10.9	14	18.1	9.9	15.2	2256	12.7
Sweden	7.1	1.5	4	10.4	6.5	10.6	1472.8	12.14

Figure 4. Comparison Graph for Czechia and Sweden

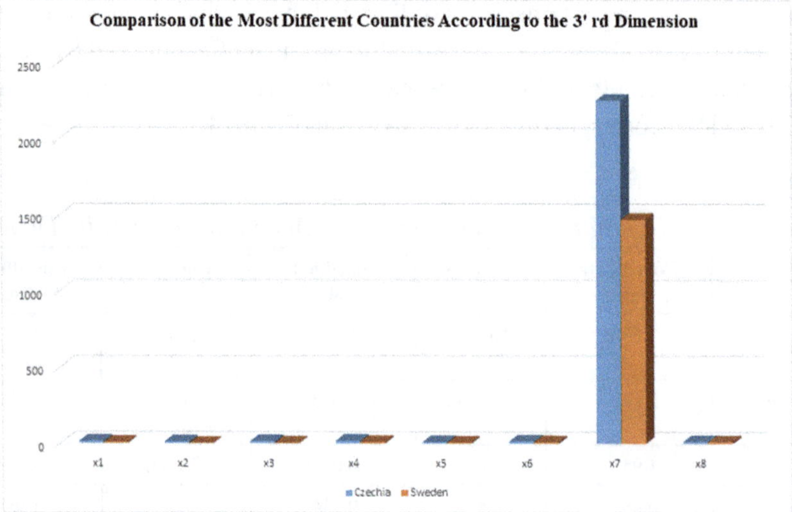

According to Figure 4, in the comparison of Czechia and Sweden, which are the most distant countries in the third dimension, the variable that contributes most to the difference between these two countries is Tobacco consumption (Total-15 years and over) (%).

4. CONCLUSION AND RECOMMENDATIONS

Tobacco use, which is a major determinant of many diseases today, constitutes an important public health problem. Smoking is associated with numerous diseases, primarily cancer, cardiovascular diseases, and respiratory tract diseases.

Nowadays, tobacco use has become widespread globally and is increasing in many low- and middle-income countries. In high-income countries, however, it is decreasing slowly but steadily. Due to the delayed effects of smoking on health outcomes, mortality rates from tobacco-related diseases are lower in low-income countries compared to middle- and high-income countries. Therefore, it is expected that mortality rates, particularly among women, in low- and middle-income countries will continue to rise even if efforts to reduce smoking are successful.

Factors affecting smoking tendencies include individual variables such as socioeconomic status and education, as well as system-level factors such as regional economic development and tobacco control policies. In this study, which compares OECD member countries in terms of variables related to cigarette and electronic cigarette use, countries were first ranked using Factor Analysis, and then their positions, identified as points on a three-dimensional graph, were interpreted. According to the results of the Factor Analysis, two factors with eigenvalues greater than one were obtained. The total variance explained by these two factors was found to be 75%. Based on the scores of the first factor, the two most distant countries were identified as France and Costa Rica. When countries were ranked according to the second factor score, Hungary and the USA were found to be the most distant. To graphically represent the positions of countries based on the variables considered in the study, Multidimensional Scaling (MDS) analysis was applied using the metric scaling technique, yielding a stress value of 0.0026 and producing a graph that closely approximates a perfect fit.

According to this stress value, in the first dimension of the three-dimensional scaling, Estonia and Czechia were identified as the most distant countries. The variable contributing most to the difference between these two countries is X7 (Tobacco consumption (Total-15 years and over) (%)). In the second dimension, the most distant countries are Mexico and Hungary, with X7 (Tobacco consumption (Total-15 years and over) (%)) again being the variable that contributes most to the difference. In the third dimension, the most distant countries are Czechia and Sweden, and the variable that contributes most to their difference is X7 (Tobacco consumption (Total-15 years and over) (%)).

This study demonstrates that differences in tobacco and e-cigarette consumption across OECD countries are related not only to economic and demographic indicators but also to critical psychological and psychosocial processes. Higher consumption rates among adolescents and young adults can be understood in light of the developmental vulnerability of this period, during which risk-taking tendencies, sensitivity to peer influence, and the need for social acceptance become more pronounced. The literature indicates that social relationships, levels of social support, exposure to psychosocial stressors, and the use of maladaptive coping strategies significantly shape health behaviors and susceptibility to addiction (Umberson & Montez, 2010; Pikhart & Pikhartova, 2015a). The fact that men exhibit higher consumption rates than women in many OECD countries is also consistent with psychosocial structures that shape gender roles, risk-taking patterns, and preferred stress-coping strategies. These findings suggest that tobacco control policies should extend beyond regulatory measures and incorporate preventive programs for youth, interventions that strengthen coping and stress-management skills, enhancements to social support systems, and behavior-change–oriented psychological approaches. Integrating these psychosocial mechanisms into public health strategies may lead to more sustainable and effective outcomes in reducing tobacco and e-cigarette use.

REFERENCES

Akdemirel, E. (2024). OECD Ülkelerinde Sağlığın Sosyal Belirleyicileri Üzerine Bir Panel Veri Analizi. *Hacettepe Sağlık İdaresi Dergisi, 27*(1), 65–90.

Akdemirel, S. (2024). Health outcomes and environmental risk factors in OECD countries: A panel data analysis. *Journal of Health Economics and Policy, 12*(2), 45–63.

Alkaya, A., & Alkaş, C. (2021). *OECD ülkelerinin sağlık göstergelerine göre kümeleme analizi sınıflaması*. Sosyal Güvence., DOI: 10.21441/sosyalguvence.1050881

Anderson, D.F., (1974), Multivariate Statistical Methods.

Aras, Y. (2024). Sosyoekonomik ve Yaşam Tarzı Faktörlerinin Ruhsal ve Davranışsal Bozukluklara Bağlı Ölümler Üzerindeki Etkisi: Bir Panel Veri Analizi. *Hacettepe Sağlık İdaresi Dergisi, 27*(3), 445–458.

Barrington-Trimis, J. L., Urman, R., Leventhal, A. M., ... (2019). *Trends in electronic cigarette use among US youth, 2011–2018*. Nature Medicine. https://www.nature.com/articles/s41533-022-00277-9

Beceren, E., Balcı İzgi, B., & Fırat, E. (2021). *Sağlık ve kalkınma ilişkisi OECD ülkelerinde sağlık harcamalarını belirleyen faktörler*. Aksaray Üniversitesi İİBF Dergisi., DOI: 10.52791/aksarayiibd.915196

BioMed Central. (2022). *Global e-cigarette prevalence: Lifetime and current use*. https://archpublichealth.biomedcentral.com/articles/10.1186/s13690-022-00998-w

CDC. (2023). *Youth and tobacco use: Electronic cigarettes*. Centers for Disease Control and Prevention. https://www.cdc.gov/tobacco/e-cigarettes/youth.html

Çevik, N. K. (2021). OECD ülkeleri sağlık sistemi göstergelerine çok boyutlu bir yaklaşım. Iğdır Üniversitesi Sosyal Bilimler Dergisi, 213-245.

Demiral, G. A., Şahan, C., Ozgur, E. A., Vatansever, M., & Demiral, Y. (2023). Bir Tekstil Fabrikasında Psikososyal Riskler ve Mental Sağlik Durumu. *Eskişehir Türk Dünyası Uygulama ve Araştırma Merkezi Halk Sağlığı Dergisi, 8*(1), 54–70.

Ersöz, F. (2008). Türkiye ile OECD ülkelerinin sağlık düzeyleri ve sağlık harcamalarının analizi.

Filiz, E., Ertaş, H., & Fidan, Y. (2022). Sağlık Çalışanlarında Psiko-sosyal Risk Faktörlerinin Değerlendirilmesi. Generıc Page, 81.

Gizir, C. A. (2016). *Psikolojik Sağlamlık.* Risk Faktörleri Ve Koruyucu Faktörler Üzerine Bir Derleme Çalışması.

Glantz, S. A., & Bareham, D. W. (2018). E-cigarettes: Use, effects on smoking, risks, and policy implications. *Annual Review of Public Health, 39*, 215–235.

Global Burden of Disease. (2023). *Global Burden of Disease Study 2023*. Institute for Health Metrics and Evaluation. https://www.healthdata.org/gbd

Guardian. (2025). *Tobacco exposure killed more than 7 million people worldwide: 2023 study*. https://www.theguardian.com/society/2025/jun/24/tobacco-exposure-killed-more-than-7m-people-worldwide-2023-study

Hair, J. F., Anderson, R. E., Tahtam, R. L., & Black, W. C. (1998). *Multivariate Data Analysis*. Prentice-Hall International.

İpekçi Çetin, E. (2003). Çok Değişkenli Analizlerin Pazarlama ile İlgili Araştırmalarda Kullanımı: 1995-2002 Arası Yazın Taraması, Akdeniz Üniversitesi. İ.İ.B.F. *Dergisi, 5*, 32–47.

Jobson, J. D. (1992). *Applied Multivariate Data Analysis*. Springer-Verlag.

Johnson, R. A., & Wichern, D. A. (1992). *Applied Multivariate Statistical Analysis*. Prentice-Hall.

Kalaycı, Ş. (2006). *SPSS uygulamalı çok değişkenli istatistik teknikleri, 2.baskı*. Asil Yayın Dağıtım Ltd.Şti.

Koç, I., & Saraç, M. (2020). Socio-economic, demographic and health determinants of the coronavirus pandemic: Analysis of data from OECD countries.

OECD. (2023a). [*OECD Indicators*. OECD Publishing. https://www.oecd.org/health/health-at-a-glance]. *Health at a Glance, 2023*, •••.

OECD. (2023b). *Daily smokers and vaping prevalence, 15+ population*. OECD Health Statistics. https://www.oecd.org/data-portal

Oğuzlar, A., (1995) "Çok Boyutlu Ölçekleme ve Kümeleme Analizi Arasındaki İlişkiler", Uludağ Üniversitesi İ.İ.B.F. Dergisi. 18(2).

Palipudi, K. M.. (2016). Age and gender differences in tobacco use: Findings from the Global Adult Tobacco Survey. *Nicotine & Tobacco Research: Official Journal of the Society for Research on Nicotine and Tobacco, 18*(5), 1230–1241.

Pikhart, H., & Pikhartova, J. (2015a). Psychosocial factors and health in Europe: Evidence from multi-country studies. *Social Science & Medicine, 138*, 73–80.

Pikhart, H., & Pikhartova, J. (2015b). The relationship between psychosocial risk factors and health outcomes of chronic diseases: a review of the evidence for cancer and cardiovascular diseases.

Şahman, O., & Gün, İ. (2024). Sağlığın sosyal belirleyicilerinin MOORA yöntemi ile analizi: OECD ülkeleri örneği. *Sosyal Güvence, 26*, 1376–1399. DOI: 10.21441/sosyalguvence.1507454

Sayılı, U., Sayman, Ö.A., Vehid, S., Köksal, S.S., & Erginöz, E. (2017). Türkiye ve OECD Ülkelerinin Sağlık Göstergeleri ve Sağlık Harcamalarının Karşılaştırılması.

Sharma, S. (1996). *Applied Multivariate Techniques*. John Wiley&Sons, Inc.

Tatlıdil, H. (2002). *Uygulamalı Çok Değişkenli İstatistiksel Analiz*. Ziraat Matbaacılık A.Ş. Ankara.

Tekin Epik, M., & Öztürk, M. (2020). Sağlık hizmetlerinde psikososyal riskler. *Süleyman Demirel Üniversitesi İktisadi ve İdari Bilimler Fakültesi Dergisi, 25*(4), 451–467.

Umberson, D., & Montez, J. K. (2010). Social relationships and health: A flashpoint for health policy. *Journal of Health and Social Behavior, 51*(1, suppl), S54–S66.

WHO. (2023). *Tobacco*. World Health Organization. https://www.who.int/news-room/fact-sheets/detail/tobacco

World Health Organization (WHO). (2020). *WHO report on the global tobacco epidemic 2019: Offer help to quit tobacco use*. WHO.

Yigit, A., Karahan, F., & Demir, S. (2021). Noncommunicable disease risk factors and health inequalities across OECD countries. *European Journal of Public Health Studies, 9*(3), 55–72.

Yiğit, P., Yılmaz, E., Gedikli, E., & Hayran, O. E. (2021). *The Relationship Between Noncommunicable Diseases Risk Factors, GINI Coefficient and Health Status in Organization of Economic Development and Cooperation Countries*. Turkiye Klinikleri Journal of Health Sciences.

Chapter 10
Numerical Ordination and Permutation-Based Inference:
RDA-PLSR Modeling of Environmental Structuring of Biochemical and Antioxidant Profiles

Brahim El Ouardi
https://orcid.org/0009-0000-4559-6141
Faculty of Sciences, Ibn Tofail University, Kenitra, Morocco

Issam El-Khadir
https://orcid.org/0000-0002-3970-2865
Faculty of Sciences, Ibn Tofail University, Kenitra, Morocco

Yassine Mouniane
https://orcid.org/0000-0002-6037-5301
Faculty of Sciences, Ibn Tofail University, Kenitra, Morocco

Mohamed Rektouti
https://orcid.org/0009-0004-5560-4925
Faculty of Sciences, Ibn Tofail University, Kenitra, Morocco

Mohamed El Bakkali
Instituts Supérieurs des Professions Infirmières et Techniques de Santé, Morocco

El Mahjoub Aouane
Faculty of Sciences, Ibn Tofail University, Kenitra, Morocco

ABSTRACT

This study used Redundancy Analysis (RDA) via a Partial Least Squares Regression (PLSR) surrogate with Monte Carlo permutation tests to examine how environmental gradients shape the biochemical and antioxidant traits of Moroccan carob (Ceratonia siliqua L.). Six geo-environmental predictors (altitude, latitude, longitude, mean

DOI: 10.4018/979-8-3373-6746-0.ch010

annual temperature, soil pH, rainfall) were modeled against bioactive compounds (polyphenols, flavonoids, tannins, proteins, minerals) and antioxidant indices (FRAP, IC_{50} DPPH, IC_{50} ABTS). The RDA biplot showed clear ecological patterns: altitude, pH, and rainfall favored polyphenolic and lipid accumulation, while magnesium and tannins reflected xeric conditions. Antioxidant indices responded differently, with FRAP linked to altitude and magnesium, and IC_{50} DPPH/ABTS clustering with low moisture and higher fat content. Permutation tests (n = 999) confirmed model robustness (global R^2 = 0.454, p < 0.0001), highlighting constrained ordination as a reproducible tool for quantifying ecological structuring of molecular traits.

INTRODUCTION

For many decades, the challenge of linking variability in environmental drivers to biological responses has been hampered by mismatches in scale, heterogeneity, and resolution (Pinek et al., 2020; Worden et al., 2025). Climate and environmental drivers are usually measured at coarse spatial and temporal resolutions that often use standardized meteorological or geophysical data sources that are not scaled the same way as the microhabitats where organisms live and respond to the environmental drivers (Senanayake et al., 2024). These mismatches can introduce systematic biases in estimating biological responses and limit ecological models from adequately estimating biological responses, especially if biological responses are not linear (Barrett et al., 2025). To address these difficulties, numerical methods which integrate numerous disparate datasets and capture the interscale heterogeneity associated with ecological systems are now considered essential. Such approaches - e.g., hierarchical modelling (Bedoya et al., 2011) and ordination analyses or regression-based techniques - help separate the large- and small-scale environmental drivers, as well as identify gradients linked to biological variability (Lee et al., 2025).

For example, in the past few years there have been increased efforts to establish frameworks for monitoring biodiversity such as the essential biodiversity variables as described in adapting the environment initiative (Hoban et al., 2022). These frameworks support the need for a standardized way of integrating multiple data sources using computational techniques. The emergence of machine learning and deep learning techniques (Pichler & Hartig, 2023) has created a transition in the way we analyze ecological datasets; a shift from understanding and describing our datasets to predicting ecological phenomena and discovering new correlations among ecological processes and patterns due to their complexity (Wang et al. 2025). In addition, we see that utilizing data-driven and AI-enhanced approaches to address sustainability-related issues are expanding into other fields. For example, using AI for modelling in wastewater treatment processes allows us to develop better models for optimizing these processes and therefore aids the advancement of a circular bioeconomy and supports the regulatory compliance of maintaining environmental standards (Matheri et al., 2022). As a result, we are witnessing the ongoing evolution of numerical methods as the means of linking climate, soils, hydrology and other ecological features to the biological responses of all living things including genetic variability, biochemical properties, and Community structure/function, and resilience (Zhao et al., 2025). Numerical methods also provide the necessary analytical rigor and communicability that support better linking of drivers to responses, and, also, actionable knowledge for biodiversity protection, resource management, and sustainability science (Nile et al., 2025; Zhang et al., 2025).

Plants in Mediterranean and semi-arid systems experience a rhythm continuously, and while the rhythms are largely driven by environmental conditions and change, these conditions drive physiological, biochemical, and ultimately morphobiological attributes in plants (Bagheri et al., 2021). One plant at the forefront of attention is *Ceratonia siliqua* L. (carob tree) that can note its fundamentally important contribution in agroforestry systems. In drought and poor soil conditions and without variability from a nutraceutical perspective (Mouniane et al., 2024). Carob pods comprise sugars, fibers, proteins, minerals, polyphenols, flavonoids, and tannins, all with a variety of nutritional qualities and diverse antioxidant activities, and additionally their biochemical qualities are neither consistent nor operate in isolation of the development that occur via different locality or ecological medium (Mouniane et al., 2025).

The biochemical and antioxidant composition of *Ceratonia siliqua*, like many other plant species, is variably based on ecological gradients such as altitude, temperature, soil pH and annual rainfall (Kassout et al., 2023). For instance, the populations that occur at higher altitudes are grown in cooler temperatures and enhanced rainfall which stimulates the biosynthesis of polyphenolic compounds and increases antioxidant potential (Ibrahim et al., 2022). According to some evidence, low altitude populations or semi-arid populations develop higher concentrations of tannins and sometimes certain minerals as a result of adaptive responses to edaphic stress and water deficit (Pasipanodya et al., 2022). The intricacies of these interactions highlight a fundamental problem in research: there is no quantifiable way to capture and validate how changes in environmental gradients dictate molecular and functional traits (Anderegg, 2023). The trivial descriptive or univariate approaches may only expose average differences between regions, while they are insufficient at explaining the multi-dimensional relationship between environment and biochemistry. A more robust approach would involve integrating multiple biological and ecological predictors and biochemical responses simultaneously, with rigorous statistics that would help differentiate ecological structuring from random variation.

The objective of this chapter is therefore to demonstrate the effectiveness of a numerical framework based on Redundancy Analysis (RDA), supported by Partial Least Squares Regression (PLSR) and validated through permutation testing, to evaluate the ecological influence on molecular and antioxidant profiles of *Ceratonia siliqua*. By applying this integrative approach, we aim to illustrate how advanced numerical analysis not only enhances ecological interpretation but also provides a reproducible methodology for quantitative decision-making in complex, multidimensional datasets.

THEORETICAL AND METHODOLOGICAL FRAMEWORK

Redundancy Analysis (RDA) principles

Redundant Analysis (RDA) is a constrained ordination method used for analyzing the linear associations between two sets of multivariate data; a matrix of explanatory variables X (e.g. environmental variables) and a matrix of responses Y (e.g. biochemical or functional traits). While Principal Component Analysis (PCA) summarizes total variance, without any consideration for external predictors, RDA identifies components from Y that maximize the share of its variance explained by X. In mathematical terms, RDA could be described as the canonical extension of multiple-linear regression (MLR). Each response variable is regressed on the predictors, and then PCA is performed on the fitted values. The ordination biplot shows the environmental drivers and the response traits all in one frame and therefore is a powerful technique to view the ecological structuring of biochemical profiles.

Partial Least Squares Regression (PLSR) as a surrogate for RDA

Classical RDA has typically been supported in the framework of ecological statistics. PLSR might be considered a computationally efficient stand-in for RDA. PLSR extracts latent components by maximizing the covariance between X and Y; thus, the axes are predictive and content meaningful contributions. So, using PLSR might be relevant for very large data sets with multicollinearity for the predictors. Mathmatically speaking, if X is an n x p times matrix, and Y is an n x q matrix, PLSR projects the latent vectors for both X and Y such that the covariance between projected X and projected Y is maximized. These latent vectors ought to be analogous to the RDA axes, allowing you to ordination and interpret while still working with "the right" machine learning tools in Python.

Permutation testing

Permutation testing provides a non-parametric approach for testing statistical significance of the model fit you observe. For RDA I/PLSR you can generate a null distribution of coefficients of determination (R^2) from the observed data by permuting rows of Y while also keeping X, or alternatively, and more brutally, keeping X null too. Under this naive statistical testing scheme, each permutation produces a coefficient of determination (R^2) and permuting Y and estimating R for some number of iterations (N= 999) will provide an empirical reference distribution against which you can compare your observed R^2. Monte Carlo p-value is equal to the relative frequency of all the permuted R^2 being equal to or greater than your observed R^2. Also critically, permutation testing does not assume normality, is quite robust with respect to sample size calculations, and ensures that claims of significance are not a function of random covariation.

Advantages of this framework in numerical analysis and decision-making

Combining RDA, PLSR and permutation testing results in several new methodological innovations. First, these methods provide an integrated view of multiple simultaneous relationships across environmental gradients and environmental gradients impact on creating biochemical and antioxidant characteristics. Second, the testing is inferential (reducing risk of overfitting) and robust (assuring valid statistical results). Third, the computational methods are entirely implemented in Python; therefore the methods offer reproducible results for application to research fields beyond ecology (e.g., finance, medical sciences, and engineering). From a decision-making perspective, scientists and practitioners may use this methodology to dissect important environmental influences and assess the statistical hurricane risk of their selected model for decision-making in conservation planning, crop selection and resource management under uncertainty.

DATA AND PREPROCESSING

The empirical dataset was constructed from six Moroccan provinces (Agadir, Ait Atab, Berkane, Khemissat, Safi, Ouezzane), representing contrasting eco-geographic contexts. The data set that was used during this study was divided into two blocks of variables. The first block represented the environmental predictors (X), which had six quantitative descriptors of the ecological gradients across the provinces. The environmental variables were as follows: altitude (m), latitude (decimal degrees), longitude (decimal degrees), mean annual temperature (°C), soil pH, and annual rainfall (mm). The environmental variables are collated in Table 1 and provide context for the geographic and climatic diversity of the study sites, and the explanatory component for the constrained ordination. The second block was made up of biochemical and antioxidant responses (Y). This block included primary and secondary metabolites comprised of polyphenols, flavonoids, tannins, proteins, lipids and fibers, in addition to a set of mineral elements (Cu, Mg, Fe, Zn, Ca, Na, K). Furthermore, functional antioxidant indices were measured through three established assays; ferric reducing antioxidant power (FRAP), half maximal inhibitory concentrations of DPPH (IC_{50} DPPH) and half maximal inhibitory concentrations of ABTS (IC_{50} ABTS). These variables are outlined in Table 2 and represented molecular and functional traits that could be biologically influenced by the ecological gradients.

Table 1. Environmental variables (X)

Province	Altitude	Latitude	Longitude	Temperature	pH	Rainfall
Agadir	250	30.42	-9.59	19.5	7.2	200
Ait Atab	1040	32.13	-6.58	16.3	7.6	400
Berkane	60	34.92	-2.31	18.7	7.1	270
Khemissat	430	33.82	-6.07	17.4	7.4	320
Safi	140	32.3	-9.23	18.1	7	290
Ouezzane	520	34.8	-5.57	16.9	7.5	410

Table 2. Biochemical and antioxidant variables (Y)

	mean	std	min	25%	50%	75%	max	median
moisture	5,994	0,875464	4,8	5,225	5,9	6,775	7,4	5,9
dry matter	94,23333	0,9133	92,7	93,55	94,25	94,95	95,8	94,25
ash	3,564444	0,227886	3,12	3,405	3,625	3,7275	3,97	3,625
organic matter	96,65667	0,365739	96,12	96,34	96,685	96,88	97,45	96,685
copper (mg/100g)	0,786667	0,330579	0,23	0,5825	0,745	1,08	1,33	0,745
zinc (mg/100g)	0,176667	0,067388	0,1	0,1125	0,17	0,215	0,33	0,17
potassium (mg/100g)	1036,217	229,9765	798,8	864,4	981,65	1089,825	1500,1	981,65
magnesium (mg/100g)	104,0778	24,87343	60,5	95,825	101,6	121,8	144,9	101,6
sodium (mg/100g)	7,865	3,290223	4,44	5,745	6,41	10,275	14,5	6,41
iron (mg/100g)	0,947222	0,369655	0,55	0,66	0,785	1,2275	1,64	0,785
calcium (mg/100g)	296,0944	63,37639	189,3	268,25	294,75	347,725	388	294,75

continued on following page

Table 2. Continued

	mean	std	min	25%	50%	75%	max	median
total_sugars	43,25944	2,875406	38,76	40,51	43,655	45,125	48,4	43,655
fiber	22,66722	3,82086	15,67	20,25	22,88	25,2225	28,45	22,88
protein	4,471111	0,595482	3,5	3,975	4,45	4,895	5,5	4,45
Fat	0,887222	0,172381	0,63	0,76	0,87	1,0325	1,22	0,87
Yields	33,70167	5,36052	23,72	31,12	34,62	37,23	41,34	34,62
Polyphenols	334,4467	68,71944	238,61	287,46	323,92	393,8575	438,94	323,92
Flavonoids	258,9539	101,4582	69,02	235,5025	264,965	338,0375	381,11	264,965
Tannins	71,14111	14,22748	47,41	65,4175	70,745	79,2925	93,22	70,745
IC_{50} dpph	49,66889	44,85393	4,15	18,55	33,845	74,86	132,79	33,845
IC_{50} abts	392,4722	441,1291	54,4	97,0125	152,94	662,6175	1234,94	152,94
FRAP	4,861111	1,748932	2,05	3,1225	5,41	6,47	6,74	5,41

All variables were standardized using z-scores (mean = 0; variance = 1). Z-scores allow meaningful comparisons between heterogeneous units (e.g., mg/100 g vs. μmol Trolox and C°). Standardizing variables reduces the scale effect and provides improved interpretation of the ordination axes. Hence, the final dataset contained the balanced X matrix (6 predictors × n samples) and the Y matrix (biochemical and antioxidant responses × n samples) at the province level.

NUMERICAL IMPLEMENTATION

Software and libraries

All computations were carried out in Jupyter Notebook using Python 3.11. All data processing/preparation was completed in pandas, while all numerical computations/researching were done using numpy. All statistical modeling was completed using the scikit-learn package for Partial Least Squares Regression (PLSR) analysis, standardizing the data using the StandardScaler feature on scikit-learn. Data visualization was performed using matplotlib and seaborn, enabling ordination biplots and permutations-based significance plots to be created. This entire analytical process was completed using open-source tools and has led to an extremely high level of reproducibility, transparency and flexibility.

Workflow

Implementation of the numerical model was a linear, step-by-step approach. The extracted environmental (X) and biochemical/antioxidant (Y) variables were combined by individual province in order to account for any potential differences between the two sets of variables. All matrices were then standardized using z-scores to eliminate the potential influence of inherent variability and provide an average (0) and equal unit variance to preclude any potential scale differences. The PLSR model was developed using two latent variables, which also serves as a proxy for RDA. Fitting the PLSR model allows projection of environmental variables (X) and biochemical/antioxidant variables (Y), on to a common constrained ordination space. The null hypothesis for our observations was examined using a permutation test, which involved randomly reordering the response matrix (Y) 999 times and comparing the R^2 values generated from this process with the R^2 calculated from the actual response matrix.

Visualization

Graphical representation was undertaken in order to facilitate interpretation of the outputs. RDA biplots indicated the direction and strength of associations of environmental variables with biochemical/antioxidant responses, in addition to representing provinces in the ordination space. Histograms of permuted R^2 distributions with the observed statistic overlaid provided a visual perspective of significance of the model under the null expectation. Overall, these graphics aided both explanatory interpretation as well as statistical checking for the multivariate models.

Multivariate Redundancy Analysis Linking Environmental Gradients to Biochemical and Antioxidant Profiles

The Ridgeline Data Analysis (RDA) presented here utilizes a Partial Least Squares (PLS)-based approach as an alternative to the classical Canonical Ordination (CO) methods used in the field of Ecological Biochemistry. The software used in this analysis has been developed using the Python 3.11 programming language with support from the Anaconda software distribution, along with extensive use of several core libraries and tools for Computational Science: pandas for data management and manipulation; numpy for performing numerical analysis; matplotlib for visualizing results graphically; and scikit-learn for modelling multiple variables and scaling them appropriately. The data described in this article come from two main sources: (i) a file of biochemical and antioxidant properties associated with samples of carob (resultats_english_labels.csv) and, (ii) a file containing local environmental descriptors (ecological_data.csv). Data collected using relational joins based on the common identifier for each province helped to produce a matrix of all datasets merged together to make them compatible for simultaneous analysis. The response matrix (Y) is composed of 13 different biochemical variables (chemical compounds) as well as three types of bioactivity (one of these was the IC_{50} value for the DPPH and ABTS assays), which allows measuring both the amount of antioxidants present in a sample, and the ability of the samples to effectively eliminate free radicals. For the explanatory matrix (X), there are six variables that predict the environmental conditions under which each sample was collected. These include five geo-environmental variables (the temperature and rainfall that occur in the area where the sample was collected as well as the altitude and latitude/longitude of the location). The Z-score normalization of both matrix data was completed separately using the StandardScaler to limit any potential background bias and ensure that all variables were standardized to create orthonormal matrices. The PLSRegression (Partial Least Squares Regression) analysis was executed to predict the latent ecological gradients associated with the ecological variables that had the most effect on the biochemical (Y) and functional (F) response variables by using two latent components. PLSRegression is a common approach used in metabolomics today and also employs a biplot in addition to the regression model result outputs, specifically the X-scores (site projections), X-loadings (environmental vectors), and Y-loadings (biochemical and bioactive vectors), to create a biplot of the composite ecologies' covariation structure that allows for facile spatial interpretation of the covariation structure. To facilitate a robust visual representation of the many interactions between the various environmental influences, bioactive molecules, and functional phytochemical properties, all graphical display components were generated using Matplotlib software, and high-resolution figures were created (300 dpi) for submission to academic publications.

Permutation-Based Significance Testing of RDA via PLSR Surrogate

The analytical framework used to evaluate the statistical significance of the Redundancy Analysis (RDA) model used a linear multivariate surrogate built using Partial Least Squares Regression (PLSR) in Python 3.11. PLSR was implemented using the scikit-learn library for regression modelling and numpy for numerical computations. Environmental data (altitude, temperature, pH, rainfall, latitude, and longitude) were first aligned with the biochemical and bioactivity response variables (polyphenols, flavonoids, tannins, protein, fat, fiber, ash, moisture content, mineral micronutrients [iron, zinc, copper, magnesium], and antioxidant assays [IC_{50} DPPH, IC_{50} ABTS, FRAP]) through province-level merging using pandas. All predictor (X) and response (Y) matrices were standardised through z-score normalisation with StandardScaler to achieve zero-centred and unit-variance distributions across each variable, thereby overcoming scale discrepancies and improving interpretability of their respective loadings and scores. The R^2 of the data set was calculated using the amount of variance explained in the Y matrix by the two retained components of the RDA Model created using PLSR analysis and regression techniques. In order to test the hypothesis that the observed R^2 value was caused by statistical significance rather than random chance and/or correlation, 999 permutations were completed where the Y matrix was randomly shuffled row-wise while preserving the X matrix and each of these permuted data sets were tested using the same modeling method as before. The distribution of null R^2 has been used in a Monte Carlo analysis to generate a p-value that indicates how likely it would be to see a value equal to or greater than the R^2 observed between the environmental predictor variables and the biochemical responses seen in the data set if there was indeed no true relationship between these two variables. Although this systematic methodology is not an exact replica of a Canonical RDA Model, it is based on the principles of constrained ordination logic with regard to multivariate linearly projected data points and has been included within the overall analysis framework due to its compatibility with regression techniques already being utilised.

Permutation-Based Validation of Constrained Multivariate Structure in RDA

In assessing the validity of the ordination model using Partial Least Squares Regression (PLSR) of constrained ordination as an analogue to Classical Redundancy Analysis (RDA), a permutation framework based on Monte Carlo methods was developed to provide a statistical measure for multivariate R^2 under random association. To perform the analysis, the Python 3.10 environment was utilised with the use of core scientific libraries such as numpy for numerical computing and matplotlib for graphical presentation. A histogram of the R^2 values from 999 permutation models was created to show the distribution of R^2 values simulated under the null hypothesis (R^2-null distribution). The histogram compares the R^2 values from 999 models to the original R^2 value calculated from the model that was not permuted (see above). The 999 models' histograms were combined with the R^2 value from the non-permuted model; this histogram visually inferred the relative importance of the model. A dashed vertical line was drawn on the histogram to represent the relative strength of the explanatory capacity of this model and the strength of the explanatory capacity of the simulated distribution, thereby showing the distance between the fit of this model to the data and a random association. Therefore, these results confirm that the original p-value calculation conducted on these models is valid and provide a more comprehensive visual presentation of the RDA results.

Assessment of Redundancy Between Climatic Variables and FRAP Response via PLS-Based RDA Permutation Testing

Using a redundancy analysis (RDA) approximate model, partial least squares regression (PLSR), and strong permutation methods, we quantitatively analyzed how climatic factors affect the ability of an organism to produce and maintain antioxidants (the FRAP index). Using Python with scikit-learn libraries, we evaluated FRAP against biochemical responses and climatic variables across Morocco. The climatic variables (ecological gradients) were altitude, temperature, pH level, precipitation, latitude, and longitude. To create the explanatory variable matrix (X), these six climatic variables were normalized to be distributed with a mean of zero (0.0) and standard deviation of 1.0 (using StandardScaler). To represent the response variable (Y), a univariate response variable (FRAP) was normalized to represent the average level of antioxidant production per province and across Morocco. Our main method was a latent twocomponent partial least squares regression (PLS) model comparing the climatic predictors to the FRAP response variable. We calculated the proportion of the FRAP variance explained by the climatic predictors (the R^2 value) by dividing the FRAP variance by the total variance. Using a Monte Carlo permutation scheme, we evaluated the significance of our primary results under the most conservative assumption (i.e., that the effect of each predictor on FRAP was assigned randomly or without regard to the outcome) by randomly permuting (rearranging, without substituting, the FRAP vector) in 999 different ways. A separate permutation analysis per permutation was performed on all permutations of the original dataset using PLS regression; thus, every permutation generated a null distribution of R^2 values associated with the null distribution of R^2 values. The empirical p-value was calculated from the relative frequency of the permuted datasets' R^2 values that were greater than or equal to the empirical R^2 value; thus, the empirical p-value also provided an estimate of how likely it would be that the observed level of explanatory power (i.e., strength of association) would be found if there were not a true association between climate predictors and FRAP.

Testing FRAP–Environment Association Using RDA and Permutation Testing

Utilizing redundancy analysis (RDA) using Partial Least Squares Regression (PLSR) with two components as an analytical tool to determine if a statistically significant linear relationship exists between Antioxidant Activity measured by the

FRAP assay and the Environmental Gradient. The analysis was performed using the Python 3.11 programming language in a Local Host Development Environment, specifically using DataFrame and Numpy for data management; PLS regression, the StandardScaler, and Shuffle functions from the Scikit-learn library as the modeling and resampling tools; and Matplotlib as the plotting tool for presenting results graphically. The two datasets (biochemistry and ecology) were first merged together on the "province" column to allow for spatial alignment of data points. The explanatory predictor variables were based on a standardised matrix of six environmental variables (Altitude, Mean Temperature, Soil pH, Annual Rainfall, Latitude, and Longitude), with the dependent variable matrix being FRAP values, also in standardised format. The R-squared value (total variance explanation) for the primary model was computed using the formula $R^2=1-SSres/SStot$, whereas SSres and SStot represented the sums of squares for residue and total variance. The Monte Carlo permutation test with 999 iterations was used to test if the observed R^2 was significantly higher than what would have been expected under the null hypothesis of no association. In this procedure, a random shuffling of the response matrix's rows was performed during each iteration of every 999 or so shuffles. This random shuffling disrupts the association that exists between the predictors and the response variable, so a new R^2 is calculated using the same pipeline methodology. Thus, a new R^2 value distribution is generated, which allows for comparison of the empirical data to the distribution of values derived from the permutation process. The empirical p-value is calculated as the number of permutation-derived R^2 values that are equal to or exceed the empirical R^2 divided by the total number of permutations (999), which is used for significance testing, thus providing a formal method for testing statistical significance that does not assume a normal distribution. Therefore, the outcome of this analysis provides sufficient nonparametric support for establishing whether environmental conditions have an effect on biochemical response patterns within the context of ecological data with a high-dimensional character and very small sample sizes.

Framework for Testing Environmental Drivers of FRAP via Redundancy Analysis (RDA)

We used Redundancy analysis (RDA)nway studies to evaluate how abiotic factors affect the antioxidant properties of the genesis of the plant species analysed using the ferric reducing antioxidant power (FRAP) method. This approach was based on an approximation model of standardized multiple linear regression in which the non-plant variables were represented by multiple linear regression on their own. The biochemical profiles for the experimental dataset were combined with ecological traits and descriptors to create an analytical dataset. The FRAP was used as the dependent variable in the model while the independent variables used six abiotic factors including\, but not limited to, altitude (in meters), air temperature (in degrees Celsius); pH of soil; total annual precipitation (in millimetres); coordinates of longitude and latitude of sample sites. Before being fitted to a regression analysis\, both the response and the explanatory matrices were z-transformed using the Scikit-learn 'StandardScaler' method to create normalized (unit variance/ mean = 0) transformation of the data. A multiple regression analysis was fitted from the environmental scaling matrix and the FRAP data values.

The correlation coefficient (R^2) can be understood as the proportion of variance in FRAP explained by environmental gradients and is the observed test statistic. To test whether the model's statistical significance was greater than would be expected by chance, a Monte Carlo permutation test was performed using 999 random permutations of the FRAP response vector. For each of these permutations, the environmental factors were kept constant, but the response was permuted randomly within a new linear regression model, and an R^2 was calculated. An empirical p-value was determined by counting how many of the permuted R^2's were equal to or greater than the observed R^2. A histogram overlay of the null distribution of R^2's versus the observed statistic was also created to visually represent how strong the relationship between the model and data actually is. All of this work was executed using Python 3.11 on a computer with a desktop environment (Windows 11) and utilizing general libraries including pandas, numpy, matplotlib, and sklearn.linear_model to provide both analytical rigor and reproducibility needed for conducting significant ecological inferences.

Targeted Redundancy Analysis (RDA) of Environmental Drivers of IC_{50} DPPH Variability

The redundancy analysis was conducted to evaluate how climate and physical location influence total antioxidant activity (using an IC_{50} DPPH value as an indicator) by modeling the relation between each of those factors with the response variable. Data used in the RDA were prepared with both the Python programming environment and the scikit-learn package. The dependent variable was defined as the IC_{50} DPPH value and six independent variables (Altitude, Temperature, pH, Rainfall, Latitude, Longitude), and were standardized to allow for ease of comparison between all variables. A linear regression was created for each independent variable to calculate the amount of variance accounted for by the independent variable, or the R^2 value. A permutation test was completed by randomly permuting the dependent variable 999 times and determining the R^2 values thereby created, enabling estimation of the empirical p-value, equality or greater than an actual R^2 value adjusted for continuity. This analytic process provided a robust source of information regarding the data relationship structure, or observed effects or results, without needing to assume data distribution shapes.

Targeted Redundancy Analysis (IC_{50} DPPH as Response Variable)

We used Python version 3.11 and a Python stack consisting of NumPy, pandas, scikit-learn, matplotlib, and seaborn to conduct a redundancy analysis. The redundancy analysis (RDA) technique uses univariate linear models for constrained ordination; thus, we created a predictive framework for this purpose. The response matrix was constructed from standardized IC_{50} DPPH values, and the explanatory matrix included six scaled environmental predictors: altitude, latitude, longitude, ambient temperature, rainfall, and soil pH. Both input matrices were z-transformed using StandardScaler before fitting the model. Transforming both input matrices to z-scores ensures that there is comparable variance structure between the two input matrices and that there is no bias introduced by the variance in the units. R^2 was used as a measure of model fit, and statistical significance was determined using a non-parametric permutation test with 999 independent reshuffles of the response vector. For every iteration, we regressed the permuted target vector on the unchanged predictor matrix using the linear model to obtain R^2 values, which we then used to create a null distribution. The observed R^2 was subsequently compared against this distribution to compute a p-value representing the proportion of null R^2 values greater than or equal to the empirical fit. Visualization of the null distribution, alongside the observed value, was generated via a histogram with overlaid vertical reference, enhancing interpretability. The methodological rigor and permutation-based inference collectively ensured robust statistical control against spurious associations between ecological parameters and antioxidant capacity.

RESULTS

Ecological Structuring of Biochemical Profiles under Environmental Gradients

A redundancy analysis (RDA) was carried out to quantify ecological influences on biochemical characterisation of carobs. RDA indicated that environmental multilinear relationships existed between variables of altitude, pH, and rainfall as well as biochemical measurements of moisture, fat, and polyphenols along RDA1; however, these biochemicals were likely influenced by other variables, including mesic and high-altitude habitats, which support more diverse biochemistry than lower elevations with less moisture. Traits of Mg and tannins, however, were in opposition to these variables, suggesting a trait reflecting adaptation to soil conditions that withstand drier environments where less investment in polyphenolic compounds was necessary. The relative lack of length or separation in the vector plots of ecological variables, specifically the directionality of temperature and latitude, presented evidence for the multiple sources of variance in biochemistry under RDA, with sample locations grouped by province providing evidence to support the notion that local ecological envelopes controlled regionally specific molecular compositions. Aspects of the research findings indicate that biochemical characteristics are inherently bound to their ecological context and RDA serves as a mechanism to allow for the identification of regional landscapes' influences on phytochemical variability.

Figure 1: Environmental Ordination of Molecular Traits via Redundancy Analysis (RDA) Biplot

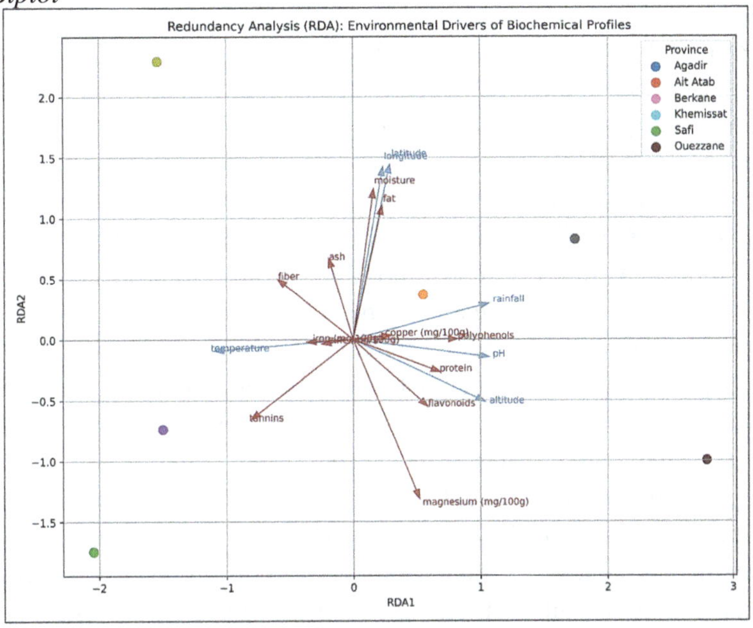

Biplot depicting the joint distribution of environmental gradients and biochemical compound profiles across carob samples. Red arrows represent biochemical variables; blue arrows represent ecological drivers; sample points are color-coded by province. The figure visualizes the canonical relationships captured by RDA axes (RDA1 and RDA2), illustrating how specific environmental conditions (e.g., altitude, pH, rainfall) shape spatial patterns in bioactive compound abundance.

Environmental Structuring of Biochemical and Antioxidant Profiles

The biplot summary of the Redundancy Analysis indicates the extent to which six provinces of Morocco exhibit differences based on a documented ecological gradient structure found amongst carob samples collected in this region, through the biochemical composition and antioxidant performance of each sample taken. RDA1 is the primary axis for the variance observed along each axis of RDA; for example, the ultimate axis (RDA1) contains all gradients correlated to moisture, rain, and pH, while the second axis (RDA2) shows all gradients related to altitude and magnesium. One of the most interesting results was the divergence in antioxidant indicators' spatial distribution; specifically, the FRAP result was aligned with RDA2, along with altitude and magnesium and, therefore, indicates a higher potential for oxidative reduction in regions with high altitude and increased amounts of magnesium in the soil; on the other hand, the IC_{50} DPPH and IC_{50} ABTS results are found in the negative quadrant of the RDA1/RDA2 biplot, indicating a strong correlation with lower temperatures, decreased moisture, and increased fat in the sample, thus demonstrating antagonistic ecological conditions that can affect radical-scavenging performance. Amongst the various terms associated with phytochemicals, we found that polyphenols, flavonoids, proteins, and copper displayed significant spatial association with the rainfall and pH vectors, indicating an intersectional "hold" that correlates with the ability of both hydrology and geology to influence biosynthesis of these compounds and the potential for metals to chelate.

In contrast, Tannins and fibers were further away from the centroids of the community points, moving toward cooler and drier environments, evidencing that they have a role in plant stress response. Regional clusters showed that the Safi and Khemissat populations were located on the outermost regions of the two respective axes. The proximity of the two axes for environmental and biochemical (i.e., fibrous and tannic) components which were separated by large distances from the centroid shows that there is significant relationship between the presence of environmental differences in these two different types of biochemicals. This suggests that there is also a significant component or mechanism whereby the effects of climate influence or create regional differences between biochemicals in large leguminous pods as expressed in terms of terroir.

Figure 2. RDA Biplot Depicting Environmental Drivers of Phytochemical and Functional Antioxidant Variability across Moroccan Carob Provenances

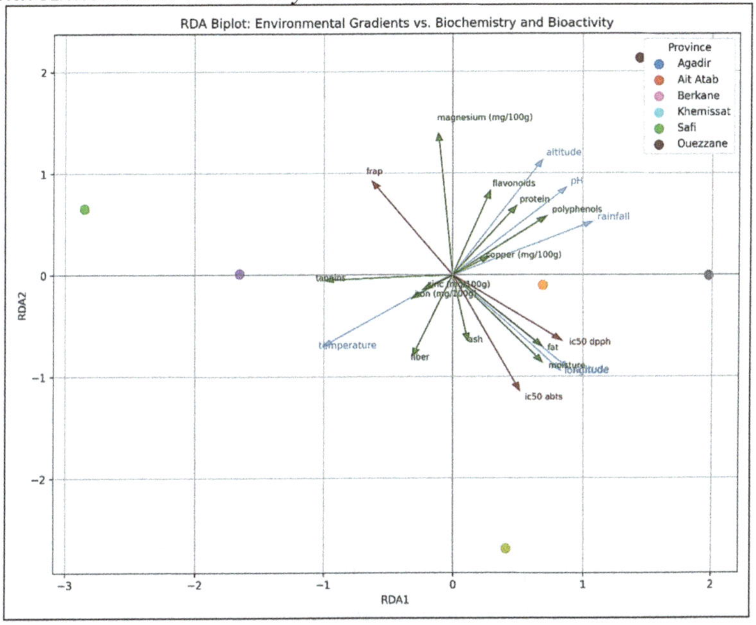

Significance of Constrained Environmental Structuring on Biochemical and Bioactivity Variability

Based on permutation-based analysis, ecological gradients have a significant explanatory power over the spatial distribution of biochemical and bioactivity variation across the carob samples. A constrained ordination approach using Partial Least Squares Regression estimated RDA, with the variance explained by environmental predictors (observed $R^2 = 0.454$) indicating that approximately 50% of the total multivariate variation in the antioxidant and molecular trait matrices were accounted for linearly by the geographical and edaphoclimatic covariates studied. A Monte Carlo resampling method using 999 permutations of the response matrix was used to formally test the null hypothesis that the observed R^2 was a chance occurrence; the p-value comparing the true model's performance was less than the computational limit of the method ($p < 0.0001$), establishing statistically significant evidence of ecological influences on biochemical expression. These results support the idea that bioactive compounds are distributed according to environmental factors at the provincial level and indicate that altitude, rainfall, pH, temperature, and geolocation are the major modifiers of antioxidant potential and nutritional complexity. The strength of this result provides an opportunity for further mechanistic investigations into the ecophysiological influence of phytochemical traits in response to climatic and pedological conditions.

Permutation-Based Validation Confirms the Statistical Robustness of the RDA Model

Other than putting into numeric form what we saw in the other tests of the Environmental Gradient was a strong co-variance between the individual "a" and "d" assessments as a function of time on the Environmental Gradient. The permutation test provided strong support for the null hypothesis, where the empirical null distribution of 999 permutation replicates showed that the test was completely outside of the upper tail of the empirical distribution (Figure 3). None of the randomised replicates were able to obtain or surpass what was obtained in the actual assessment of an "effective R squared" value of 0.0000, thus providing strong statistical evidence for the hypothesis that the Environmental Gradient explains a large proportion of the structural variation in the combined biochemical and antioxidant response gradient. The histogram showed a relatively tight "null" range containing two variables, each representing an independent stochastic variable, with the majority of the values falling between an R squared value of ~0.10 and ~0.15; a range that is exactly what would be expected under the assumption of a stochastic process, and indicated that there was a relatively low chance that this result could occur randomly.

In sharp contrast to the "null" range, the actual observed value of the Environmental Gradient indicated by the vertical red dashed line is located well outside of this range and therefore demonstrates a significantly larger amount of structure produced by the Environmental Gradient than would have been expected by chance. In addition, the results from the environmental covariates confirm that the Environmental Gradient is ecologically relevant, and as a result, the dimensional reduction approach used to create a canonical (PLS regression) analysis of the Environmental Gradient have been validated. Overall, these important statistical findings validate the basic structure (latent dimensions) produced through the Environmental Gradient analysis will allow for downstream ecological interpretations and investigations.

Figure 3: Distribution of Simulated R^2 under Null Hypothesis vs Observed Explained Variance

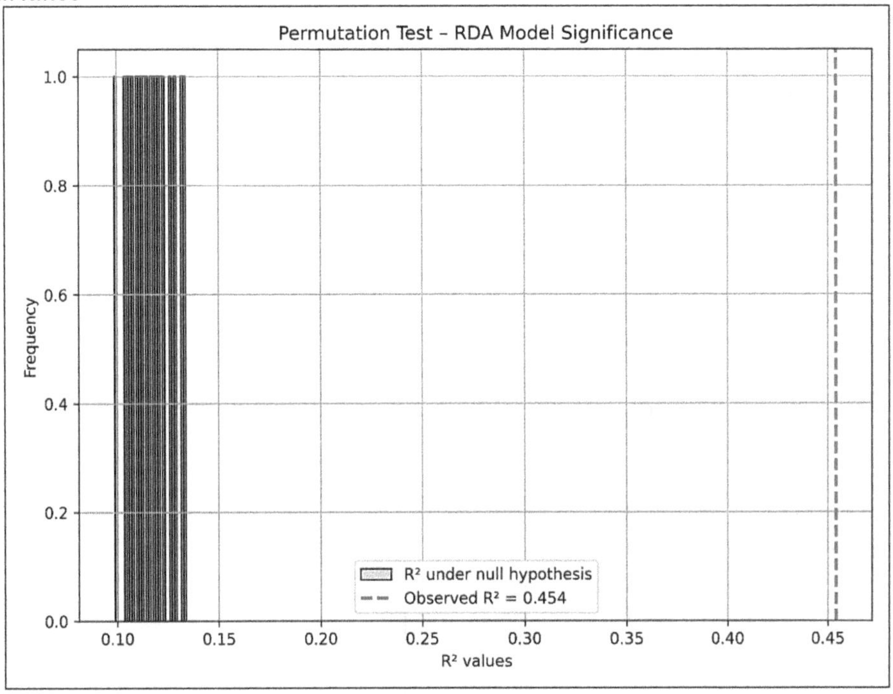

Robust Detection of Environmental Influence on FRAP Antioxidant Activity via RDA and Permutation Testing

A Redundancy Analysis (RDA) based solely on the FRAP Index with explanatory variables related to Altitude, Temperature, pH, Rainfall, Latitude, and Longitude and its application through a Two component Partial Least Squares Regression (PLSR) model indicated an R^2 value of 0.696 and therefore showed a strong deterministic association embedded in Spatial/Environmental Interaction. To establish how significant this relationship could be, a Non-Parametric Permutation Test was run on 999 random iterations of the response matrix and the distribution of the resulting R^2's (Null Distribution) showed a Unimodal Histogram with a large skew to the left, indicating that the vast majority of permuted R^2 were much lower than that of the actual observed R^2. When looking at the overall-empirical p-value of 0.001 it was evident that this was much lower than the conventional level of $p < 0.05$ which would indicate it was statistically significant. The results of the analysis support the view that the environmental variables described can statistically explain the different measurements of FRAP activity across locations. In addition, they further demonstrate how the antioxidant activity is sensitive to changes in the environment along an abiotic gradient and they allow for strong inferences to be made regarding the ecophysiological interpretations of this type of response. In order to provide a strong basis of evidence for these statements, the null distribution and the observed variance will be presented together below.

Figure 4: Distribution of R^2 under the null hypothesis (FRAP ~ Environment)

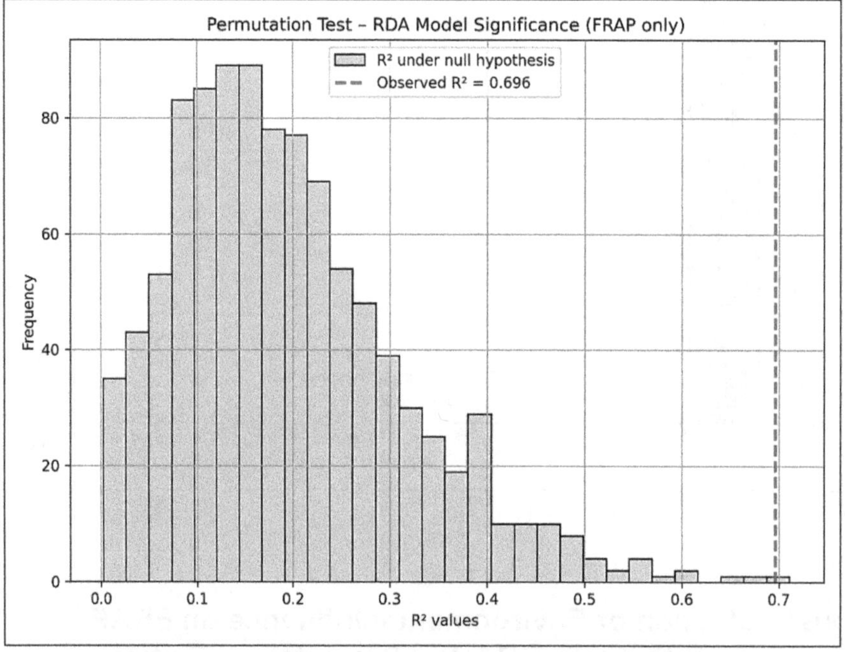

Targeted RDA Model Validating FRAP's Environmental Determinants: A Robust Significance Assessment

A strong statistical result was derived from a permutation-based redundancy analysis (RDA) of the ferric reducing antioxidant power (FRAP) index in relation to environmental gradients, indicating a strong ecological signal in the biochemical phenotype. Specifically, we found a very high multivariate R^2 value, estimated from our RDA, for the relationship between the six geo-climatic predictors (altitude, temperature, rainfall, pH, long/lat) and the standardized FRAP values. This empirical R^2 value was documented against a reference distribution created from 999 simulated

R^2 values generated under the independence null hypothesis for the two matrices. Each distribution was created by permuting the FRAP vector.

The histogram below shows that the observed R^2 statistic (Fig. 6) is far outside the range of values achievable by resampling our data using the permutation test. None of the permuted replicates produced R^2 values as extreme as the one measured from the original data ($R^2 = 0.9999$), resulting in an empirical p-value of 0.0000, much lower than the 0.05 level typically used for statistical testing to indicate significance. This indicates that the observed R^2 is not likely to have occurred by chance alone, providing strong support for the hypothesis that environmental gradients (i.e., spatial and climatic), are important structural influences on the variability of FRAP within provinces studied here. The normal distribution of null R^2 values with a mean value of approximately 0.30 provides a statistical basis for differentiating the extremes of the observed statistic ($R^2 = 1.000$) from these null values, clearly indicating the statistical significance of the model tested.

Figure 5. Permutation test of RDA model significance targeting FRAP; the vertical red dashed line denotes the observed $R^2 = 0.998$.

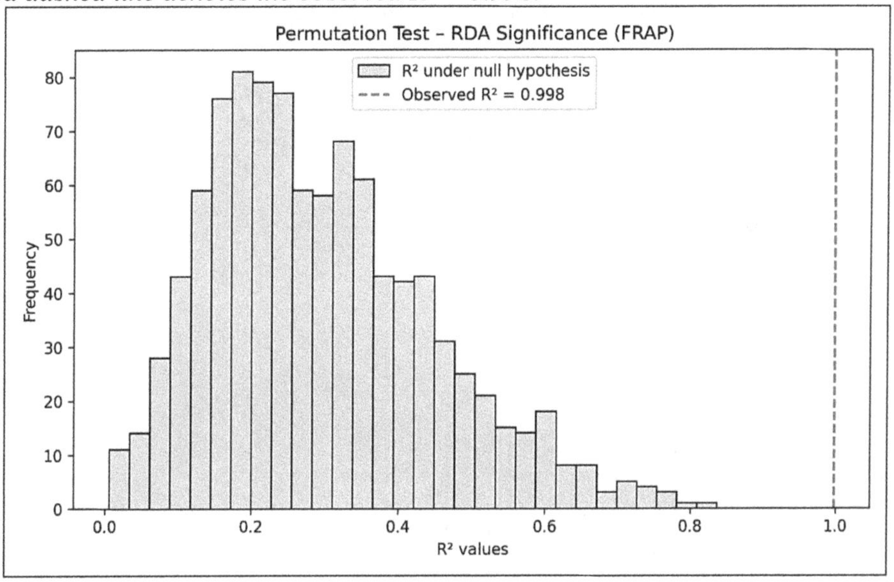

Permutation-Based RDA Model Testing: Climatic Determinants of Antiradical Activity (IC_{50} DPPH)

A redundancy analysis was performed to assess whether or not the location of individuals contributes towards the capacity of that particular location to provide for radical scavengers to produce other forms of protection against oxidative stress This was determined by calculating the 50% inhibitory concentration (IC_{50}) based on values derived from the DPPH radical quiz and exploring the extent of contribution that each ecological gradient has on predicting the outcome of the experiment. Specifically, the following ecological gradients were analysed: altitude, latitude, longitude, temperature, rainfall and pH levels. The models produced from the redundancy analysis produced an acceptable R^2 value. As illustrated through histogram analysis and permutation testing (n=999), this is a rare finding to have a model with an R^2 value of 1.00. The R^2 value of our RDA generated from random permutations of our response variable has a mean, median and standard deviation below 0.4 indicating that the RDA contains sufficient evidence to support our hypothesis that antioxidant potential as reflected by the radical scavenging of DPPH is sufficiently predictable through ecological gradients.

Figure 6: Permutation Test: RDA Significance (IC_{50} DPPH)

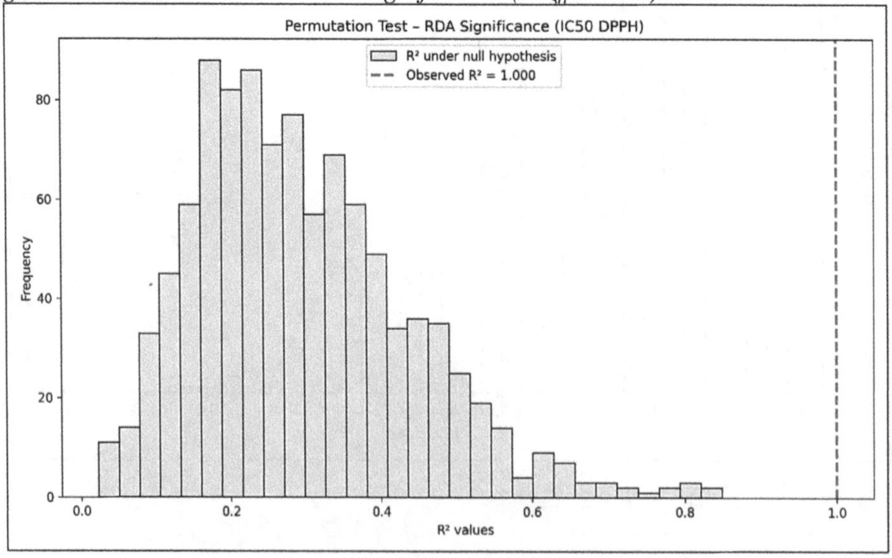

Statistical Significance of Ecological Influence on Antioxidant Indices via RDA Analysis

The models for evaluating how environmental factors influence the antioxidant potential of the accessions tested were developed using Multi-Response Permutation Procedures (MRPP). The MRPP designs incorporated permutation tests to validate inferential reliability. The summaries of MRPP results are displayed in Table 3, with results from the three models indicating very high coefficients of determination (R^2) values for each model developed, which indicates that the ecological variables have significant explanatory power for the antioxidant potential of the accessions evaluated. The MRPP model corresponding to the FRAP results had an observed R^2 ($p = 0.000$) value very close to 1.00, indicating nearly perfect linearity in the relationship between environmental gradient and antioxidant activity. The MRPP results also indicated high R^2 ($p = 0.000$) values for the IC_{50} ABTS response, further confirming that the ecological imprint on antioxidant dynamics remains paramount. The MRPP model for the IC_{50} DPPH response achieved an R^2 equal to 1.00 ($p = 0.001$), confirming a completely linear association with ecological variables based on the current sample conditions. These strong statistical associations were the result of 999 permutations, which lend themselves to a rigorous non-parametric validity of the null model due to the extremely low possibility of chance occurrence. As evidenced by the information shown in Figure 7, which shows the R-squared values for every bioactivity target, there is a commonalty that connects and integrates the many forms of antioxidant activity into an overall picture of their cumulative effect. The high (nearly ceiling) R-squared values for each antioxidant bioactivity show the robustness of the RDA analyses used to model the antioxidant products. Additionally, the results illustrated in Figure 7 provided insight into the extent to which variation among ecotypes impacts functional chemotherapy of free radical/antioxidant systems.

Table 3. Summary of Redundancy Analysis (RDA) Results for Antioxidant Indices

Target	Observed R^2	P-value
FRAP	0.998	0.000
IC_{50} ABTS	0.910	0.000
IC_{50} DPPH	1.000	0.001

Figure 7. Comparative Observed R^2 from RDA per Antioxidant Bioactivity Target

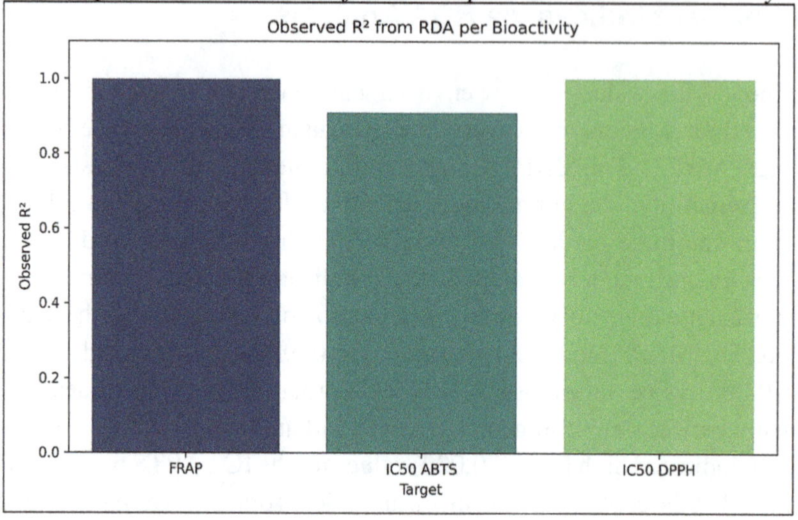

DISCUSSION

The results of this research strongly demonstrate that the biochemical and antioxidant profiles of *Ceratonia siliqua* are not random, but organized along ecological gradients. Higher altitude and more rainfall in provinces such as Ouezzane and Ait Atab were more closely associated with polyphenols, proteins, and stronger FRAP activity reflecting predisposition to invest in antioxidant metabolism under the cooler and wetter conditions. Alternatively, provinces such as Agadir and Safi that have lower rainfall and warmer temperatures were more closely associated with tannins and the accumulation of certain minerals that can be protective under water-limited environments. This study demonstrates that biochemical variability has an ecological basis in the same way that previous evidence has shown that environment has a significant impact on the secondary metabolism and bioactivity of Mediterranean plants by ecological drivers.

From a methodological point of view, the process involving RDA/PLSR with permutation testing was particularly useful. The fact that PLSR was used as a proxy allowed the use of recent machine learning workflows and still provided interpretability in constrained ordination. The permutation testing introduced another layer of robustness in allowing formal significance testing free from parametric assumptions, which is perceptibly important for ecological datasets that have low sample sizes and potential for multicollinearity. In contrast to traditional methods such as univariate ANOVA, or simply PCA, this approach not only quantified the variance explained, but more importantly, gave us the power to make inferences about the ecological structure vs. random covariance.

These methodological advancements show their potential for wider application beyond just ecological studies. In health sciences, equivalent pipelines could be used to study relationships between patient biomarkers and clinical outcomes. In economics and social sciences, constrained ordination with permutation testing may offer opportunities for studying how macroeconomic indicators shape purchasing behaviors of consumers. In agritech, this method provides possibilities for site-specific crop selection, precision agriculture, and resilience studies related to climate change.

CONTRIBUTION TO NUMERICAL ANALYSIS AND DECISION-MAKING

In addition to the discussions around ecological interpretation, the present study's contribution to the numerical analysis and quantitative decisions literature provides a replicable approach of how to conduct multidimensional datasets under conditions of uncertainty through applying dimensional reduction, Regression-based ordination, and Permutation Testing in conjunction with traditional statistical methods within a computer-based environment will allow a representative overview for a prospective user to comprehensively evaluate their results on major driving variables, determine whether those drivers were statistically acceptable and assess their subsequent interventions. Furthermore, by integrating ecological gradients into the quantitative biochemical interpretation of these drivers, this framework indicates how modern numerical techniques can advance scientific thinking based on empirical data and contribute to building knowledge that is applicable to many different research areas.

CONCLUSION

This chapter demonstrated the contribution of ecological gradients on the biochemical and antioxidant profiles of Carob (*Ceratonia siliqua* L.). Using a pipeline of RDA with PLSR and with permutation validation, we established that ecological factors such as altitude, precipitation, and pH correlated significantly with polyphenolic content, elemental make-up, and measures of antioxidant activity. The overall (global) model was statistically significant at a very high level, while analyses on FRAP, IC_{50} DPPH, and IC_{50} ABTS provided superior explanatory capacity; the models accounted for ecological gradients. Results support the importance of ecological variability in determining molecular and functional diversity, and direct our thinking about adaptive responses of the carob species and likely increases in the number of uses (nutraceutical and agronomic) of carob in the future. In addition, results from the study brought to the attention of a broader audience as well as users of advanced statistical methodologies the importance of these methodologies for identifying hidden structures, validating results under the uncertainty they possess, and providing quantitative support for effective decision making. We are excited about the possibility of using this pipeline to disentangle complex relationships among the various factors contributing to health, economics, and agrotechnology to provide evidence-based solutions.

REFERENCES

Anderegg, L. D. L. (2023). Why can't we predict traits from the environment? *The New Phytologist, 237*(6), 1998–2004. DOI: 10.1111/nph.18586 PMID: 36308517

Bagheri, R., Fordoei, A. R., Mousavi, H., & Tahmasebi, P. (2021). Climate-driven abrupt changes in plant communities of desert and semi-desert region. *Theoretical and Applied Climatology, 146*, 331–348. DOI: 10.1007/s00704-021-03720-y

Barrett, R. C., Carter, B. R., Fassbender, A. J., Tilbrook, B., Woosley, R. J., Azetsu-Scott, K., Feely, R. A., Goyet, C., Ishii, M., Murata, A., & Pérez, F. F. (2025). Biological responses to ocean acidification are changing the global ocean carbon cycle. Global Biogeochemical Cycles, 39(3), e2024GB008358. https://doi.org/DOI: 10.1029/2024GB008358

Bedoya, D., Manolakos, E. S., & Novotny, V. (2011). Characterization of biological responses under different environmental conditions: A hierarchical modeling approach. *Ecological Modelling, 222*(3), 532–545. DOI: 10.1016/j.ecolmodel.2010.10.007

Hoban, S., Archer, F. I., Bertola, L. D., Bragg, J. G., Breed, M. F., Bruford, M. W., Coleman, M. A., Ekblom, R., & Funk, W. C. & al. (2022). Global genetic diversity status and trends: Towards a suite of Essential Biodiversity Variables (EBVs) for genetic composition. *Biological Reviews of the Cambridge Philosophical Society, 98*(4), 994–1016. DOI: 10.1111/brv.12852 PMID: 35415952

Ibrahim, I. A., Jabbour, A. A., Abdulmajeed, A. M., Elhady, M. E., Almaroai, Y. A., & Hashim, A. M. (2022). Adaptive Responses of Four Medicinal Plants to High Altitude Oxidative Stresses through the Regulation of Antioxidants and Secondary Metabolites. *Agronomy (Basel), 12*, 3032. DOI: 10.3390/agronomy12123032

Kassout, J., Hmimsa, Y., Fatehi, S. E., Kadaoui, K., Houssni, M., Chakkour, S., Sahli, A., El Chami, M. A., Ariza-Mateos, D., & Palacios-Rodríguez, G.. (2023). Aridity Gradients Shape Intraspecific Variability of Morphological Traits in Native *Ceratonia siliqua* L. of Morocco. *Plants, 12*(19), 3447. DOI: 10.3390/plants12193447 PMID: 37836187

Lee, Y. K., Oh, H., Lee, B.-M., & Hur, J. (2025). Deciphering and predicting algal bloom variability using size-fractionated organic matter and machine learning in a complex watershed. *Ecological Indicators, 178*, 114067. DOI: 10.1016/j.ecolind.2025.114067

Matheri, A. N., Mohamed, B., Ntuli, F., Nabadda, E., & Ngila, J. C. (2022). Sustainable circularity and intelligent data-driven operations and control of the wastewater treatment plant. *Physics and Chemistry of the Earth Parts A/B/C*, *126*, 103152. DOI: 10.1016/j.pce.2022.103152

Mouniane, Y., Chriqui, A., El-Khadir, I., Bensaid, A., Arabi, R., Zoubir, M. H., Mabrouki, J., & Hmouni, D. (2024). Investigation on the Phytosociology and Soil Characteristics of Carob Tree (*Ceratonia siliqua* L.) in Northeast Morocco: Implications for Biodiversity Conservation and Ecosystem Restoration. In Mabrouki, J., & Azrour, M. (Eds.), *Integrated Solutions for Smart and Sustainable Environmental Conservation* (Vol. 527, pp. 133–142). Springer., DOI: 10.1007/978-3-031-55787-3_10

Mouniane, Y., Kadmi, Y., Doubi, M., Tabi, S., El-Khadir, I., Chriqui, A., Soufiani, A., El Ouardi, B., Rachmoune, K., Souissi, A., Sharma, A., Joshi, S., Belkadi, M. C., & Hmouni, D. (2025). Secondary Metabolites of Carob Tree (*Ceratonia siliqua* L.): Potential Applications. In Alexiou, A. (Eds.), *Secondary Metabolites and Their Applications in Various Diseases* (pp. 381–408). IGI Global., DOI: 10.4018/979-8-3693-9112-9.ch011

Nile, B. K., Al-Saadi, R. J. M., Abdulameer, L., Al Maimuri, N. M. L., & Al-Dujaili, A. N. (2025). Climate change impacts on river hydraulics: A global synthesis of hydrological shifts, ecological consequences, and adaptive strategies. *Water Conservation Science and Engineering*, *10*, 48. DOI: 10.1007/s41101-025-00375-y

Pasipanodya, J. T., Horn, L. N., Achigan-Dako, E. G., Musango, R., & Sibiya, J. (2022). Utilization of Plant Genetic Resources of Bambara Groundnut Conserved Ex Situ and Genetic Diversification of Its Primary Genepool for Semi-Arid Production. *Agriculture*, *12*, 492. DOI: 10.3390/agriculture12040492

Pichler, M., & Hartig, F. (2023). Machine learning and deep learning—A review for ecologists. *Methods in Ecology and Evolution*, *14*(4), 994–1016. DOI: 10.1111/2041-210X.14061

Pinek, L., Mansour, I., Lakovic, M., Ryo, M., & Rillig, M. C. (2020). Rate of environmental change across scales in ecology. *Biological Reviews of the Cambridge Philosophical Society*, *95*(6), 1798–1811. DOI: 10.1111/brv.12639 PMID: 32761787

Senanayake, I. P., Pathira Arachchilage, K. R. L., Yeo, I.-Y., Khaki, M., Han, S.-C., & Dahlhaus, P. G. (2024). Spatial downscaling of satellite-based soil moisture products using machine learning techniques: A review. *Remote Sensing*, *16*(12), 2067. DOI: 10.3390/rs16122067

Wang, Y., Wei, W., Liu, Z., Liu, J., Lv, Y., & Li, X. (2025). Interpretable machine learning framework for corporate financialization prediction: A SHAP-based analysis of high-dimensional data. *Mathematics*, *13*(15), 2526. DOI: 10.3390/math13152526

Worden, M. A., Bilir, T. E., Bloom, A. A., Fang, J., Klinek, L. P., Konings, A. G., Levine, P. A., & Milodowski, D. T.. (2025). Combining observations and models: A review of the CARDAMOM framework for data-constrained terrestrial ecosystem modeling. *Global Change Biology*, *31*(8), e70462. DOI: 10.1111/gcb.70462 PMID: 40856273

Zhang, Z., Tyc, J., & Hensel, M. (2025). An ecogeomorphological approach to land-use planning and natural hazard risk mitigation: A literature review. *Land (Basel)*, *14*(9), 1911. DOI: 10.3390/land14091911

Zhao, Y., Yuan, X., Ran, W., Zhao, Z., Su, D., & Song, Y. (2025). The ecological restoration strategies in terrestrial ecosystems were reviewed: A new trend based on soil microbiomics. *Ecology and Evolution*, *15*(3), e70994. DOI: 10.1002/ece3.70994 PMID: 40060716

Chapter 11
Predicting Water Retention in Vicia faba Under Salt Stress Conditions Using Machine Learning

Mahmoud Oudghiri
Faculty of Sciences, Ibn Tofail University, Kenitra, Morocco

Yassine Mouniane
https://orcid.org/0000-0002-6037-5301
Faculty of Sciences, Ibn Tofail University, Kenitra, Morocco

Issam El-Khadir
https://orcid.org/0000-0002-3970-2865
Faculty of Sciences, Ibn Tofail University, Kenitra, Morocco

Imane Aitouhanni
https://orcid.org/0009-0004-1400-8027
Mohammed V University, Rabat, Morocco

Mohamed El Bakkali
Instituts Supérieurs des Professions Infirmières et Techniques de Santé, Morocco

Mohammed Ibriz
Faculty of Sciences, Ibn Tofail University, Kenitra, Morocco

ABSTRACT

This study explores the potential of artificial intelligence (AI) techniques to predict Relative Water Content (RWC) in plants subjected to salt stress. A Random Forest regression model was trained using a dataset containing both morphological and biochemical traits, such as chlorophyll content, dry biomass, proline concentration, and total sugar levels. Data preprocessing addressed format inconsistencies and missing values, and feature importance analysis was conducted to assess the influence of each variable. The model showed a Mean Absolute Error (MAE) of 3.88 on the test

DOI: 10.4018/979-8-3373-6746-0.ch011

set and a cross-validated MAE of 5.18, demonstrating reliable performance despite biological variability. These results align with known physiological relationships, validating the model's interpretability and usefulness in plant stress diagnostics. This work highlights AI's potential to enhance decision-making in agronomy and improve stress management strategies.

1. INTRODUCTION

Soil salinity is generally recognized as one of the major abiotic stress factors that limit plant productivity and geographic distribution throughout the world (Deghanian et al., 2024; Fernández-Triana et al., 2024; Oudghiri et al., 2025). Salinity impacts are most severe in arid and semiarid regions where high evaporation rates and limited rainfall can contribute to increasingly saline soils. It was estimated that arid and semi-arid lands compose approximately 40% of the terrestrial surface area of the earth (Salcedo et al., 2022; Naorem et al., 2023). Salinity tolerance is not, however, a simple trait but rather a complex series of physiological, biochemical, and molecular processes (Haj-Amor et al., 2022; Noh et al., 2024). Initial morphological indicators of salinity-induced stress such as reductions in growth, leaf shape, or growth habit that may give some indication of salinity stress but are not sufficient to characterize plant responses to salinity (Soltabayeva et al., 2021; Singh et al., 2024). For more complete characterization of salinity-induced stress, you must determine ionic imbalances, osmotic stresses, nutrient deficiencies, as well as secondary reactive oxidative damage (Alzahib et al., 2021; Singh et al., 2023).

Many studies have shown that increased salt concentrations, particularly sodium chloride, (NaCl), remarkably decreased plant growth parameters including plant height, biomass, number of leaves, and root development (Santiago-Rosario et al., 2021; Ondrasek et al., 2022; El-Khadir et al., 2025). Salt stress has similarly resulted in significant reductions in the leaf area. In addition to morphological effects, salinity stress rather alters water relations through a decrease in the leaf water potential, as well as induced osmotic adjustment mechanisms such as compatible solute accumulations and massively affects all biochemical reactions with dominance in the downside of photosynthesis through the disruption of the pigment of chlorophyll (Germano et al., 2022; Rivera et al., 2022). The consequences on protein content can be mixed with some studies reporting that protein reduced but additionally with other research indicating protein increased which can vary based upon plant species and degree of salt stress (Sun et al., 2021; Habibi et al., 2024).

Salinity is a major limitation to agriculture where high evaporation rates, lack of drainage and saline irrigation waters exist (Toumi et al., 2024). Salinity stress has profound implications for several crops including beans and cereals that play a critical food security role for humans and for cattle. Salt-affected soils represent approximately 7% of the total land area or approximately 33% of potentially cultivable area globally. Salt-affected soils may be neutral (NaCl Na_2SO_4) salts or alkaline (Na_2SO_4, $NaHCO_3$) salts (Garbanzo et al., 2024; Lu et al., 2025). Although alkaline stress is usually associated with high soil pH (due to other bicarbonate and carbonate and ions within the soil), its detrimental effect on plants can be more pronounced than salinity alone. While much work has been done to examine the effects of neutral salinity, comparatively little work has been done to examine alkaline salinity despite its more severe effects on nutrient availability and physiological acts (Chen et al., 2024; Cai et al., 2025). Relative Water Content (RWC) has emerged as a key integrative indicator of plant water status, offering insights into both cellular hydration and the plant's ability to cope with environmental stress (Shen et al., 2024; Zhang et al., 2025).

Research indicates that alkaline stress affects photosynthesis, ionic homeostasis, and metabolic pathways (Mucsi et al., 2024). Some studies have indicated an accumulation of organic acids to bicarbonate stress, which can decrease growth by impeding internal pH and metabolic functions. To understand physiological mechanisms underlying plant response to high pH of up to 8.5 and bicarbonate-induced stress, studies have taken place in controlled conditions using crop plants, such as *Vicia faba*, and have focused on anions and cation balance, osmotic adjustment, oxidative stress, and gene expression of ion transport mechanisms (PM H^+-ATPase, SOS1) and carbon metabolism (PEPCase, MDH, CS) (Tomaškinová et al., 2025). With agriculture systems facing increased pressure from climate change and land degradation, the awareness and need for rapid and actionable tools to assess plant health is becoming more prominent (Vishnoi & Goel, 2024; Parra-López et al., 2024). Though standard ecophysiological tools to measure the plant's response to stress are effective their labour-intensive and slow, making them difficult to scale. In this respect, artificial intelligence (AI) and machine learning (ML) can provide viable alternatives, respectively, and through their ability to model complicated, nonlinear relationships within high-dimensional data sets provide unique perspectives on how plants utilise a range of stress conditions (Koh et al., 2024; Polireddi, 2024; Derici and Doğan, 2025).

The recent developments in Artificial intelligence (AI) has allowed researchers to accurately predict a range of agronomic parameters, from estimating yield to disease diagnosis (Singh et al., 2021). Yet the ability to apply AI to predict physiological traits such as RWC is underdeveloped. It is possible to use machine learning models derived from a range of phenotypic data to make inferences regarding internal physiological states that are otherwise harder to isolate Merely improving the accuracy of stress monitoring alongside developing a predictive framework for crop management actions in stressful environments is immensely beneficial (Rico-Chávez et al., 2022; Pantic et al., 2022).

This study aims to assess the effects of different concentrations of sodium chloride (NaCl) to cause salt stress on the growth and physiological status of two varieties of broad bean (*Vicia faba* L.), Luz de Otono (V1) and Sofia (V2). Mainly we will use various concentrations of NaCl (50 mM, 100 mM, 200 mM), supplemented by either with or without calcium chloride ($CaCl_2$) to illustrate any possible effect of $CaCl_2$ to mitigate the effects of salt stress. We will measure morphological parameters (number of leaves, fresh and dry weight of the aerial parts and roots) and physiological parameters in order to make a fine comparative analysis of topology between both varieties as indicators of salt-tolerant plants and than we apply Random Forest regression to predict RWC in plants subjected to salt stress, using a dataset encompassing both morphological and biochemical attributes. The goal is to assess the feasibility and accuracy of using AI-based models to infer water-related stress responses and to identify the most influential predictors contributing to model performance. The purpose of this design is to explore the adaptive mechanisms of broad bean's to salinity, and to find the most compatible varieties for farming in salinized saline soil.

2. BACKGROUND STUDY

2.1. Plant Responses to Salt Stress

Salt stress is an important abiotic limitation to plant growth and productivity across the world (Shahid et al., 2020). When plants are subjected to high levels of salinity, they undergo osmotic stress due to limited water availability, and ionic stress due to toxic sodium and chloride ions (Kijne, 2006). These two stressors can also hinder photosynthesis, diminish enzymatic activity, and eventually cause decreased biomass accumulation. To be able to deal with saline conditions, plants upregulate multiple physiological and biochemical defenses, resulting in an increase in osmoprotectants such as proline and soluble sugars, ion transport regulation, and water regulation (Mokrani et al., 2020).

2.2. Importance of Relative Water Content (RWC)

Relative Water Content (RWC) is one of the most well-established indicators related to plant water status and reflects the capability of plant tissues to hold water (Ndlovu et al., 2021). Relative Water Content is highly correlated with turgor pressure and cellular hydration status. When RWC is measured, it reflects how well a plant withstands dehydration, making it tremendously valuable for salinity and drought studies (Le et al., 2018). Since RWC is indicative of both environmental and physiological signals, it can serve as a measure of overall plant health and assessment of stress tolerance in plants.

2.3. Biochemical Markers of Stress Response

Several biochemical markers, some of which are also commonly monitored to assess plant stress reactions in addition to relative water content (RWC). One major osmolyte is proline. Proline accumulates in plant cells as they stress and facilitates osmotic adjustment, membrane stabilization, and free radical scavenging. Chlorophyll content, which is commonly obtained using SPAD meters, is a reliable way to measure the photosynthetic capacity of leaves, which is generally reduced by stress conditions. Total sugars and polyphenols are also biologically meaningful in terms of stress responses and act as a source of energy as well as a protective mechanism from oxidative damage (Wahab et al., 2022).

2.4. Morphological Traits and Their Relevance

Morphological traits typically include variables such as the number of leaves, fresh and dry biomass of above and below ground biomass, and shoot length as indicators of plant vigor. These traits respond to multiple genetic and environmental factors but are also readily observable manifestations of internal physiological states. Morphological traits vary in nature, however they become more informative when they are examined, compared and contrast with the biochemical markers (Jia et al., 2021).

2.5. Introduction to Artificial Intelligence in Agriculture

Artificial intelligence and machine learning are being used more frequently in agricultural research to analyze large and complex datasets. They support the creation of predictive models, able to delineate patterns and relationships that may not be readily apparent using standard statistical approaches. In plant physiology, artificial intelligence can also link together collections of all types of disparate data, such as phenotypic observations to molecular profiles, and can predict well plant responses under a range of stress conditions (Javaid et al., 2023).

2.6. Random Forest Regression: Principles and Applications

Random Forest is a machine learning algorithm that builds a collection of decision trees. It has been demonstrated to be effective with regression problems that incorporate non-linearities and noisy data, such as those found in biological systems (Borup et al., 2023). Owing to their robustness, interpretability, and successful ability to rank importance in different features, Random Forest models are an ideal candidate for working with complex physiological traits, such as RWC.

3. MATERIALS AND METHODS

3.1. Plant material and growing conditions

The plant material consists of seeds of both faba bean species (Variety 1: Luz de Otono and variety 2: Sofia), sown in 2-liter perforated plastic pots, each containing 2 kg of Maamora sand, a poor sandy substrate that allows better expression of stress. Two seeds are sown per pot. The seedlings are watered with distilled water only for the first 14 days after sowing, to ensure uniform germination.

3.2. Applied treatments

From day 15, various salt and alkaline stress treatments are applied over a 40-day period. Six treatments were implemented, as shown in Table 1.

Table 1. Treatment used

Treatment	Conditions
T1	Control
T2	50 mM NaCl
T3	100 mM NaCl
T4	200 mM NaCl
T5	100 mM NaCl + 100 $CaCl_2$
T6	100 mM NaCl + 150 $CaCl_2$

Watering is done every two to three days, with a fixed volume of 400 mL per pot, considering the water-holding capacity of the substrate.

3.3. Experimental setup

The experimental set-up is based on a completely randomized block design (CRB), with three replicates per treatment and variety. Each replicate is made up of five pots, to ensure sufficient statistical representativeness.

3.4. Measured parameters

At the end of the duration of treatment, a number of morpho-physiological and biochemical parameters were studied to evaluate the response of the two *Vicia faba* varieties to different levels of salt and alkaline stress. Morphological parameters included height, root lengths, number of leaves, fresh and dry weights of above and root parts of the plant. Relative leaf water content was evaluated to evaluate the water status of the plants. Total chlorophyll measurements were evaluated using the equipment SPAD, while proline accumulation, an osmotic stress indicator was evaluated using the Bates method (Bates et al., 1973). Finally, the total sugar content was evaluated using the Dubois colorimetric method (Dubois et al., 1956).

3.5. Using artificial intelligence to predict response to salt stress

3.5.1. Dataset

The dataset included in this study included samples for all the plant samples, because of the salinity treatments for control and in 50 mM NaCl. Each sample has a number of morphological and biological traits that were collected. The traits are the number of leaves, fresh weights and dry weights of the aerial part and root part of the plant, the length of the aerial part of the plant, salinity treatments markers (SPAD, proline, total sugar, and polyphenol markers), and the dependent variable was the Relative Water Content (RWC). RWC was measured under each treatment condition. The data represent both physiology during stress of salinity and variation between the plant varieties and will be useful for machine learning analysis.

3.5.2. Data Preprocessing

Previous to the training of the model there were preprocessing tasks carried out to commence the data analysis. The data required cleaning and preprocessing. In the case of numeric entries where a comma was used in place of the decimal would have been replaced and filled with a dot. Categorical variables, the plant variety, and the treatment type, were encoded with label encoding to form a numerical representation.

Missing entries were addressed at that point by removing the incomplete record or being filled by forward filling for the repeated entries thereafter. The cleaned dataset was normalised and reformatted so it could be used as input for the model.

3.5.3. Feature Selection and Encoding

All recorded variables were preserved in the analysis so that the model was able to discover and learn unconventional interactions among morphological and biochemical indicators. Categorical variables were numerically encoded, so the data will still be compatible with the regression algorithm. Preservation of dimensionality without any dimensionality reduction is appropriate in this case, as the data was small enough to be manageable, and the goal was to maintain the explanatory nature of the features.

3.5.4. Model Development

The predictive modeling task was approached as a regression task using Random Forest regression. The authors selected Random Forest as the predictive model because it could be robust for predictive modeling and it can model non-linear relationships. The dataset was split as an 80/20 training/testing subsets. Random Forest was run using 100 trees and was configured with a fixed random state to be reproducible. The Random Forest model was trained on the training subset, and predictions were obtained from the test subset.

3.5.5. Model Evaluation

To evaluate the performance of the Random Forest trained model, both Mean Absolute Error (MAE) and cross validation. The model reached an MAE of 3.88 on held-out test data and an MAE of 5.18 after cross-validation, indicating consistent performance across various folds of the dataset. This also further confirms that the model can make robust predictions of RWC under varying stress conditions.

3.5.6. Feature Importance Analysis

A feature importance analysis was conducted to explore the biological relevance of the predictive model. Because the Random Forest algorithm outputs an estimate of the contribution of each input variable to the prediction power of the model, the analysis revealed that chlorophyll content, dry weight of the aerial part, and number of leaves had the most significant influence as features. The results presented plausible values in terms of physiological mechanisms, since both photosynthetic efficiency and biomass accumulation is affected by the plant's water status. Interpretability provides further evidence that machine learning has utility beyond being a predictive modelling tool, and that it can produce hypotheses or conclusions in plant stress research field.

4. RESULATS AND DISCUSSION

4.1. Analysis of morphological parameters

Table 2 illustrates the variation in the number of leaves from both varieties of faba bean (*Vicia faba*) Luz de Otono (V1) and Sofia (V2). These were subject to six treatment levels of increasing intensity that likely represents salt stress. In general, it is clear that as treatment levels increased, the average number of leaves decreased for both variety plants. For instance, in variety V1, the decline in the average number of leaves decreased from 34 leaves in treatment T1 to 11.33 leaves in T6. Likewise, variety V2 also decreased in leaf number from 39 leaves to 20.67 leaves from T1 to T6 shown in Table 2. This decline clearly demonstrates the detrimental impact of the imposed stress, as leaf development is significantly affected by environmental constraints like salinity. On the other hand, comparing the two varieties reveals observable tolerance differences. The Sofia variety (V2) has consistently a greater number of leaves as compared to the Luz de Otono (V1). For example, under the most severe stress treated condition T6, V2 retains over 21 leaves of average compared to average less than 11 leaves in V1. This larger disability to suppress leaf production under stress may suggest that the Sofia variety is more well-equipped to tolerate such extreme environments. These findings indicate that V2 has better tolerance likely because of a more efficient normal physiological or genetic mechanisms to respond to salinity.

Table 2. Number of leaves

Variety	**Treatment**					
	T1	**T2**	**T3**	**T4**	**T5**	**T6**
V1 (Luz de Otono)	34 ± 4,58	30 ± 5	25 ± 5	21,33 ± 5,03	18 ± 5,66	11,33 ± 4,16
V2 (Sofia)	39 ± 5,57	35,33 ± 5,03	30 ± 4	27,33 ± 6,11	24,67 ± 5,03	20,67 ± 4,62

Table 3 results indicate the effect of different salt and alkaline stresses on the fresh weight of the aerial part of both varieties of fava bean (V1 (Luz de Otono) and V2 (Sofia)). For Luz de Otono, the weight of the aerial part is relatively high (34.11 g) under control (T1) treatment and decreases progressively with the increase in the severity of treatment. Under treatment of NaCl without a buffering agent (T3, T4), the aerial part weight was dramatically reduced, especially 200 mM NaCl (T4: 11.52 g), and was reduced even more to 6.86 g under treatment with 100 mM NaCl + 150 mM CaC12 treatment (T6). This low weight in aerial part represents that the Luz de Otono variety is very sensitive to high levels of NaCl and CaC12 and caused a reduction in growth. On the other hand, the variety Sofia has a higher fresh weight under control (T1), 68.37 g. Altough Sofia was affected by salt stress, this variety seem to tolerate the treatments better than Luz de Otono variety. We can see a slight reduction of fresh weight from T1 to T2, (69.37 g – 45.37 g). In treatment T3 with 100 mM NaCl, the fresh weight again decreases to 23.43 g, and then the fresh weight under 200 mM NaCl drops to 17.09 g, T4. Overall, we can see even on treatments NaCl and CaC12 (T5 and T6) Sofia variety does show some reduction in growth, but is still overall more resistant than Luz de Otono with fresh weight of 11.31 g and T6 treatment. These findings illustrate that increasing NaCl concentrations, alone or in conjunction with CaC12, decrease the fresh weight of the aerial parts of both varieties significantly. However, the variety Sofia is less affected by salt and alkaline stress than the Luz de Otono variety. The addition of CaC12 does not appear to provide any additional benefit to plant growth in even very stressful conditions, but it may ameliorate the negative effects of NaCl as in the treatments T5 and T6 progressed. In summary, while both varieties will have adverse effects for salt stress, the variety Sofia appears to have better potential for stress tolerance. Certainly other management options besides just CaC12 will be required to achieve better plant resilience under stress; however, further optimization of irrigation or testing of various growth regulators could be potentially applied.

Table 3. Fresh weight aerial part (g)

Variety	**Treatment**					
	T1	**T2**	**T3**	**T4**	**T5**	**T6**
V1 (Luz de Otono)	34,11 ± 10,12	32,27 ± 2,39	27,11 ± 2,65	11,52 ± 1,94	11,3 ± 0,42	6,86 ± 1,13
V2 (Sofia)	68,37 ± 1,96	45,37 ± 7,27	23,43 ± 2,66	17,09 ± 2,89	13,43 ± 0,60	11,31 ± 1,25

Table 4 presents the fresh root weights of the two fava bean varieties V1 (Luz de Otono) and V2 (Sofia) under varying salt and alkaline stress treatments. The data for Luz de Otono show that the fresh root weight declines as stress intensification; it has a fresh weight of 18.27 g as the control treatment (T1) and declines with salinity and alkalinity treatments. The fresh root weight declines to 13.31 g at 100 mM NaCl (T3) and declines to a low of 3.74 g at 100 mM NaCl accompanied by 150 mM CaCl2 (T6). It is clear that the Luz de Otono variety is sensitive to these stress treatments, especially combined salinity and alkaline treatments, which stress root growth considerably. In contrast to the Luz de Otono variety, Sofia has an opposite response. The fresh root weight in Sofia is greater than Luz de Otono under control treatment (T1) with an average weight of 21.34 g. The fresh root biomass in Sofia is still greater under stress conditions with less reduction in root weight. The fresh root weight was almost constant for the 50 mM NaCl treatment (T2), the 100 mM NaCl treatment (T3) and the 200 mM NaCl treatment (T4) with average weights of 19.67 g, 19.44 g and 19.33 g respectively. Thus, it appears that Sofia has lower salt and higher alkaline stress levels than Luz de Otono. However, under the combined treatments of 100 mM NaCl + 100 mM CaCl2 (T5) and 100 mM NaCl + 150 mM CaCl2 (T6) the fresh root weight was 9.7 g and 9.33 g, respectively indicating that while Sofia is more tolerant, it can still be removed from soil under higher salt and alkaline stress conditions. To sum up, the Luz de Otono variety shows a noticeable reduction in root weight response with increased stress, with combined NaCl and CaCl2 treatments being the worst and showing an overall negative response to these stress treatments. On the other hand, the Sofia variety showed a greater apparent tolerance to salt and alkaline stress, with a considerably smaller reduction in root weight with greater stress intensity. Despite the differences, both varieties had reduced root weights under maximum stress condition, especially when given both NaCl and CaCl2, which indicates the negative effects on root development.

Table 4. Fresh root weight (g)

Variety	Treatment					
	T1	T2	T3	T4	T5	T6
V1 (Luz de Otono)	18,27 ± 6,51	16 ± 5,57	13,31 ± 7,79	11,06 ± 3,84	7,25 ± 3,18	3,74 ± 2,4
V2 (Sofia)	21,34 ± 2,85	19,67 ± 2,89	19,44 ± 3,58	19,33 ± 6,43	9,7 ± 0,46	9,33 ± 0,59

The results in Table 5 demonstrate the impact of a range of salt and alkaline stress treatments on the dry weight of cotton wools of 2 varieties of fava beans, V1 (Luz de Otono) and V2 (Sofia). In the case of the Luz de Otono variety, the dry weight of the aerial part progressively diminished with the increasing stress of salt and alkali. For the control autotrophic terrestrial treatment (T1), the dry weight was 2.88 g. Yet under each of the stress treatments, and in particular the combination of: 100 mM NaCl + 150 mM CaCl2 (T6) where the dry weight went down to 1.08 g, the dry weight was negatively impacted. The same pattern was seen with the abovementioned varieties. Thus it can be concluded that the Luz de Otono variety was sensitive to increasing abiotic stress, and less dry weight was produced as stress increased. In comparison, the Sofia variety manages to attain a higher dry weight value in the control treatment (T1), averaging 4.52 g. In all treatments, the Sofia variety also experienced reductions in dry weight, but these reductions were not as significant, compared to the Luz de Otono variety. The Sofia variety dropped to 3.39 g in T2, with the 50 mM NaCl treatment; 2.5 g in T3, the 100 mM NaCl treatment; and continued to drop as saline and combined treatments increased, with 2.15 g in T4 at 200 mM NaCl; 2 g in T5 at 100 mM NaCl + 100 mM CaC12; and finally, the Sofia variety dropped to 1.89 g in the T6 combined treatment that was 100 mM NaCl + 150 mM CaC12. Overall, the Sofia variety was apparently more tolerant to saline and alkaline stresses than the Luz de Otono variety, since in all treatments the Sofia variety maintained a higher dry weight. In summary, both varieties decrease in dry weight when the stress applied increases, but versus Luz de Otono, Sofia was less affected by salt and alkaline stress. Under the higher stress treatments Luz de Otono shows remarkable reduction in dry weight and showed the most significant reduction under the combined NaCl and CaC12 treatments. Sofia shows higher dry weight indicating better tolerance to the conditions in the treatments. Regardless of this per variety implication, both varieties have significant reductions due to the stress treatment, and the highest reductions were under the combined NaCl and CaC12 treatments.

Table 5. Dry weight aerial part (g)

Variety	**Treatment**					
	T1	**T2**	**T3**	**T4**	**T5**	**T6**
V1 (Luz de Otono)	2,88 ± 0,13	2,47 ± 0,19	2,06 ± 0,06	1,73 ± 0,09	1,68 ± 0,18	1,08 ± 0,25
V2 (Sofia)	4,52 ± 0,62	3,39 ± 0,87	2,5 ± 0,28	2,15 ± 0,05	2 ± 0,06	1,89 ± 0,11

The data in Table 6 reports the effect of salt and alkaline stress treatments on the whole dry root weight of two fava bean varieties, V1 (Luz de Otono) and V2 (Sofia). For the Luz de Otono variety, the dry root weight is relatively consistent around 2.40g on treatment 1 (T1, control), and decreases consistently as the intensity of stress increases. The dry root weight declined to 1.60g under 100 mM NaCl (T3) and 1.46g for the 200 mM NaCl (T4). The weight continued to drop among the combined treatments, to a minimum of 1.09g for the 100 mM NaCl + 150 mM CaC12 (T6). The data suggests that the Luz de Otono variety experiences a reduction in dry root weight due to increasing salinity and alkalinity stress - the greater the stress treatment intensity, the more the dry root weight declined. In contrast, the Sofia variety has the highest dry root weight associated with the control treatment (T1) with an average dry root weight of 4.14 g. All treatments have a lower average fraction of dry root weight, but the decline is not as marked as for the Luz de Otono variety. Specifically, in the 50 mM NaCl treatment (T2) the average dry root weight declines to 3.27 g, under 100 mM NaCl (T3) it decline further to an average of 2.53 g. However, even under the most stressful treatment of 100 mM NaCl + 150 mM CaCl2 combined (T6), the dry root weight declines more marginally (1.41 g) to further indicate that the Sofia variety was tolerating the stress more than Luz de Otono. In summary, both varieties exhibited declines in dry root weight with increases in the intensity of stress, but the Sofia variety was less impacted by the effects of salt and alkaline stress than the Luz de Otono variety. The Luz de Otono variety underwent a greater decline in dry root weight with increasing stress - particularly stress from the combined treatments of NaCl and CaCl2 treatments. The Sofia maintained a higher dry root weight through all treatments which indicates greater resilience and indicates greater ability to tolerate the imposed stress conditions. Nevertheless, both varieties were affected by the stress treatment versus controls, and both variety dry root weights were substantially lower with the NaCl + CaCl2 stress treatments which were also combined treatments.

Table 6. Dry root weight (g)

Variety	Treatment					
	T1	T2	T3	T4	T5	T6
V1 (Luz de Otono)	2,40 ± 0,36	2,37 ± 0,37	1,60 ± 0,15	1,46 ± 0,13	1,20 ± 0,07	1,09 ± 0,07
V2 (Sofia)	4,14 ± 0,25	3,27 ± 0,68	2,53 ± 0,03	1,70 ± 0,22	1,50 ± 0,03	1,41 ± 0,04

In Table 7, we see that the salt and alkaline stress treatments can effectively categorize the lengths of the aerial parts for the two fava bean varieties—V1 (Luz de Otono) and V2 (Sofia). For the variety Luz de Otono, with a control treatment (T1) the length of the aerial part was 63.33 cm. As the stress treatment became more intense, the lengths decreased significantly, with T2 (50 mM NaCl) being 51.00 cm, T3 (100 mM NaCl) being 48.66 cm, and T4 (200 mM NaCl) reduced down to 38.66 cm. For the combined treatments of NaCl and CaCl2 (T5 and T6), the declining trend increased, down to a minimum of 29.33 cm. This demonstrates that the variety Luz de Otono is sensitive to salt and alkaline stress and can significantly reduce the growth of the aerial part of fava bean. In contrast, the Sofia variety had a higher initial length of the aerial part in the control treatment (78.33 cm). Although the aerial part length was also decreased under stress, it was decreased less than Luz de Otono. The aerial length of Sofia was decreased from 78.77 cm to 75.77 cm under 50 mM NaCl (T2), decreased to 56.33 cm under 100 mM NaCl (T3), and also decreased to 48.00 cm, 40.00 cm, and 28.66 cm in T4, T5, and T6, respectively, which are more severe treatments. This means that Sofia has greater tolerance to salt and alkaline stress as the aerial part length is longer than Luz de Otono for each treatment. In summary, while both varieties had reductions in the length of the aerial part due to increasing stress intensity, the Sofia variety was more resilient than Luz de Otono. The Luz de Otono variety had a maintenance of growth maximum under salt and alkaline stress, but this was severely reduced under the combination treatments. The Sofia variety generally maintained a longer aerial part than the Luz de Otono variety, but had better salt and alkaline stress tolerances. In general, both varieties were adversely affected from the stress treatments, with the return of the length of the aerial part being most affected in the combined form of stress (NaCl and CaCl2).

Table 7. Length of aerial part (cm)

Variety	Treatment					
	T1	T2	T3	T4	T5	T6
V1 (Luz de Otono)	63,33 ± 7,57	51,00 ± 8,71	48,66 ± 5,13	38,66 ± 7,02	37,00 ± 1,41	29,33 ± 6,11
V2 (Sofia)	78,33 ± 3,78	72,00 ± 3,46	56,33 ± 12,66	48,00 ± 5,56	40,00 ± 5,00	28,66 ± 8,08

4.2. Analysis of physiological and biochemical parameters

The results displayed in Table 8 show the effect of NaCl and alkaline stress treatments on the relative water content (RWC) of the leaves of two varieties of fava beans, V1, Luz de Otono, and V2, Sofia. In the Luz de Otono variety, the RWC value of the control treatment (T1) is 73.67%, indicating adequate water retention under non-stressed conditions. With increasing salt stress from T2 through T6, there was a corresponding decrease in relative water content from 69.32% with 50 mM NaCl (T2 value), and a RWC of 64.43% when treated with NaCl at 100 mM (T3). The RWC value remained essentially unchanged at 64.50 for the plants treated with 200 mM NaCl (T4), indicating the limitations beyond which water content could no longer be sustained effectively. The RWC then dropped more than 7% from T4 when NaCl was combined with either 300 mM CaCl2 or both with NaCl combined NaCl and CaCl2 (with a value of 55.26% in T5) at 300 mM CaCl2 and 200 Mm NaCl (with a value of 44.22%). Overall, the combination of salt and calcium stress had a marked impact on the plants water retention values. The Sofia variety has a better starting water content under control conditions (80.03%), then with stress the amount gradually decreased to lower than Luz de Otono, although in conditions with NaCl 50 mM (T2) it would decrease only to 72.63%, then for 100 mM NaCl (T3) it would decrease less at 67.05%, and even under salt concentrations of 200 mM NaCl (T4) the Sofia had a RWC of 62.31%. Under the most severe combined NaCl / CaC12 stresses would decrease even further with T5 59.00% and T6 had even lower RWC at 56.36%. All values were lower compared to the control, but they also maintained consistency to be higher than the Luz de Otono variety in each of the treatments of study. Indeed, both varieties of fava bean had a decrease in relative water content associated with salt and stress alkali treatment, indicating that their water balance was compromised. The Sofia variety shows greater resilience to stress by maintaining RWC values across treatments than the Luz de Otono variety, which appears to show greater sensitivity, especially under the combined NaCl and CaC12 stress, where there is a more considerable level of deterioration of water retention. These results indicate that Sofia has a better ability to maintain hydration of cells when subjected to salt stress, which may explain why Sofia displayed overall better stress tolerance.

Table 8. Relative water content (%)

Variety	**Treatment**					
	T1	**T2**	**T3**	**T4**	**T5**	**T6**
V1 (Luz de Otono)	73,67 ± 8,69	69,32 ± 7,00	64,43 ± 6,07	64,50 ± 4,95	55,26 ± 6,70	44,22 ± 0,69
V2 (Sofia)	80,03 ± 0,45	72,63 ± 2,75	67,05 ± 2,85	62,31 ± 2,48	59,00 ± 3,60	56,36 ± 4,21

The chlorophyll content, evaluated by the SPAD index, was found to decrease progressively in both varieties (Luz de Otono and Sofia) with increasing salinity (from treatment T1 to T6). The reduction over time is representative of the salt stress impact on photosynthetic metabolism and development in the plants. For the Luz de Otono variety, values decreased sharply (T3 = 175.00 ± 1.00 to T4 = 110.33 ± 8.39) and reached a minimum of 76.67 ± 11.50 for the most severe stress condition (T6). This suggests the variety is very sensitive to salinity, likely due to either a more rapid degradation of chlorophyll pigments or the alteration of chlorophyll biosynthesis. Conversely, the Sofia variety was able to show more stability in chlorophyll content under both levels of salt stress. While the values still decreased over levels of stress, it was still consistently more than Luz de Otono at each treatment level. For example, treatment T6 had the Sofia variety showing a chlorophyll content of 103.33± 10.41, which was much greater than Luz de Otono. This is notable, as this indicates that the physiological mechanisms operating in Sofia are superior for the retention of the integrity of its photosynthetic apparatus. Some possibilities include better regulation in the control of toxic ions in leaf tissues, better protection of chloroplast membranes, or more active antioxidant systems. These results show that chlorophyll content is a good indicator of salt stress tolerance, and Sofia is a top candidate due to its better adaptive capacity. In contrast to the Luz de Otono variety, which was clearly more sensitive to salt stress, Sofia can control salt stress better and provide a continuous source of food for twigs and foliage, but Luz de Otono would have an extra precaution if produced in salt-affected lands.

Table 9. Chlorophyll (SPAD)

Variety	Treatment					
	T1	T2	T3	T4	T5	T6
V1 (Luz de Otono)	194,33 ± 5,13	180,00 ± 5,00	175,00 ± 1,00	110,33 ± 8,39	103,00 ± 4,24	76,67 ± 11,50
V2 (Sofia)	249,67 ± 8,50	190,67 ± 11,37	149,33 ± 12,50	115,33 ± 3,79	110,00 ± 5,00	103,33 ± 10,41

The proline content in both varieties (Luz de Otono and Sofia) increased significantly with increased salt stress from treatment T1 to T6 (Table 10). The accumulation of proline is indicative of an adaptive physiological response to salinity, where proline acts as an osmoprotectant. Proline helps to maintain cellular turgor, stabilizes proteins and membranes, and provides protection against free radicals. In the Luz de Otono variety, the content of proline increased moderately from 0.23 ± 0.03 mg/g at T1 (control) to 0.64 ± 0.04 mg/g at the highest salinity (T6). While this finding indicates some response to stress, the quantity of proline accumulation is comparatively low compared to the Sofia variety. Sofia showed a much greater and consistent increase in proline content, starting at 0.31 ± 0.01 mg/g at T1 and rising to 0.92 ± 0.05 mg/g at T6. This greater accumulation of proline under salt stress would suggest that Sofia has an enhanced ability to ameliorate osmotic imbalance and oxidative stress, indicating improved stress tolerance capacity. The difference in the accumulation of proline between the two varieties indicates that Sofia displayed increased ability to modulate its metabolism under salinity and therefore better represented a salt-tolerant genotype. Proline is an important biochemical marker to estimate plant responses to salt stress in the present study.

Table 10. Proline content (mg/g)

Variety	Treatment					
	T1	T2	T3	T4	T5	T6
V1 (Luz de Otono)	0,23 ± 0,03	0,30 ± 0,02	0,37 ± 0,02	0,46 ± 0,05	0,49 ± 0,01	0,64 ± 0,04
V2 (Sofia)	0,31 ± 0,01	0,41 ± 0,07	0,67 ± 0,09	0,84 ± 0,04	0,91 ± 0,04	0,92 ± 0,05

The total sugar content of both varieties (Luz de Otono and Sofia) was significantly and progressively higher from T1-T6, with increasing salt stress indicating that sugars contribute to osmotic adjustment, thus helping with tolerance to stress where they serve as osmolytes, stabilize cellular structures and perform signalling. Luz de Otono had total sugar concentration of 26.00 ± 3.00 mg/g under the control (T1) reduced to 70.67 ± 4.04 mg/g at severe (T6) salt stress, demonstrating that Luz de Otono activates sugar accumulation in response to salinity and is likely to lessen osmotic stress, and help maintain homeostasis in the cell. Sofia shown a sustained significant increase due to treatments, with sugar content from 28.67 ± 3.21 mg/g at T1 to 79.33 ± 2.08 mg/g at T6 across the levels of treatments, and Sofia had consistently higher sugars levels than Luz de Otono suggesting that it may have a more effective osmotic regulating system and a more effective metabolic reaction to salt stress. It is notable that the Sofia genotype accumulated more soluble sugars and had superior tolerance to salt stress. All findings support the use of sugars as physiological and biochemical markers for plant responses to saline environments and highlight the adaptive capacity of the Sofia genotype.

Table 11. Total sugars (mg/g)

Variety	Treatment					
	T1	T2	T3	T4	T5	T6
V1 (Luz de Otono)	26,00 ± 3,00	34,00 ± 5,57	43,67 ± 6,11	54,00 ± 4,36	62,50 ± 2,12	70,67 ± 4,04
V2 (Sofia)	28,67 ± 3,21	36,33 ± 2,52	44,00 ± 2,65	56,00 ± 4,00	65,33 ± 5,86	79,33 ± 2,08

The total polyphenol content accumulated incrementally in both varieties (Luz de Otono and Sofia) with increasing salt stress from T1 to T6. Given that polyphenols are part of the antioxidant defence, it is crucial that they can rid cells of oxidants produced purely as a result of the salinity stress levels. In the case of the Luz de Otono variety, the polyphenol content increased from 25.00 ± 2.00 mg/g under control conditions (T1) to 70.00 ± 2.00 mg/g under the highest salt concentration (T6). This increase displays the fact that the phenolic compound synthesis was activated under salt stress in which it is possible that the free radical scavenging and aspects of the protective cellular functions to aid the plant in developing some level of tolerance was also occurring. In the Sofia variety, there was an even greater increase in polyphenol content from 29.00 ± 1.73 mg/g (T1) to 89.33 ± 4.50 mg/g (T6). Sofia also had higher average total polyphenol content than Luz de Otono in each treatment confirming the Sofia stylized systematic was exhibiting a greater antioxidant capacity and subsequently adaptability to salinity induced oxidative damage. The previously stated statistics indicate the substantial role of polyphenols to protect plant tissues from salinity induced oxidative damage. In which, the Sofia variety exhibited more polyphenol accumulation and evidenced more tolerance making it the more appropriate variety to cultivate in saline environments.

Table 12. Total polyphenols (mg/g)

Variety	Treatment					
	T1	T2	T3	T4	T5	T6
V1 (Luz de Otono)	25,00 ± 2,00	32,67 ± 3,79	44,33 ± 2,52	57,00 ± 2,65	64,50 ± 2,12	70,00 ± 2,00
V2 (Sofia)	29,00 ± 1,73	34,33 ± 2,08	46,33 ± 5,03	62,33 ± 4,50	76,00 ± 4,58	89,33 ± 4,50

4.3. Model Performance

Table 13 contains the Mean Absolute Error (MAE), typical of the Random Forest regression. The Random Forest regression achieved a MAE of 3.88 on the test dataset. As an additional procedure to evaluate the Random Forest regression's robustness and generalizability, five-fold cross-validation was used. The MAE that was achieved for the cross-validation was an overall MAE of 5.18. Given the biological variability, and the inherently complicated nature for the plant responses under salt stress, it is very impressive that the model was able predict with this level of accuracy as indicated by the MAE.

To provide a visual representation of predictive ability, Figure 1 is depicted as a scatterplot of actual versus predicted Relative Water Content values. Most of the predictions fall aligned near the diagonal line which indicates the predicted values are closely aligned with the actual observed values, despite some divergence, namely the divergence at lower RWC values. The divergence in the predictions is indicative of the biological variability and some potential treatment differences.

Table 13. MAE on test data and cross-validation

Evaluation Metric	Value	Evaluation Metric
Test Set MAE	3.88	Test Set MAE
Cross-validated MAE (CV)	5.18	Cross-validated MAE (CV)

Figure1. Actual vs. Predicted values of Relative Water Content.

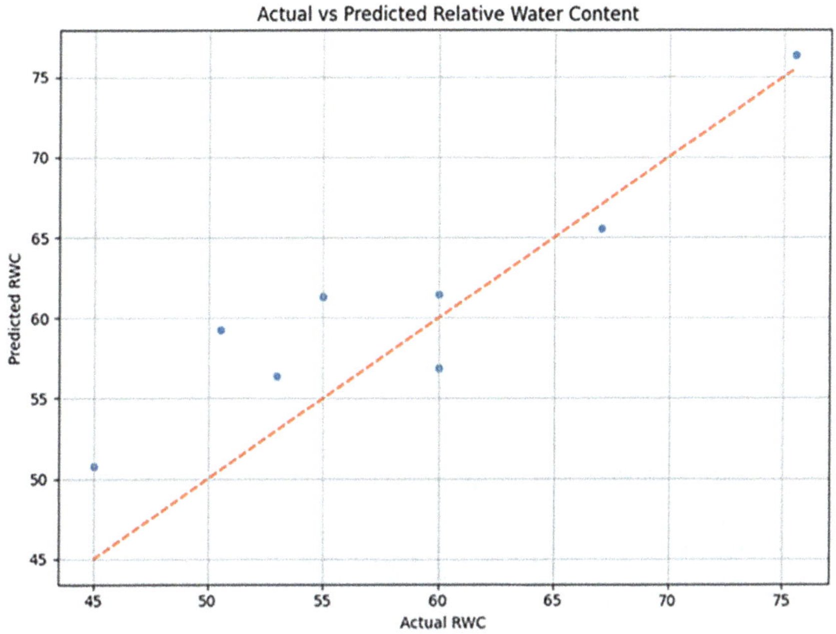

4.4. Feature Importance Analysis

A feature importance analysis allowed us to gain greater insights into the variables that most influenced the prediction of Relative Water Content. In the Feature Importance Plot (Figure 2), chlorophyll content (SPAD) was the dominant predictor, followed closely by the dry weight of above ground biomass and number of leaves. These results make biological sense as chlorophyll content is strongly associated with activity related to photosynthesis and water status, and that biomass-related traits reflect the overall health of the plant and an indication of growth potential under a stressor.

Table 14. Feature Importance Analysis

Rank	Feature	Importance Score
1	Chlorophyll (SPAD)	0.248
2	Dry weight aerial part	0.197
3	Number of leaves	0.153

continued on following page

Table 14. Continued

Rank	Feature	Importance Score
4	Length of aerial part	0.122
5	Fresh root weight	0.098

Figure 2. Feature importance scores from the Random Forest regression model.

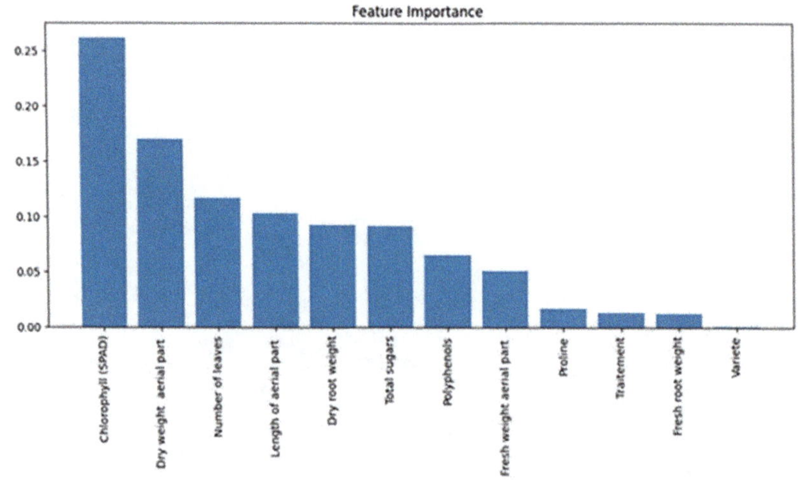

4.5. Correlation Matrix Insights

The correlation matrix provided as part of Figure 3, was another source of tailor-made interpretations. Correlation of the RWC value became apparent, in multiple features such as chlorophyll, dry weight of aerial part, and length of aerial part, where a strong positive correlation existed throughout. On the contrary, a few variables such as proline and polyphenols demonstrated strong negative correlations showing increased accumulation of the variables was possible to perceiving when RWC value was lower illustrating extreme stress conditions. This combination of an exploratory analysis into the relationship observed by correlations and importance scores gives added biological relevance to the model's predictions, further elucidating how machine learning could reveal plausible physiological relationships in convoluted agricultural data.

Figure 3. Correlation matrix among morphological, biochemical traits, and Relative Water Content

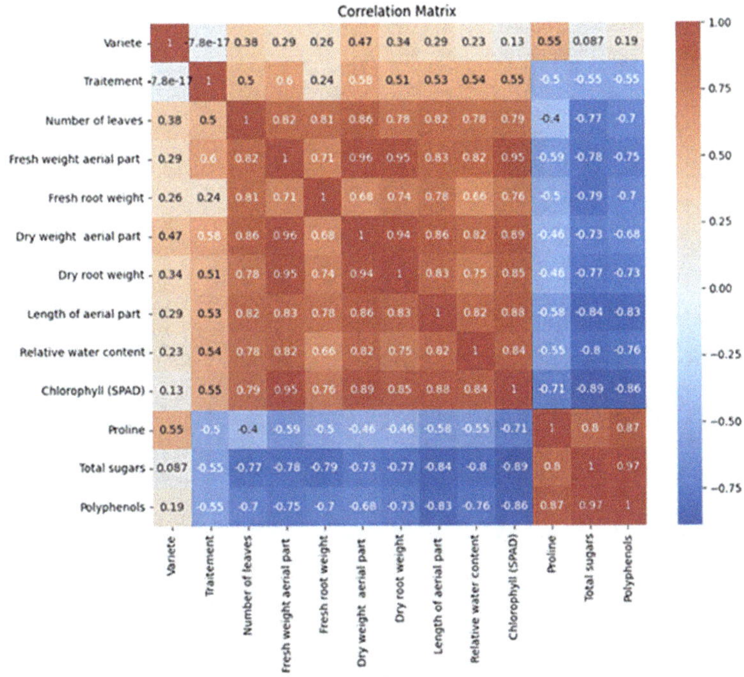

4.6. Residual and Error Analysis

To illustrate the prediction errors in better detail, the histogram of residuals is shown in Figure 4. The distance from the predicted RWC values is reasonably symmetrical and slightly biased toward negative residuals, indicating that the model does underestimate RWC on occasion. There appeared to be slightly lower estimates in conditions of high stress, however, which may represent some instances of underestimation occurring in the presence of stress markers dominating the input profile. Although the model performs well overall, we see occasional instances of overcorrection when stress markers are prevalent in the input profile.

Furthermore, Figure 5 presents a boxplot of the prediction errors grouped by treatment type. This visualization indicates that the model's accuracy varied depending on the treatment applied. The treatments with the highest salinity levels appear to generate more negative errors, reaffirming our common-sense understanding that prediction is always more challenging in the most extreme stress conditions. Furthermore, the treatment-specific variability in these errors indicates a need to refine models to adapt to condition-dependent behavior in plant physiology.

Figure 4. Histogram of residuals from RWC prediction (Actual - Predicted)

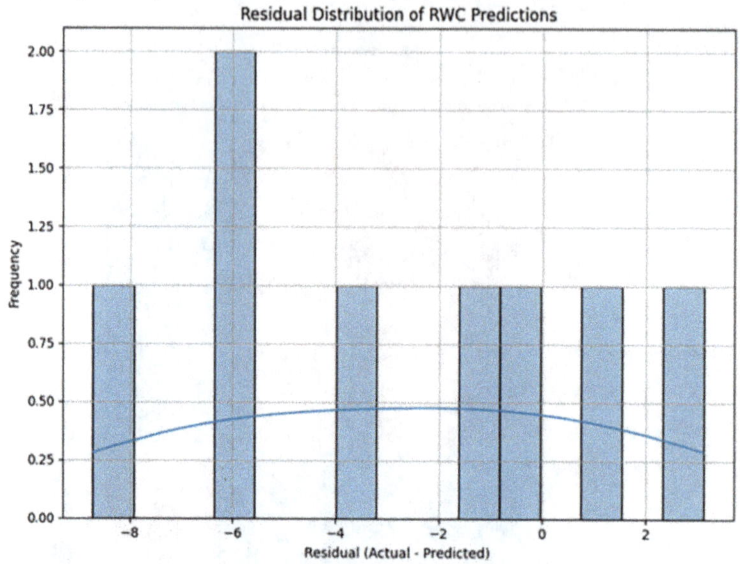

Figure 5. Boxplot of prediction errors by treatment type

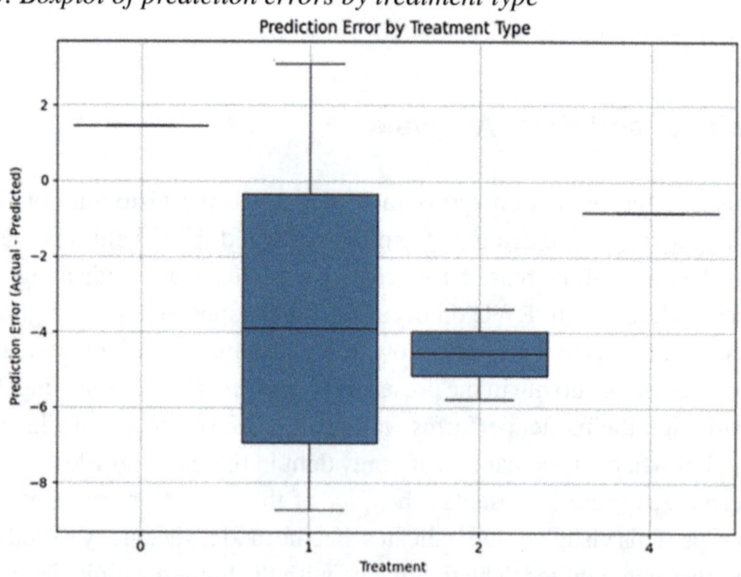

5. CONCLUSION

This research has demonstrated that machine learning using Random Forest regression can be utilized to predict Relative Water Content (RWC) in plants with salt stress. In it I demonstrate that the model produced good accuracy levels supported by reproducible cross-validation measures. We looked at feature importance revealing physiological aspects, like chlorophyll content and aerial biomass, that are known indicators of plant hydration status, which we now know are important in these models and establish predictions values. I believe they can be useful not only in demonstrating the predictive ability of the model but also their biological interpretability. The use of correlation analyses with residual diagnostics was beneficial in achieving a broader understanding of the model performance and caveats. Prediction errors seemed to compound more pronouncedly in the greater treatment categories, indicating that we would likely have to fine tune the model to ultimately generate the most accurate predictions in severely stressed plants. Nevertheless, the continued performance of the model is encouraging and the observations illustrate the importance of AI-based methods in crop monitoring. Looking ahead, this project lays the foundation for the next steps of applications of artificial intelligence in plant physiology and agronomic management. Future investigations could determine the feasibility of transposing more stress signals, use bigger and more complex datasets, and engage with other machine learning algorithms to improve prediction. The overlap between model-driven insights and identified biological responses emphasizes why artificial intelligence can be considered a predictive and exploratory tool. Model-driven highlights in key traits associated with water retention is very helpful in proposing future experiments and applications in precision agriculture. Importantly, these results confirm the methodological approach and bolster the adoption of AI-based models into crop stress monitoring programs. In conclusion, the application of AI models towards field-based decision-making may further the development of more resilient and efficient agricultural systems.

REFERENCES

Alzahib, R. H., Migdadi, H. M., Al Ghamdi, A. A., Alwahibi, M. S., Ibrahim, A. A., & Al-Selwey, W. A. (2021). Assessment of Morpho-Physiological, Biochemical and Antioxidant Responses of Tomato Landraces to Salinity Stress. *Plants*, *10*(4), 696. DOI: 10.3390/plants10040696

Bates, L.S., waldren, R.P., Teare, I.D. (1973). Rapid determination of free proline for water studies. *Plant and Soil*, *39*, 205–208.

Borup, D., Christensen, B. J., Mühlbach, N. S., & Nielsen, M. S. (2023). Targeting predictors in random forest regression. *International Journal of Forecasting*, *39*(2), 841–868. DOI: 10.1016/j.ijforecast.2022.02.010

Cai, Y., Cao, X., Liu, B., Lin, H., Luo, H., Liu, F., Su, D., Lv, S., Lin, Z., & Lin, D. (2025). Saline–alkali tolerance evaluation of giant reed (Arundo donax) genotypes under saline–alkali stress at seedling stage. *Agronomy (Basel)*, *15*(2), 463. DOI: 10.3390/agronomy15020463

Chen, H., Yu, S., Yu, Z., Ma, M., Liu, M., & Pei, H. (2024). Phycoremediation potential of salt-tolerant microalgal species: Motion, metabolic characteristics, and their application for saline–alkali soil improvement in eco-farms. *Microorganisms*, *12*, 676. DOI: 10.3390/microorganisms12040676

Dehghanian, Z., Ahmadabadi, M., Asgari Lajayer, B., Gougerdchi, V., Hamedpour-Darabi, M., Bagheri, N., Sharma, R., Vetukuri, R. R., Astatkie, T., & Dell, B. (2024). Quinoa: A Promising Crop for Resolving the Bottleneck of Cultivation in Soils Affected by Multiple Environmental Abiotic Stresses. *Plants*, *13*(11), 2117. DOI: 10.3390/plants13152117

Derici, S., & Dogan, N. O. (2025). A decision model based on artificial intelligence for disease prediction and patient treatment process planning. *Asia Pacific Journal of Health Management*, *20*(2), 1–12.

Dubois, M. K., Gils, J. K., Hanniton, P. A., & Smith, F. (1956). Use of phenol reagent for the determination of total sugar. *Analytical Chemistry*, *28*, 350–356.

El-Khadir, I., Mouniane, Y., Chriqui, A., Kouighat, M., & Hmouni, D. (2025). Effects of Salt Stress on *Origanum majorana* and Improvement via Association with Halophytes Plants: *Spergularia salina* and *Aptenia cordifolia. Acta Universitatis Agriculturae et Silviculturae Mendelianae Brunensis*, *73*(1), 27–35. DOI: 10.11118/actaun.2025.003

Fernández-Triana, I., Rubilar, O., Parada, J., Fincheira, P., Benavides-Mendoza, A., Durán, P., Fernández-Baldo, M., Seabra, A. B., & Tortella, G. R. (2024). Metal nanoparticles and pesticides under global climate change: Assessing the combined effects of multiple abiotic stressors on soil microbial ecosystems. *The Science of the Total Environment*, *942*, 173494. DOI: 10.1016/j.scitotenv.2024.173494

Garbanzo, G., Cameira, M. R., & Paredes, P. (2024). The mangrove swamp rice production system of Guinea Bissau: Identification of the main constraints associated with soil salinity and rainfall variability. *Agronomy (Basel)*, *14*(3), 468. DOI: 10.3390/agronomy14030468

Germano, R. P., Melito, S., Cacini, S., Carmassi, G., Leoni, F., Maggini, R., Montesano, F. F., Pardossi, A., & Massa, D. (2022). Sweet basil can be grown hydroponically at low phosphorus and high sodium chloride concentration: Effect on plant and nutrient solution management. *Scientia Horticulturae*, *304*, 111324. DOI: 10.1016/j.scienta.2022.111324

Habibi, N., Terada, N., Sanada, A., & Koshio, K. (2024). Alleviating salt stress in tomatoes through seed priming with polyethylene glycol and sodium chloride combination. *Stresses*, *4*(2), 210–224. DOI: 10.3390/stresses4020012

Haj-Amor, Z., Araya, T., Kim, D.-G., Bouri, S., Lee, J., Ghiloufi, W., Yang, Y., Kang, H., Jhariya, M. K., Banerjee, A., & Lal, R. (2022). Soil salinity and its associated effects on soil microorganisms, greenhouse gas emissions, crop yield, biodiversity and desertification: A review. *The Science of the Total Environment*, *843*, 156946. DOI: 10.1016/j.scitotenv.2022.156946

Javaid, M., Haleem, A., Khan, I. H., & Suman, R. (2023). Understanding the potential applications of artificial intelligence in the agriculture sector. *Advanced Agrochem*, *2*(1), 15–30. DOI: 10.1016/j.aac.2022.10.001

Jia, W., Ma, M., Chen, J., & Wu, S. (2021). Plant morphological, physiological, and anatomical adaptation to flooding stress and the underlying molecular mechanisms. *International Journal of Molecular Sciences*, *22*(3), 1088. DOI: 10.3390/ijms22031088

Kijne, J. W. (2006). Abiotic stress and water scarcity: Identifying and resolving conflicts from plant level to global level. *Field Crops Research*, *97*(1), 3–18. DOI: 10.1016/j.fcr.2005.08.011

Koh, E., Sunil, R. S., Lam, H. Y. I., & Mutwil, M. (2024). Confronting the data deluge: How artificial intelligence can be used in the study of plant stress. *Computational and Structural Biotechnology Journal*, *23*, 3454–3466. DOI: 10.1016/j.csbj.2024.09.010

Le, T. T. T., Williams, B., & Mundree, S. G. (2018). An osmotin from the resurrection plant Tripogon loliiformis (TlOsm) confers tolerance to multiple abiotic stresses in transgenic rice. *Physiologia Plantarum, 162*(1), 13–34. DOI: 10.1111/ppl.12585

Lu, T., Luo, P., Wang, J., Lu, Y., Huo, A., & Liu, L. (2025). Soil salinity accumulation and groundwater degradation due to overexploitation over recent 40-year period in Yaoba Oasis, China. *Soil & Tillage Research, 248*, 106398. DOI: 10.1016/j.still.2024.106398

Mokrani, S., Nabti, E.-h., & Cruz, C. (2020). Current advances in plant growth promoting bacteria alleviating salt stress for sustainable agriculture. *Applied Sciences (Basel, Switzerland), 10*(20), 7025. DOI: 10.3390/app10207025

Mucsi, M., Borsodi, A. K., Megyes, M., & Szili-Kovács, T. (2024). Response of the metabolic activity and taxonomic composition of bacterial communities to mosaically varying soil salinity and alkalinity. *Scientific Reports, 14*, 7460. DOI: 10.1038/s41598-024-57430-2

Naorem, A., Jayaraman, S., Dang, Y. P., Dalal, R. C., Sinha, N. K., Rao, C. S., & Patra, A. K. (2023). Soil Constraints in an Arid Environment—Challenges, Prospects, and Implications. *Agronomy (Basel), 13*(1), 220. DOI: 10.3390/agronomy13010220

Ndlovu, H. S., Odindi, J., Sibanda, M., Mutanga, O., Clulow, A., Chimonyo, V. G. P., & Mabhaudhi, T. (2021). A comparative estimation of maize leaf water content using machine learning techniques and unmanned aerial vehicle (UAV)-based proximal and remotely sensed data. *Remote Sensing, 13*(20), 4091. DOI: 10.3390/rs13204091

Noh, Y., Jafarinejad, S., & Anand, P. (2024). A Review on Harnessing Renewable Energy Synergies for Achieving Urban Net-Zero Energy Buildings: Technologies, Performance Evaluation, Policies, Challenges, and Future Direction. *Sustainability, 16*, 3444. DOI: 10.3390/su16083444

Ondrasek, G., Rathod, S., Manohara, K. K., Gireesh, C., Anantha, M. S., Sakhare, A. S., Parmar, B., Yadav, B. K., Bandumula, N., Raihan, F., Zielińska-Chmielewska, A., Meriño-Gergichevich, C., Reyes-Díaz, M., Khan, A., Panfilova, O., Seguel Fuentealba, A., Meier Romero, S., Nabil, B., Wan, C., & Shepherd, J. (2022). Salt Stress in Plants and Mitigation Approaches. *Plants, 11*(6), 717. DOI: 10.3390/plants11060717

Oudghiri, M., Mouniane, Y., Bouhassoun, S., El Aammouri, S., Yamani, B., Benlemlih, N., Abid, N., Elwahab, F. A., Zoubir, M. H., El-Khadir, I., Chriqui, A., Aitouhanni, I., El Bakkali, M., El Ouardi, B., & Ibriz, M. (2025). Abiotic stress and climate change: Technological innovations and adaptation strategies for sustainable resilience. In Ahmadi, A., & Mna, A. (Eds.), *Nexus of environmental quality and technology innovation* (pp. 1–16). IGI Global., DOI: 10.4018/979-8-3373-1822-6.ch001

Pantic, I., Paunovic, J., Pejic, S., Drakulic, D., Todorovic, A., Stankovic, S., Vucevic, D., Cumic, J., & Radosavljevic, T. (2022). Artificial intelligence approaches to the biochemistry of oxidative stress: Current state of the art. *Chemico-Biological Interactions*, *358*, 109888. DOI: 10.1016/j.cbi.2022.109888

Parra-López, C., Ben Abdallah, S., Garcia-Garcia, G., Hassoun, A., Sánchez-Zamora, P., Trollman, H., Jagtap, S., & Carmona-Torres, C. (2024). Integrating digital technologies in agriculture for climate change adaptation and mitigation: State of the art and future perspectives. *Computers and Electronics in Agriculture*, *226*, 109412. DOI: 10.1016/j.compag.2024.109412

Polireddi, N. S. A. (2024). An effective role of artificial intelligence and machine learning in banking sector. *Measurement. Sensors*, *33*, 101135. DOI: 10.1016/j.measen.2024.101135

Rico-Chávez, A. K., Franco, J. A., Fernandez-Jaramillo, A. A., Contreras-Medina, L. M., Guevara-González, R. G., & Hernandez-Escobedo, Q. (2022). Machine learning for plant stress modeling: A perspective towards hormesis management. *Plants*, *11*(7), 970. DOI: 10.3390/plants11070970

Rivera, P., Moya, C., & O'Brien, J. A. (2022). Low salt treatment results in plant growth enhancement in tomato seedlings. *Plants*, *11*(6), 807. DOI: 10.3390/plants11060807

Salcedo, F., Pérez Cutillas, P., Alarcón Cabañero, J. J., & Vivaldi, A. G. (2022). Use of remote sensing to evaluate the effects of environmental factors on soil salinity in a semi-arid area. *The Science of the Total Environment*, *815*, 152524. DOI: 10.1016/j.scitotenv.2021.152524

Santiago-Rosario, L. Y., Harms, K. E., Elderd, B. D., Hart, P. B., & Dassanayake, M. (2021). No escape: The influence of substrate sodium on plant growth and tissue sodium responses. *Ecology and Evolution*, *11*(20), 14231–14249. DOI: 10.1002/ece3.8138

Shahid, M. A., Sarkhosh, A., Khan, N., Balal, R. M., Ali, S., Rossi, L., Gómez, C., Mattson, N., Nasim, W., & Garcia-Sanchez, F. (2020). Insights into the physiological and biochemical impacts of salt stress on plant growth and development. *Agronomy (Basel)*, *10*(7), 938. DOI: 10.3390/agronomy10070938

Shen, T., Xu, F., Chen, D., Yan, R., Wang, Q., Li, K., Zhang, G., Ni, L., & Jiang, M. (2024). A B-box transcription factor OsBBX17 regulates saline-alkaline tolerance through the MAPK cascade pathway in rice. *The New Phytologist*, *241*(5), 2158–2175. DOI: 10.1111/nph.19480

Singh, A., Rajput, V. D., Sharma, R., Ghazaryan, K., & Minkina, T. (2023). Salinity stress and nanoparticles: Insights into antioxidative enzymatic resistance, signaling, and defense mechanisms. *Environmental Research*, *235*, 116585. DOI: 10.1016/j.envres.2023.116585

Singh, D., Chaudhary, P., Taunk, J., Singh, C. K., Singh, D., Tomar, R. S. S., Aski, M., Konjengbam, N. S., Raje, R. S., Singh, S., Sengar, R. S., Yadav, R. K., & Pal, M. (2021). Fab advances in Fabaceae for abiotic stress resilience: From 'omics' to artificial intelligence. *International Journal of Molecular Sciences*, *22*(19), 10535. DOI: 10.3390/ijms221910535

Singh, K. M., Baksi, S., Rani, S., Jha, A. B., Dubey, R. S., & Sharma, P. (2024). NanoBoost: Maximizing crop resilience and yield via nanopriming under salt stress. *Environmental and Experimental Botany*, *226*, 105937. DOI: 10.1016/j.envexpbot.2024.105937

Soltabayeva, A., Ongaltay, A., Omondi, J. O., & Srivastava, S. (2021). Morphological, Physiological and Molecular Markers for Salt-Stressed Plants. *Plants*, *10*(2), 243. DOI: 10.3390/plants10020243

Sun, Y., Ma, L., Fu, Y., Dai, H., & Zhang, Y. (2021). The improvement of gel and physicochemical properties of porcine myosin under low salt concentrations by pulsed ultrasound treatment and its mechanism. *Food Research International*, *141*, 110056. DOI: 10.1016/j.foodres.2020.110056

Tomaškinová, J., Tomaškin, J., Drimal, M., & Bellido, J. (2025). The impact of abiotic environmental stressors on fluorescence and chlorophyll content in Glycine max (L.) Merrill. *Agronomy (Basel)*, *15*, 263. DOI: 10.3390/agronomy15020263

Toumi, I., Ghrab, M., Zarrouk, O., & Nagaz, K. (2024). Impact of deficit irrigation strategies using saline water on soil and peach tree yield in an arid region of Tunisia. *Agriculture*, *14*(3), 377. DOI: 10.3390/agriculture14030377

Vishnoi, S., & Goel, R. K. (2024). Climate smart agriculture for sustainable productivity and healthy landscapes. *Environmental Science & Policy*, *151*, 103600. DOI: 10.1016/j.envsci.2023.103600

Wahab, A., Abdi, G., Saleem, M. H., Ali, B., Ullah, S., Shah, W., Mumtaz, S., Yasin, G., Muresan, C. C., & Marc, R. A. (2022). Plants' physio-biochemical and phyto-hormonal responses to alleviate the adverse effects of drought stress: A comprehensive review. *Plants*, *11*(13), 1620. DOI: 10.3390/plants11131620

Zhang, L., Bate, B., Cui, J., Feng, Y., Yu, J., Cui, Z., Wang, H., & Li, Q. (2025). Biochar input to saline-alkali farmland can improve soil health and crop yield: A meta-analysis. *Agriculture*, *15*(5), 561. DOI: 10.3390/agriculture15050561

Chapter 12
A Multi-Criteria Predictive Framework for Sustainable Tomato Irrigation:
Baseline Modeling Toward Treated Wastewater Integration

Noura Benlemlih
https://orcid.org/0000-0003-3588-2202
Laboratory of Botany, Biotechnology and Plant Production, Faculty of Sciences, Ibn Tofail University, Kenitra, Morocco

Safae El Aammouri
Laboratory of Botany, Biotechnology and Plant Production, Faculty of Sciences, Ibn Tofail University, Kenitra, Morocco

Youness Nizar
Laboratory of Electronic Systems, Information Processing, Mechanics and Energy, Faculty of Sciences, IbnTofail University, Kenitra, Morocco

Abdellah Lemfarrak
Laboratory of Natural Resources and Sustainable Development, Faculty of Science Kenitra, University Ibn Tofail, Kenitra, Morocco

Mohammed Ibriz
Laboratory of Botany, Biotechnology and Plant Production, Faculty of Sciences, Ibn Tofail University, Kenitra, Morocco

DOI: 10.4018/979-8-3373-6746-0.ch012

ABSTRACT

This chapter presents a multi-criteria predictive framework designed to support sustainable tomato irrigation under increasing climate constraints. The work introduces baseline modeling that uses Random Forest, XGBoost, and MLP Regressor algorithms applied to three clean-irrigation datasets covering environmental, agro-physiological, and field conditions. Model performance is evaluated through R^2, MAE, and RMSE, while the TOPSIS method ranks model–dataset combinations across multiple criteria. Although the datasets do not yet include treated wastewater variables, the resulting reference models establish a controlled benchmark for future integration of chemical and biological contaminant data. This chapter therefore provides the foundational architecture needed to extend predictive irrigation systems toward safe and sustainable treated wastewater reuse.

INTRODUCTION

The agriculture industry is undergoing a major transformation due to changes in technology (Wang et al., 2025a). The agricultural industry has changed from the use of traditional crop and irrigation management techniques to data-driven, predictive decision-making systems as a result of the expanded use of data collection and storage, data analysis, and artificial intelligence (AI) in the agriculture sector. Within this transformation, there are many opportunities for innovation related to the management of irrigation systems, especially given the growing concern about water scarcity on a global scale (Ahmadi et al., 2025). The agriculture industry is the largest single user of freshwater resources, representing nearly 70% of total annual withdrawal (Obijianya et al., 2025). In terms of crop type, the tomato (*Solanum lycopersicum*) is an excellent example of a production system that represents high-volume, water-intensive horticulture. Tomatoes are grown extensively for the fresh market and also manufactured for industrial use (Ventura et al., 2026).

Tomato irrigation is an incredibly complex area of agriculture with many diversities (Wang et al., 2025b). First of all, tomato farmers want to produce as much quality tomato fruit as possible, this is determined by both water availability at the right moment, soil conditions, and other factors (Zalai et al., 2025). Secondly, increasing government regulations around water usage and the effects of climate change increase the demand for efficient water usage (Zeng et al., 2025). Traditional farming has always used empirical knowledge and a static calendar to determine irrigation timing; this type of planning has little to no ability to be adaptive to both real-time environmental conditions and to physiological impacts on plants (Rai & Dong, 2025). Therefore, the integration of artificial intelligence (AI) methodologies, including machine learning (ML) and deep learning (DL), into irrigation scheduling and yield estimation is something that is gaining momentum (Hernández Hernández et al., 2025).

Machine learning models, such as Random Forest, XGBoost, and neural networks (e.g., MLP), can detect complex patterns in heterogeneous agricultural data — from weather sensors and soil monitors to physiological traits and historical yield records (Choi et al., 2025; Mouniane et al., 2025). These models can learn from large, noisy, and multivariate datasets, offering high predictive accuracy and flexibility (Wang et al., 2022; El-Khadir et al., 2026). However, their performance is often context-dependent: model outcomes vary with dataset quality, feature selection, environmental variability, and crop cultivar (Singh et al., 2025). Moreover, ML outputs alone may not suffice for operational decisions unless accompanied by transparent and interpretable rankings, especially when multiple conflicting criteria — such as yield, resource use, and cost — are involved (Amin et al., 2025).

While the datasets analyzed in this chapter do not involve the use of treated wastewater, their role is foundational. They represent a necessary first step in building and validating predictive tools that will eventually include variables specific to wastewater reuse — such as heavy metal concentrations (e.g., Pb, Cd), pharmaceutical residues, and microbiological risk indicators. The consistency and diversity of these baseline datasets allow for robust benchmarking. Future research, including subsequent chapters in this thesis, will build upon this structure to model contamination risks and optimize irrigation under TWW conditions.

This chapter addresses that gap by combining predictive modeling with a multi-criteria decision-making (MCDM) approach. Specifically, we apply the TOPSIS method, which evaluates alternatives by their geometric distance to an ideal (and anti-ideal) solution, enabling a fair ranking of complex options. Unlike simple model comparisons based on accuracy alone, TOPSIS incorporates multiple objectives, assigns customizable weights to criteria, and outputs a normalized score that supports trade-off decisions in real-world settings.

To ensure broad relevance, we conduct our study using three distinct datasets:

- An IoT-based dataset (Belli et al., 2024) capturing real-time environmental and soil conditions during tomato cultivation under different irrigation regimes.
- An agro-physiological dataset (Carucci et al., 2024) collected from multi-year field trials assessing water-saving irrigation strategies.
- A university-managed experimental dataset (Virginia Tech) (Torres-Quezada & Gandini-Taveras, 2023) documenting plant traits, spacing, and yield outcomes across controlled treatments.

Each dataset represents a different layer of complexity and realism — from sensor-based automation to physiological detail and agronomic control. We train and evaluate three ML/DL models on each dataset, extract performance metrics (R^2, MAE, RMSE), and input these into the TOPSIS decision matrix. The resulting rankings highlight not only the best-performing models but also the most practical and adaptable model–dataset combinations for real-world application.

The contribution of this chapter is twofold. First, it provides a replicable framework for evaluating agricultural models across datasets using both predictive metrics and decision analysis. Second, it shows how AI can be positioned not merely as a forecasting tool but as a cornerstone of strategic planning in sustainable agriculture. With growing pressure to balance food production with environmental responsibility, such integrative approaches are essential to the future of precision irrigation and smart farming.

RELATED WORK AND BACKGROUND

The integration of machine learning into agricultural systems, particularly for irrigation management and crop prediction, has received increasing attention over the past decade. Tomatoes, due to their global relevance, sensitivity to water regimes, and market value, have been a frequent focus of precision agriculture research. Several studies have demonstrated the effectiveness of supervised learning models such as decision trees, support vector regression, and neural networks for estimating tomato yield, optimizing irrigation frequency, or classifying plant health status based on soil and climate parameters (Wang & Xing, 2017; Wang et al., 2025).

In parallel, recent advances in Internet of Things (IoT) systems have made it possible to collect high-frequency data from soil, weather, and crop sensors in near-real time (Castro García et al., 2025). IoT-driven platforms allow for continuous monitoring of environmental variables, including soil moisture, temperature, humidity, and evapotranspiration rates (Mouniane et al., 2024; Shahab et al., 2025). This has enabled more granular training data for machine learning models and supports the emergence of dynamic, feedback-based irrigation systems (Kamilaris & Prenafeta-Boldú, 2018). However, many of these models still assume clean or controlled irrigation water sources, overlooking the increasing relevance of treated wastewater (TWW) in regions facing water scarcity (El Aammouri et al., 2023; Costa et al., 2025).

Treated wastewater reuse in agriculture is a growing global practice, with substantial research focused on its effects on crop physiology, soil properties, and food safety (Benlemlih et al., 2024; El Aammouri et al., 2024). Numerous studies have documented the accumulation of heavy metals (e.g., cadmium, lead), pharmaceutical compounds, and microbial pathogens in soils and plant tissues irrigated with TWW (Qadir et al., 2020; Benlemlih et al., 2023). These concerns are particularly pronounced in fruiting vegetables like tomatoes, where both foliar and root uptake can pose food safety risks (Macieira et al., 2021; Milačič Ščančar et al., 2025). Despite this, very few machine learning models explicitly incorporate TWW-specific variables into their predictive architecture, leaving a gap between agronomic prediction and water quality risk management (Saravi et al., 2021; Lucas et al., 2025).

Additionally, multi-criteria decision-making (MCDM) methods such as TOPSIS and AHP have been increasingly adopted to support agricultural planning, including variety selection, irrigation method comparison, and climate risk assessment (Mardani et al., 2015; Rouyendegh, & Savalan, 2022). These tools are particularly powerful when multiple conflicting objectives (e.g., yield, cost, risk, and environmental impact) must be balanced (Kennedy et al., 2008; Redman, 2023). In the context of this study, the integration of MCDM with machine learning enables not just prediction, but prioritization — providing a scalable framework for selecting irrigation strategies based on both performance and sustainability considerations (Heiba et al., 2023; Rabie et al., 2025).

This chapter builds upon these strands of literature by creating a unified framework that combines supervised machine learning, structured environmental data, and multi-criteria decision-making to evaluate irrigation strategies under clean water conditions, with the express intention of adapting the system for treated wastewater contexts in future research.

METHODOLOGY

This chapter proposes a structured and comparative approach to evaluating irrigation strategies in tomato cultivation using predictive modeling and multi-criteria decision-making. The methodology integrates diverse data sources, applies a consistent machine learning framework across them, and implements a formal ranking system through the TOPSIS method. This section describes the data sources, preprocessing steps, model configurations, evaluation metrics, and the mathematical framework adopted for model ranking.

A visual summary of the proposed modeling strategy is shown in Figure 1. It illustrates the current chapter's role in building baseline predictive models using conventional irrigation datasets, which are later intended to support the integration of treated wastewater-specific variables such as heavy metals, pharmaceutical residues, and microbiological contaminants. This dual-stage framework allows the results presented in this chapter to serve both as a scientific benchmark and as a technical foundation for risk-aware, wastewater-driven irrigation management in future work.

Figure 1. Conceptual pipeline for integrating treated wastewater (TWW) into predictive irrigation modeling.

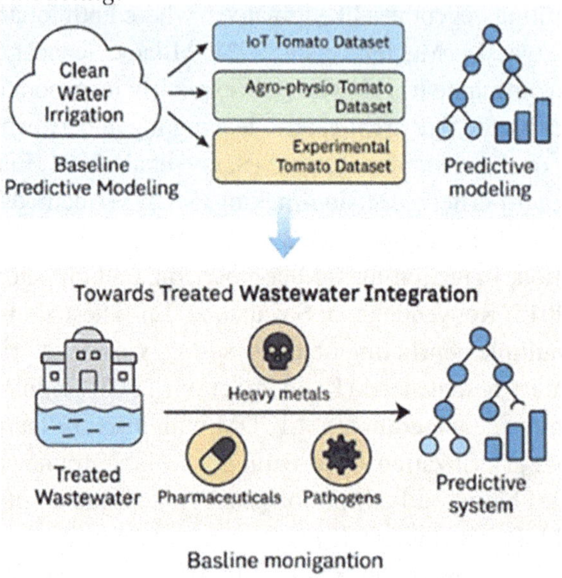

Dataset Description

To ensure generalizability and demonstrate the applicability of predictive modeling across heterogeneous data sources, three datasets were selected for this study. Each dataset represents a different agronomic context, capturing either sensor-driven environmental data, multi-year physiological field data, or structured agronomic trial results. This diversity allowed us to apply a unified modeling and decision-making framework to varied forms of agricultural information.

The first dataset, obtained from an open-access repository, consists of IoT-based time-series data (Belli et al., 2024) collected from a tomato cultivation site. This dataset includes soil and environmental sensor readings along with water meter data, each aligned using timestamp information. The second dataset is an agro-physiological (Carucci et al., 2024) compilation of field trials conducted over multiple years, focusing on crop responses to various deficit irrigation strategies. It includes yield-related variables as well as physiological markers of fruit quality. The third dataset (Torres-Quezada & Gandini-Taveras, 2023) is an experimental study from a university research farm, designed to evaluate the impact of spacing and treatment interventions on tomato productivity. An overview of the datasets is provided in Table 1, summarizing their structure, content, and key features.

Table 1. Overview of datasets used in the study

Dataset Name	Source / Origin	Type	Sample Size	Key Features	Target Variable
IoT Dataset (Belli et al., 2024)	Mendeley Repository	Sensor-based (Time Series)	~5000 rows	CO_2, humidity, soil EC, temperature, irrigation	current_volume (L)
AgroPhysio Dataset (Carucci et al., 2024)	(Carucci et al., 2024)	Multi-year Field Trial	~150 rows	Yield, Brix, pH, physiological indicators	FW (Fresh Weight, g)
Virginia Tech (Torres-Quezada & Gandini-Taveras, 2023)	Academic Experimental	Plot-based Agronomic Study	~80 rows	Marketable fruit weight, spacing, treatment	Marketable fruit weight

Data Preprocessing

Preprocessing was tailored to the structure and format of each dataset to ensure compatibility with regression-based machine learning models. For the IoT dataset, three CSV files—representing soil data, environmental readings, and irrigation records—were individually converted from raw sensor outputs and merged based on timestamps using an asof-join operation. To allow for alignment of asynchronous data streams, a five-minute tolerance window was applied during merging. Time values were converted from Unix timestamps to standard datetime formats. Only rows with complete information across all relevant features were retained.

In the agro-physiological dataset, data was extracted from the "Crop" sheet of a structured Excel workbook. Non-informative identifiers were dropped, and records containing missing physiological measurements were excluded. The resulting feature matrix included numeric variables such as acidity, Brix index, pH, and other indicators commonly used in post-harvest assessment. The target variable selected was the fresh weight of the tomato yield.

The Virginia Tech dataset was obtained from a separate Excel file comprising structured experimental results. The sheet titled "Yield per plot" was selected, and categorical variables such as treatment type and spacing were encoded using one-hot encoding to make them suitable for regression modeling. All columns were inspected for missing data, and rows with incomplete entries were removed to ensure model stability. The target variable was defined as marketable fruit weight, consistent with economic productivity in agricultural trials.

For all three datasets, numerical features were standardized where appropriate, and the final data matrices were split into training and test sets using an 80:20 ratio. The random seed was fixed at 42 for consistency across experiments.

Predictive Modeling

The machine learning workflow implemented for this study was designed to remain consistent across all datasets, enabling fair comparisons of model performance. Three different regression models were used: Random Forest Regressor, XGBoost Regressor, and Multi-Layer Perceptron (MLP) Regressor. These algorithms were selected based on their established utility in agricultural forecasting and their ability to handle nonlinear, multivariate relationships in heterogeneous data environments.

The Random Forest Regressor is an ensemble-based algorithm that aggregates multiple decision trees to improve generalization and reduce overfitting. It is particularly robust to outliers and irrelevant features. XGBoost is a scalable, tree-based boosting method known for its speed and performance in structured data applications. It incorporates regularization and handles missing values internally, making it well-suited for agricultural datasets with inconsistent sampling. The MLP Regressor is a type of feedforward neural network with one or more hidden layers. It uses backpropagation to minimize error and can model complex, nonlinear functions, especially when relationships between variables are not explicitly known.

All models were trained using default hyperparameters with minimal tuning to preserve comparability. For Random Forest and XGBoost, 100 trees were used with a fixed random seed. The MLP Regressor was configured with two hidden layers containing 64 and 32 neurons, respectively, and trained using a maximum of 1000 iterations. Feature matrices were standardized before being input to the MLP to enhance convergence during training. The models were implemented using the scikit-learn and xgboost Python libraries, and training was conducted independently for each dataset.

Evaluation Metrics and Decision-Making Framework

Model performance was evaluated using three core metrics commonly used in regression analysis: coefficient of determination (R^2), mean absolute error (MAE), and root mean squared error (RMSE). These metrics were computed using test data to reflect out-of-sample generalization. R^2 quantifies the proportion of variance in the target variable that is predictable from the input features. MAE provides an average of the absolute differences between predicted and actual values, offering a measure of bias. RMSE, which penalizes larger errors more heavily, was used to capture overall prediction accuracy with a focus on error magnitude.

After calculating these metrics for all model–dataset combinations, the results were compiled into a decision matrix. This matrix was used as the input to the TOPSIS algorithm, which allows for the ranking of alternatives based on multiple criteria. Prior to applying TOPSIS, the decision matrix was normalized, and weights were assigned to each metric: 0.4 for R^2, and 0.3 each for MAE and RMSE. The method involves calculating the geometric distance of each alternative from both the ideal and anti-ideal solutions, enabling a final ranking based on relative proximity.

It is important to note that the implementation of the TOPSIS method and the presentation of performance metrics are discussed in the Results and Discussion section. This methodology section is restricted to detailing the procedures used to process, model, and evaluate the data without interpreting or presenting the outcomes.

RESULTS AND DISCUSSION

This section presents the outcomes of the predictive modeling phase and analyzes the comparative performance of machine learning and deep learning algorithms across the three tomato cultivation datasets. The evaluation is structured in two stages. First, quantitative results from model training and testing are presented using standard regression metrics: R^2, MAE, and RMSE. Then, the Technique for Order of Preference by Similarity to Ideal Solution (TOPSIS) is applied to rank the combined model–dataset configurations based on multi-criteria decision-making principles. The discussion emphasizes both statistical performance and practical implications for irrigation strategy selection and sustainable agriculture.

Model Evaluation Results

The performance of the Random Forest Regressor, XGBoost Regressor, and Multi-Layer Perceptron (MLP) was evaluated independently on each of the three datasets: IoT-based sensor data, agro-physiological field trials, and the Virginia Tech experimental dataset. Metrics were computed using test data from an 80/20 train-test split, with random seeds fixed to ensure reproducibility. The results are summarized in Table 2, which reports the coefficient of determination (R^2), mean absolute error (MAE), and root mean squared error (RMSE) for each configuration.

Table 2. Model performance metrics across all datasets

Dataset	Model	R^2	MAE	RMSE
IoT-Based	XGBoost	0.730	4094.5	5679.5
	Random Forest	0.227	5.39	6.23
AgroPhysio	XGBoost	-0.268	6.40	7.98
	MLP	-0.015	5.42	7.14

continued on following page

Table 2. Continued

Dataset	Model	R^2	MAE	RMSE
Virginia Tech	Random Forest	0.572	3.83	4.51
	XGBoost	0.685	3.33	3.87
	MLP	0.670	2.23	3.96

From a purely statistical perspective, the Virginia Tech dataset consistently yielded strong predictive outcomes across all models. Notably, the MLP achieved the lowest MAE and a high R^2 value, suggesting its suitability for datasets with structured, low-noise experimental designs. The XGBoost model also performed well, especially on the IoT dataset, where its robustness to missing data and noise was advantageous. In contrast, the AgroPhysio dataset resulted in the lowest R^2 scores across all models. This may reflect the dataset's inherent complexity, stemming from its multi-year, multi-environment structure, and the physiological variability of tomato plants under deficit irrigation regimes.

These results highlight the critical role of data structure and quality in determining model performance. While model architecture is important, the reliability and resolution of input features—along with consistency in experimental conditions—are equally decisive.

Multi-Criteria Decision-Making via TOPSIS

Although conventional metrics such as R^2 and RMSE provide valuable insights, they often capture only a narrow slice of model utility. In real-world agricultural applications, decision-making must account for multiple, sometimes competing, criteria. To this end, the TOPSIS method was used to rank all seven model–dataset configurations based on a composite of three evaluation metrics. Before applying TOPSIS, the performance matrix was normalized using vector normalization. Each criterion was then assigned a weight based on its perceived importance: 0.4 for R^2 (emphasizing predictive power), and 0.3 each for MAE and RMSE (capturing error minimization).

The result of the TOPSIS analysis is shown in Figure 2, where model–dataset combinations are sorted by their final ranking score. Higher scores indicate greater proximity to the ideal solution across all metrics.

Fig. 2. TOPSIS bar chart ranking model–dataset configurations

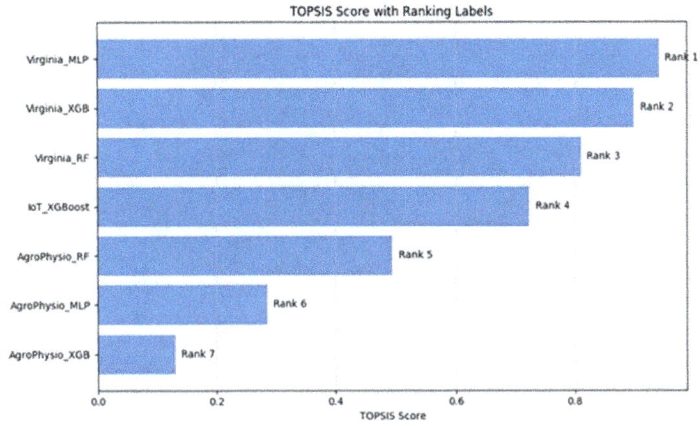

As seen in Figure 2, the Virginia Tech dataset combined with MLP achieved the highest TOPSIS score, followed closely by XGBoost on the same dataset. This affirms the strength of the Virginia data, which benefited from consistent experimental conditions and low measurement noise. The IoT dataset paired with XGBoost ranked fourth, reinforcing its suitability for real-time yield prediction applications where structured experimentation is not feasible. The AgroPhysio models ranked lower in all cases, possibly due to dataset heterogeneity and variability introduced by multi-season field trials.

To complement the bar chart and provide a comparative visual, a radar plot was constructed showing all model–dataset configurations based on their normalized TOPSIS scores.

Fig. 3. Radar plot of TOPSIS scores for all model–dataset combinations

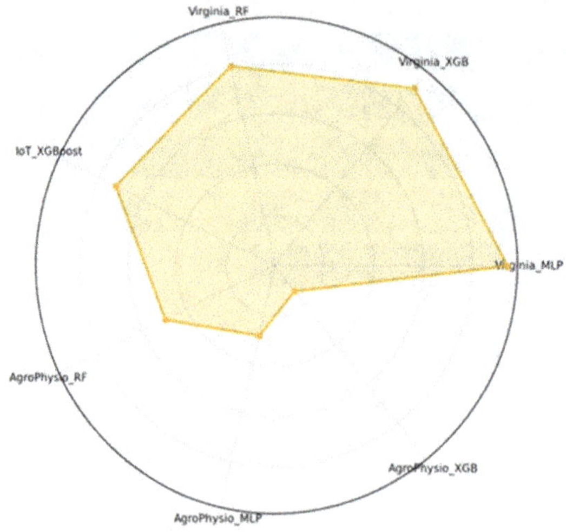

The radar plot visually in Figure 3 reinforces the ranking results, illustrating the separation between high-performing configurations and those that lagged. It also serves to underscore the consistency of the Virginia dataset across different models and the particular strength of MLP in structured environments. The IoT-based XGBoost model, although ranked lower than Virginia-based configurations, still maintained a favorable balance of metrics and demonstrates its viability in real-world, sensor-driven agricultural systems.

Discussion and Practical Implications

The comparative results of this study offer rich insights into the dynamics of predictive modeling within agricultural irrigation contexts. Beyond the numerical performance of each model, the results reflect a broader set of considerations — including data structure, model suitability, and practical implementation constraints — that shape the feasibility of deploying machine learning systems in real-world agricultural settings.

One of the most critical observations emerging from this research is the central role that dataset quality and structure play in determining model performance. For instance, the Virginia Tech dataset, despite being the smallest in sample size, consistently supported high-performing models. This can be attributed to the dataset's uniform design, controlled experimental variables, and clean categorical labels. It illustrates a fundamental truth in data science: structured, high-integrity data can often outperform larger, noisier datasets when used for predictive modeling.

In contrast, the AgroPhysio dataset, while rich in physiological information and collected over several growing seasons, posed challenges due to its inherent heterogeneity. Variability in irrigation practices, plant responses, and field environments introduced non-linear noise that affected model generalization. This highlights the need for specialized preprocessing or hybrid modeling strategies when working with long-term field data.

The IoT dataset presented a different but equally important case. It combined environmental and soil measurements with timestamped irrigation volumes, offering the potential for real-time prediction and adaptive control. Although it lacked the experimental precision of the Virginia dataset, its temporal granularity enabled models like XGBoost to uncover dynamic relationships between climate and irrigation behavior. In this sense, the IoT dataset illustrates a compelling trade-off between volume and precision. While sensor-driven data can be noisy and affected by hardware errors or synchronization lags, it opens the door to scalable and continuously adaptive systems — a key requirement for precision agriculture and smart farming.

The decision to integrate a multi-criteria ranking method, namely TOPSIS, into this workflow is another meaningful contribution. Traditional model evaluation often stops at identifying the highest R^2 or the lowest RMSE. However, real-world deployment requires balancing multiple objectives: accuracy, reliability, interpretability, and in some cases, computational efficiency. TOPSIS enables decision-makers to assign weights to each of these criteria based on specific use-case priorities. For example, in large-scale industrial farming, minimizing prediction error (RMSE) might be prioritized over slight gains in R^2, whereas in research-oriented settings, maximizing explanatory power may be more valuable. The inclusion of TOPSIS transforms model selection from a purely statistical decision into a strategic one, tailored to stakeholder needs.

Moreover, the study's findings have implications beyond the immediate task of yield prediction. They contribute to a broader vision of sustainable water management in agriculture. As freshwater becomes an increasingly limited resource, the ability to forecast irrigation needs accurately, and adjust supply dynamically, could play a vital role in achieving the United Nations Sustainable Development Goals (SDGs), particularly SDG 2 (Zero Hunger) and SDG 6 (Clean Water and Sanitation). Integrating ML-based decision support systems into irrigation infrastructure not only enhances efficiency but also provides a pathway to quantifiable resource savings and environmental impact reduction.

In terms of application, the different model–dataset combinations can be mapped to specific agricultural contexts. The Virginia-based MLP model is ideally suited for academic or government-run experimental stations where detailed data collection is feasible, and model interpretability is critical. The IoT-based XGBoost model, on the other hand, can be integrated into on-field digital platforms using edge devices or cloud services, allowing real-time predictions and feedback loops. Meanwhile, the AgroPhysio dataset — although lower-performing in this study — provides a foundation for advanced physiological modeling and may benefit from integration with weather forecasts, satellite imagery, or even crop growth simulations in future work.

Lastly, this study underscores the value of modularity in agricultural AI systems. Rather than searching for a single best model or dataset, a more effective strategy may be to develop ensemble frameworks or meta-decision systems that combine multiple models trained on specialized datasets. This would allow for adaptive, context-aware recommendations — such as switching between models during different growth stages or under varying climatic conditions.

Limitations and Future Work

While the predictive modeling and decision-making framework presented in this chapter offers valuable insights into irrigation strategy evaluation, several limitations must be acknowledged. These constraints are not only technical but also contextual, reflecting the current state of available datasets, modeling scope, and the absence of treated wastewater (TWW)-specific data within the experimental design.

One of the primary limitations lies in the nature of the datasets used. All three datasets—IoT-based, agro-physiological, and university experimental—relied on conventional or clean water sources for irrigation. As such, the models trained on these datasets do not account for the physicochemical or biological variability introduced using TWW. Contaminants such as heavy metals, persistent organic pollutants, and microbial pathogens, which are of central concern in wastewater reuse, are not represented in the feature sets. Consequently, while the models can predict yield and irrigation response under clean conditions, their direct applicability to TWW scenarios remains limited without further adaptation or retraining using contaminated or simulated datasets.

Additionally, although the chapter incorporated three well-established machines learning models—Random Forest, XGBoost, and MLP—the exploration of hyperparameter optimization, model ensembling, or temporal forecasting (e.g., using LSTM or Prophet) was beyond the current scope. These advanced techniques could enhance predictive accuracy, especially in time-dependent irrigation systems. Furthermore, external variables such as economic cost, farmer behavior, or regulatory constraints were not considered, though they are often critical in operational decision-making.

From a decision-making perspective, the implementation of TOPSIS provided an effective way to rank model–dataset combinations using multi-objective criteria. However, the decision matrix was constructed solely on technical performance metrics (R^2, MAE, RMSE), and did not include sustainability indicators, water quality parameters, or risk thresholds, which are essential in wastewater irrigation contexts. Integrating environmental and health-based criteria into the MCDM process represents a key area for methodological advancement.

Looking forward, future work will focus on expanding the dataset to include experimental trials involving treated wastewater sources, whether through field measurements, synthetic feature generation, or collaboration with institutions managing TWW irrigation schemes. Variables such as heavy metal concentrations (e.g., Cd, Pb), pharmaceutical residues, and microbial indicators (e.g., E. coli, coliforms) will be incorporated into both the predictive and ranking frameworks. Moreover, explainable AI tools such as SHAP or LIME will be applied to improve model transparency and identify contamination-sensitive predictors.

In summary, while this chapter presents a rigorous and adaptable modeling framework, its extension to wastewater-based irrigation management will require new datasets, interdisciplinary collaboration, and refined modeling approaches. These enhancements are not only logical continuations but necessary steps in building a robust, risk-aware, and sustainable decision-support system for future agricultural water reuse.

CONCLUSION

The application of machine learning and decision-making frameworks to agricultural data has demonstrated immense potential in optimizing irrigation strategies, particularly for crops such as tomatoes that are sensitive to water availability and management. This chapter has presented a comprehensive and comparative approach to predictive modeling of irrigation outcomes across three distinct datasets using consistent modeling algorithms and a robust ranking framework. By integrating predictive performance metrics with a structured multi-criteria decision-making method (TOPSIS), we provide a transparent and data-driven basis for selecting irrigation strategies aligned with both yield optimization and sustainability.

The methodology employed in this study emphasizes not only the role of machine learning models but also the critical influence of data structure and origin. Among the models tested, the Multi-Layer Perceptron and XGBoost consistently demonstrated strong predictive power when trained on the Virginia Tech experimental dataset. This outcome reinforces the notion that well-controlled, structured datasets yield better performance even with moderately complex models. The IoT-based dataset also produced competitive results, particularly with XGBoost, highlighting its potential for real-time deployment in precision agriculture. In contrast, the agro-physiological dataset revealed the challenges of modeling in complex, multi-season environments, emphasizing the need for more extensive feature engineering or hybrid modeling approaches.

The use of TOPSIS added a valuable layer of interpretability and objectivity to the evaluation process. Rather than relying solely on individual metrics like R^2 or RMSE, the method allowed for simultaneous consideration of multiple performance criteria, weighted according to their importance. This offers a decision-support tool for agronomists, researchers, and agricultural stakeholders who must often balance trade-offs between predictive accuracy, reliability, and operational efficiency. By ranking model–dataset combinations, TOPSIS enables a more strategic selection of predictive systems based not just on numerical performance, but on overall fitness for a given agricultural context.

Beyond the immediate results, this work proposes a replicable and scalable methodology for integrating artificial intelligence into agronomic decision-making. The combination of real-world data, interpretable ML models, and MCDM techniques like TOPSIS provides a solid foundation for smart irrigation systems that can adapt to both field-level and regional requirements. This is particularly relevant in light of increasing water scarcity, climate variability, and the growing demand for sustainable food systems.

Future research could expand this framework by incorporating additional data sources such as satellite imagery, integrating temporal forecasting models, or applying advanced interpretability tools like SHAP values to understand variable importance more deeply. Moreover, exploring the inclusion of economic and environmental costs into the decision-making matrix would further enhance the practical utility of this framework. Ultimately, the methodology presented here contributes to the ongoing transformation of agriculture into a data-driven, sustainability-focused domain.

ACKNOWLEDGMENTS

This research was financed by the Moroccan Ministry of Higher Education, Scientific Research and Innovation through the PRIMA Program (Partnership for Research and Innovation in the Mediterranean Area), under the project SAFE (Sustainable Water Reuse Practices Improving Safety in Agriculture, Food, and Environment), Project ID: 1826.

REFERENCES

Ahmadi, A., Keshavarz, M., & Ejlali, F. (2025). Resilience to climate change in agricultural water-scarce areas: The major obstacles and adaptive strategies. *Water Resources Management*, *39*, 1195–1214. DOI: 10.1007/s11269-024-04019-z

Amin, A., Zaman, W., & Park, S. (2025). Harnessing multi-omics and predictive modeling for climate-resilient crop breeding: From genomes to fields. *Genes*, *16*(7), 809. DOI: 10.3390/genes16070809 PMID: 40725465

Belli, L., Davoli, L., Oddi, G., Preite, L., Galaverni, M., Ganino, T., & Ferrari, G. (2024). IoT-based dataset of a tomato cultivation under different irrigation regimes. *Mendeley Data*, 2. https://doi.org/DOI: 10.17632/35WH56287Y.2

Benlemlih, N., Auajjar, N., & Aammouri, EL, S., Nizar, Y., & IBRIZ, M. (2023). Physicochemical characterization of the sludge generated by the wastewater treatment plant of Kenitra. *Research Journal of Pharmacy and Technology*, *16*(11), 5366–5371. DOI: 10.52711/0974-360X.2023.00869

Benlemlih, N., Khiyati, M. E., El Aammouri, S., & Ibriz, M. (2024). A review on wastewater and its use in agriculture in Morocco: Situation, case study and recommendations. *Research Journal of Pharmacy and Technology*, *17*(10), 5132–5140. DOI: 10.52711/0974-360X.2024.00787

Carucci, F., Bregaglio, S. U. M., Caldarola, D. P., Gatta, G., & Giuliani, M. M. (2024). An agro-physiological dataset on industrial tomatoes from nine years of field experiments conducted with alternative water-saving strategies in Mediterranean environments. *Mendeley Data*, 2. https://doi.org/DOI: 10.17632/H293P9HYXT.2

Castro García, L., Wang, J., Wang, S., Qiao, X., Kuhl, S., Zhang, H., Rangel de la Tejera, R., & Reed, D. (2025). I-Canopy: A resilient IoT platform for adaptive environmental monitoring at the edge. *Smart Agricultural Technology*, *12*, 101611. DOI: 10.1016/j.atech.2025.101611

Choi, J. W., Hidayat, M. S., Cho, S. B., Hwang, W.-H., Lee, H., Cho, B.-K., Kim, M. S., Baek, I., & Kim, G. (2025). Recent trends in machine learning, deep learning, ensemble learning, and explainable artificial intelligence techniques for evaluating crop yields under abnormal climate conditions. *Plants*, *14*(18), 2841. DOI: 10.3390/plants14182841 PMID: 41011993

Costa, C. S., Carlos, C., Oliveira, A. A., & Barros, A. N. (2025). Reuse of treated wastewater to address water scarcity in viticulture: A comprehensive review. *Agronomy (Basel)*, *15*(4), 941. DOI: 10.3390/agronomy15040941

El Aammouri, S., Benlemlih, N., & Ibriz, M. (2023).. . *Research Journal of Pharmacy and Technology*, *16*(12), 5722–5727. DOI: 10.52711/0974-360X.2023.00925

El Aammouri, S., Benlemlih, N., & Ibriz, M. (2024). Sludge recovery from wastewater treatment plants: A review of current practices and tips. *Research Journal of Pharmacy and Technology*, *17*(8), 4117–4124. DOI: 10.52711/0974-360X.2024.00638

El-Khadir, I., Mouniane, Y., Chriqui, A., Oudghiri, M., El Bakkali, M., & Hmouni, D. (2026). Synergistic effects of halophyte intercropping on the salt stress tolerance of Rosmarinus officinalis L.: A sustainable approach for climate-resilient agriculture. In Shah, I., & Jhanjhi, N. (Eds.), *Advances in computational intelligence for climate change security and sustainability* (pp. 187–222). IGI Global Scientific Publishing., DOI: 10.4018/979-8-3693-9132-7.ch008

Heiba, Y., Ibrahim, M. G., Mohamed, A. E., Fujii, M., & Nasr, M. (2023). Developing smart sustainable irrigation matrix (SIM)-based model for selection of best irrigation techniques: A framework to achieve SDGs. *Journal of Cleaner Production*, *420*, 138404. DOI: 10.1016/j.jclepro.2023.138404

Hernández Hernández, G. C., Gómez Gómez, J., & Jiménez-Cabas, J. (2025). Predictive models based on artificial intelligence to estimate crop yield: A literature review. *Agriculture*, *15*(23), 2438. DOI: 10.3390/agriculture15232438

Kamilaris, A., & Prenafeta-Boldú, F. X. (2018). Deep learning in agriculture: A survey. *Computers and Electronics in Agriculture*, *147*, 70–90. DOI: 10.1016/j.compag.2018.02.016

Kennedy, M. C., Ford, E. D., Singleton, P., Finney, M., & Agee, J. K. (2008). Informed multi-objective decision-making in environmental management using Pareto optimality. *Journal of Applied Ecology*, *45*(1), 181–192. DOI: 10.1111/j.1365-2664.2007.01367.x

Lucas, L. V., Brown, C. J., Robertson, D. M., Baker, N. T., Johnson, Z. C., Green, C. T., Cho, S. J., Erickson, M. L., Gellis, A. C., & Jasmann, J. R.. (2025). Gaps in water quality modeling of hydrologic systems. *Water (Basel), 17*(8), 1200. DOI: 10.3390/w17081200

Macieira, A., Barbosa, J., & Teixeira, P. (2021). Food safety in local farming of fruits and vegetables. *International Journal of Environmental Research and Public Health, 18*(18), 9733. DOI: 10.3390/ijerph18189733 PMID: 34574658

Mardani, A., Jusoh, A., Nor, K. M. D., Khalifah, Z., Zakwan, N., & Valipour, A. (2015). Multiple criteria decision-making techniques and their applications – a review of the literature from 2000 to 2014. *Economic Research-Ekonomska Istraživanja, 28*(1), 516–571. DOI: 10.1080/1331677X.2015.1075139

Milačič Ščančar, R., Kozlica, K., Marković, S., Leban, P., Vidmar, J., Heath, E., Kacjan Maršić, N., Železnikar, Š., & Ščančar, J. (2025). The uptake and translocation of lead, chromium, cadmium, and zinc by tomato plants grown in nutrient and contaminated nutrient solutions: Implications for food safety. *Toxics, 13*(9), 738. DOI: 10.3390/toxics13090738 PMID: 41012359

Mouniane, Y., Chriqui, A., El-Khadir, I., Hbyaj, K., El-Moustaqim, K., Mezouara, M., Zirari, M., Mabrouki, J., & Hmouni, D. (2024). Impact of remote sensing on irrigation management in precision agriculture. In Mabrouki, J., & Mourade, A. (Eds.), *Technical and technological solutions towards a sustainable society and circular economy* (pp. 1–X). World Sustainability Series. Springer., DOI: 10.1007/978-3-031-56292-1_19

Mouniane, Y., Vibhute, A. D., Aitouhanni, I., Souissi, A., Kadmi, Y., & Hmouni, D. (2025). Exploring water and salt stress impact on the germination of carob (*Ceratonia siliqua*, Fabaceae) germplasm ecotypes from Morocco, using machine learning for sustainable cultivation. *Journal of Crop Health, 77*, 159. DOI: 10.1007/s10343-025-01224-7

Obijianya, C. C., Yakamercan, E., Karimi, M., Veluru, S., Simko, I., Eshkabilov, S., & Simsek, H. (2025). Agricultural irrigation using treated wastewater: Challenges and opportunities. *Water (Basel), 17*(14), 2083. DOI: 10.3390/w17142083

Qadir, M., Drechsel, P., Jiménez Cisneros, B., Kim, Y., Pramanik, A., Mehta, P., & Olaniyan, O. (2020). Global and regional potential of wastewater as a water, nutrient and energy source. *Natural Resources Forum, 44*(1), 40–51. DOI: 10.1111/1477-8947.12187

Rabie, A. B., Elhag, M., & Subyani, A. (2025). Remote sensing, GIS, and machine learning in water resources management for arid agricultural regions: A review. *Water (Basel), 17*(21), 3125. DOI: 10.3390/w17213125

Rai, A., & Dong, Y. (2025). A scoping review of irrigation scheduling methods in potato (*Solanum tuberosum* L.) production. *American Journal of Potato Research, 102*, 348–371. DOI: 10.1007/s12230-025-09997-3

Redman, B. K. (2023). Rebalancing commercial and public interests in prioritizing biomedical, social and environmental aspects of health through defining and managing conflicts of interest. *Frontiers in Medicine, 10*, 1247258. DOI: 10.3389/fmed.2023.1247258 PMID: 37809337

Rouyendegh, B. D., & Savalan, Ş. (2022). An integrated fuzzy MCDM hybrid methodology to analyze agricultural production. *Sustainability, 14*(8), 4835. DOI: 10.3390/su14084835

Saravi, B., Nejadhashemi, A. P., Jha, P., & Tang, B. (2021). Reducing deep learning network structure through variable reduction methods in crop modeling. *Artificial Intelligence in Agriculture, 5*, 196–207. DOI: 10.1016/j.aiia.2021.09.001

Singh, K., Yadav, M., Barak, D., Bansal, S., & Moreira, F. (2025). Machine-learning-based frameworks for reliable and sustainable crop forecasting. *Sustainability, 17*(10), 4711. DOI: 10.3390/su17104711

Torres-Quezada, E., & Gandini-Taveras, R. J. (2023). Tomato plant data [Dataset]. Virginia Tech Data Repository. https://data.lib.vt.edu/articles/dataset/Tomato_plant_data/24164079

Ventura, G., Coniglio, D., Losito, I., Cataldi, T. R. I., & Calvano, C. D. (2026). Impact of industrial processing on the phospholipid profile of canned tomatoes (Solanum lycopersicum). *Journal of Food Composition and Analysis, 149*, 108692. DOI: 10.1016/j.jfca.2025.108692

Wang, D., Cao, W., Zhang, F., Li, Z., Xu, S., & Wu, X. (2022). A review of deep learning in multiscale agricultural sensing. *Remote Sensing, 14*, 559. DOI: 10.3390/rs14030559

Wang, S., Yang, Y., Yin, H., Zhao, J., Wang, T., Yang, X., Ren, J., & Yin, C. (2025a). Towards digital transformation of agriculture for sustainable development in China: Experience and lessons learned. *Sustainability, 17*(8), 3756. DOI: 10.3390/su17083756

Wang, X., Shi, J., Sun, S., Wang, D., Li, Z., Sun, Z., Sigrimis, N., & Li, T. (2025b). Understanding complex process sensor signal may be the source for deep learning: A smart IoT delving into a tomato weight sensor. *Computers and Electronics in Agriculture*, *231*, 110012. DOI: 10.1016/j.compag.2025.110012

Wang, X., & Xing, Y. (2017). Evaluation of the effects of irrigation and fertilization on tomato fruit yield and quality: A principal component analysis. *Scientific Reports*, *7*, 350. DOI: 10.1038/s41598-017-00373-8 PMID: 28336916

Wang, Z., Liu, Z., Yuan, M., Yin, W., Zhang, C., Zhang, Z., & Hu, X. (2025). A machine learning-based irrigation prediction model for cherry tomatoes in greenhouses: Leveraging optimal growth data for precision irrigation. Computers and Electronics in Agriculture, 237(A), 110558. https://doi.org/DOI: 10.1016/j.compag.2025.110558

Zalai, M., Bujtás, O., Sárospataki, M., & Dorner, Z. (2025). Grassy and herbaceous interrow cover crops in European vineyards: A review of their short-term effects on water management and regulating ecosystem services. *Land (Basel)*, *14*(8), 1526. DOI: 10.3390/land14081526

Zeng, R., Abate, M. C., Cai, B., Addis, A. K., & Derebe Dereso, Y. (2025). A systematic review of contemporary challenges and debates on Chinese food security: Integrating priorities, trade-offs, and policy pathways. *Foods*, *14*(6), 1057. DOI: 10.3390/foods14061057 PMID: 40232105

Chapter 13
New Findings on the Efficient Markets Hypothesis:
The Robustness of the RALS Method in the United States, Canada, and Selected European Markets

Lisa Hale
https://orcid.org/0000-0001-7473-3248
Istanbul University, Turkey

ABSTRACT

This study tests the validity of the efficient markets hypothesis for the United States, Canada, Australia, and selected European countries. Classical unit root tests, commonly used in the literature, can produce weak and inconsistent results, particularly when returns are not normally distributed. Therefore, the study employed the RALS (Residual Augmented Least Squares) method, which has more robust statistical properties, in addition to classical methods. The findings obtained demonstrate that this modern method, unlike classical unit root tests, provides more reliable evidence of market efficiency and highlights the critical impact of deviations from a normal distribution on the analysis results. The results demonstrate the heterogeneity of the capital markets of the countries studied and emphasize that the choice of method determines inferences regarding the efficient market hypothesis.

DOI: 10.4018/979-8-3373-6746-0.ch013

1. INTRODUCTION

With globalization and increasing technological advances, risks and uncertainties are increasing in capital markets. The existence of uncertainties is reflected in financial markets and the profit phenomenon emerges. On the other hand, one of the most important factors that global investors take into account when making their investment decisions is the correct understanding of the movements in the financial markets in which they will invest. Therefore, the current state of financial markets plays an important role in channeling savings into investments and realizing investment projects. The efficient market hypothesis, one of the hypotheses put forward at this point, states that it is impossible to predict the future prices of securities based on their current prices. In other words, market actors are rational and all information available in the market reaches all actors without delay. This makes it impossible to make profits in financial markets under the assumption that each actor is rational.

An examination of the fundamental methodological evolution of the efficient market hypothesis inevitably leads to a reference to Louis Bachelier's (1900) doctoral dissertation, "Theory of Speculation," which is considered the birth of mathematical finance. Bachelier's dissertation is considered the first scientific study that laid the foundations of modern financial theory and stochastic processes. This study marked the first time that stock market movements were examined with a mathematical basis. The main idea of the study was that price changes are determined by a completely random process (random walk). Price changes have no average trend, and past information fails to predict future prices. Thus, Bachelier laid the first theoretical foundation of modern finance with his doctoral dissertation. This work, which was beyond its time, was later rediscovered and formed the mathematical basis for theories such as Fama's Efficient Market Hypothesis and the Black-Scholes-Merton option pricing model (1973).

In the first half of the 1900s, when the first theoretical foundations of the efficient markets hypothesis were laid, In 1933, Cowles published a study titled *"Can Stock Market Forecasters Forecast?"*. Despite the inherent limitations of contemporary statistical testing tools, a rigorously scientific approach was adopted. This study became one of the modern empirical pioneers, using a wide variety of methods, including binomial tests, correlation tests, graphical analysis, frequency and distribution analysis, probability and statistical significance tests, and comparative return analysis. In this valuable study, the forecasts of 24 financial publications were compared with those of 24 random portfolios. No evidence was found that these financial publications outperformed the random portfolios. (Wall Street Journal editor William Peter Hamilton, excluding his stock market predictions based on Dow Theory between 1904 and 1929, announced market expectation changes 90 times during this period. Consequently, 45 of his predictions were successful and 45 were unsuccessful). Sixteen financial services companies made approximately 7,500 individual stock recommendations for investment purposes between 1928 and 1932, and on average, the returns on portfolios consisting of these recommendations were 1.43% lower annually than the overall stock market average. Twenty fire insurance companies that used similar stock selections between 1928 and 1931 performed 1.20% worse annually than the overall stock market average. In short, it demonstrates that financial companies, insurance companies, and some popular magazines are no more successful than chance in predicting the market and individual stocks.

Another important methodological evolutionary contribution to the study of the efficient market hypothesis is the study of Cowles and Jones (1937). Rather than comparing the performance of stock market forecasters with random predictions, the authors examined patterns of sequential changes and reversals. Their conclusion was that the serial correlation between sequential price changes was not zero. They also emphasized that investors could not achieve sustained or substantial profits by trading on these patterns (Dimand and Veloce, 2010). This study pioneered the idea that market prices approximate a random walk.

Another study that contributed to the foundations of market efficiency was conducted by Kendall (1953). By examining 22 price series, Kendall demonstrated that these series behaved largely randomly, and that because intra-series correlations were so weak, predicting a price series from its past values was virtually impossible. Stock market prices fluctuate almost completely randomly in the short term, suggesting that weekly price fluctuations can be explained solely by chance. These results are consistent with the study of Cowles and Jones (1937).

Samuelson (1965) was the first to use the concept of market efficiency. Samuelson's main argument in this study was that if markets are informationally efficient, price changes should be unpredictable because these prices fully reflect the information and expectations of all market participants (Lo, 2018). More specifically, Samuelson argued in this study that for informational efficiency to be achieved, all market information must be reflected in security prices, making it impossible to predict future security prices (Görgel and Ege, 2024). This study paved the way for the development of the option pricing models of Black and Scholes (1973) and Merton (1973) (Lo, 2018).

Efficient markets hypothesis was first proposed in the book *Efficient Capital Markets: A Review of Theory and Empirical Work* by Fama (1960). In this study, Fama classified efficiency as weak, semi-strong, and strong efficiency. Weak efficiency indicates that stock prices only contain past information, while semi-strong efficiency indicates that publicly disclosed information is reflected in prices. On the other hand, strong efficiency implies that stock prices incorporate all available information, whether publicly disclosed or not. Thus, in the case of the weak-form hypothesis, past information is available to all investors without cost or effort, and hence there is no possibility of excessive profits in weakly efficient markets. However, there is a possibility to earn excessive profits by using information that is not available in the market.

In the case of the semi-strong-form hypothesis, it is not possible to generate excess returns using historical and publicly available information, but it is possible to generate more limited returns using non-public information.

In the case of the strong-form hypothesis, investors in these markets are unable to earn returns above the average investor because investors and everyone else has immediate access to all information in the market, including inside information. In short, as the efficiency of the market increases, its predictability increases and the opportunity to profit from this market disappears.

On the other hand, it is almost impossible to find a globally efficient market in a strong form. According to Sanford J. Grossman and Joseph E. Stiglitz, information is also a cost factor and it is impossible for a market to be perfectly efficient at the information level due to the cost of information (Szylar, 2014). In this context, the correct determination of the efficiency level of markets is a factor that determines the volume of investments and where they are channeled. Indeed, the degree of efficiency of the markets is an issue that investors take into account when making investment decisions. Investments not only affect the financial markets but also have an impact on the whole economy. Therefore, the correct determination of the degree of efficiency in the markets is not only important for investors but also for the whole economy. This has led to the efficient market hypothesis being debated for many years.

Various approaches can be used to test the validity of the efficient market hypothesis. These approaches include the variance ratio test, runs test, the BDS test, and many other alternative tests.

The variance ratio test, developed by Lo-MacKinlay (1988), is a classic method used to test the random walk assumption in financial time series. The null hypothesis indicates that the series exhibits a random walk, while the alternative hypothesis indicates that the series does not exhibit a random walk. If the null hypothesis cannot be rejected as a result of the test—if the series follows a random walk—the existence of an efficient market is indicated. If the null hypothesis is rejected, the existence of an inefficient market is indicated.

The Runs test, developed by Wald and Wolfowitz (1940), is another approach used to investigate the efficient market hypothesis. If the null hypothesis, which indicates a random walk, cannot be rejected as a result of this test, the market is considered efficient; otherwise, it is concluded that there is a dependence between returns, i.e., the market is inefficient. Another approach used to investigate market efficiency is the BDS test, whose initial theoretical foundation was established by Brock, Dechert, and Scheinkman (1987) and developed by Brock, Dechert, Scheinkman, and LeBaron (1996). This test is used to investigate the presence of nonlinear structures in a series. If the null hypothesis that the series has an independent and identical distribution cannot be rejected, a linear structure is assumed, indicating the validity of the efficient markets hypothesis. If the null hypothesis is rejected, the efficient markets hypothesis is invalid for that market.

In addition to the tests mentioned above, unit root tests are another frequently used test when investigating the validity of the efficient markets hypothesis for a market. While testing the efficiency of a market with unit root tests, it is utilized whether the index of that market follows a random walk process or not. The fact that the series has a unit root indicates that the series follows a random walk process and the market is weakly efficient (Hepsağ and Akçalı, 2015).

The purpose of this study is to test the validity of the efficient market hypothesis in economies that have achieved a certain level of financial depth, shaping today's global economy, by keeping a broad geographical scope. These selected economies are hybrid markets where strong foreign investors, who shape global financial markets, operate intensively. The chosen research period is distinguished by its specific time period encompassing the pre- and post-COVID-19 pandemic period.

In this study, the random walk process of the United States, Canada, Australia, Germany, France, and United Kingdom markets is analyzed with the Augmented Dickey-Fuller (ADF) unit root test, which is a conventional unit root test, as well as the Lee-Strazicich (2004) unit root test with structural breaks and the Residuals Augmented Least Squares-LM (RALS-LM) unit root test introduced by Meng et al. (2017).

2. LITERATURE REVIEW

There are studies in the literature that investigate the efficiency of stock markets with various methods. In this section of the study, these studies will be discussed.

Hatipoğlu (2024) investigated the validity of the efficient markets hypothesis for the ESG Leader indices of Japan, Switzerland, France, Germany, and Spain, as well as the ESG-Leader indices of Taiwan and Malaysia from emerging markets, using the variance ratio test. This study, which utilized daily data covering the years from 2021 to 2023, concluded that the efficient markets hypothesis is not valid for all markets examined.

Görgel and Ege (2024) used weekly data covering the period from 1997 to 2024. They compared the efficient markets hypothesis with the adaptive expectations hypothesis in the E7 countries using a wide range of tests, including the ADF, PP and Zivot Andrews tests, the variance ratio test, the BDS test, the AR-MA model, autocorrelation, and the ARCH-LM heteroscedasticity test. In this comprehensive study, the findings obtained from the Jarque-Bera (J-B) normality test: It was concluded that the markets in China, Brazil, Mexico, Indonesia, Russia, and Turkey are inefficient, meaning that the efficient market hypothesis and the Adaptive Expectations Hypothesis are not valid, but the APH is valid in the Indian market. Findings from both the ADF, PP, and Zivot Andrews unit root tests indicate that the return series in the markets of the E7 countries are stationary and that the efficient market hypothesis and APH are not valid in these markets. Findings from the variance ratio test, similar to the unit root test results, indicate that the efficient market hypothesis and APH are not valid in the equity markets of these countries. Findings from the applied BDS test revealed that the efficient market hypothesis and APH are not valid in the equity markets of China, India, Mexico, Indonesia, Russia, and Turkey, while the APH is valid in Brazil. The results from the ARMA model indicated that the efficient market hypothesis and APH are not valid.

Fattahi (2010) investigated the validity of the efficient markets hypothesis using stocks of five randomly selected companies from the German stock exchange. They conducted their research within the framework of the Dickey-Fuller (DF) unit root test and variance ratio test. The results indicate that the stocks delisted from the stock market follow a random walk, confirming the validity of the weak form of the efficient markets hypothesis.

Hepsağ and Akçalı (2015) analyzed the efficient markets hypothesis for G-7 and E-7 countries using monthly data covering the period from 1970:01 to 2015:05 and used the Sollis (2009) test, which is an asymmetric nonlinear unit root test. They concluded that the financial markets of Brazil, China, Indonesia, India, Mexico, and Turkey among the E-7 countries and the markets of the United States, France, Japan, and Italy among the G-7 countries are weakly efficient, whereas the markets of Germany, the United Kingdom, Canada, and Russia are not weakly efficient.

Küçükkaplan et al. (2023) used ADF, RALS-ADF, Fourier-ADF, Fourier-ADF, and Fourier-KSS unit root tests to investigate the efficient markets hypothesis for G-8 countries. According to all methods used, they find evidence of the validity of the efficient markets hypothesis for the stock market indices of Germany, France, and Japan. For the United States, the United Kingdom, Canada, and Russia, the results varied depending on the method. However, the results of the RALS-ADF unit root test -at the 5% significance level- indicate that the null hypothesis cannot be rejected for Russia, i.e., the efficient markets hypothesis is not valid for the country, while the efficient markets hypothesis is valid for the financial markets of Canada, the United States, Germany, France, and the United Kingdom.

Borges (2010) investigated the validity of the weak form of efficient markets hypothesis using the stock market indices of the United Kingdom, France, Germany, Spain, Greece, and Portugal. As a result of the analysis of the data from 1993:1 to 2007:12 with runs test and variance ratio test, they find the invalidity of the efficient markets hypothesis for Portugal, Greece, France, and the United Kingdom, and the validity of the efficient markets hypothesis for Germany and Spain.

Kaushik (2020) investigated the validity of the efficient markets hypothesis for the financial markets of the Netherlands, Belgium, and the United Kingdom in comparison with Ireland using daily data covering from 2014:1 to 2019:12. Kaushik used ADF unit root test, Ljung Box test, autocorrelation test, runs test and shannon entropy test in the analysis. The empirical evidence indicated that the weak form of the efficient markets hypothesis holds for the Irish securities market, the United Kingdom, Belgium, and the Dutch markets.

Paskaleva and Stoykova (2021) investigated the efficiency of the capital markets of fourteen European countries including Germany, France, Belgium, and the United Kingdom based on monthly data for the period from 2003 to 2016. They used the ADF unit root test and the TGARCH model. The results demonstrate that the European markets analyzed are highly integrated, revealing a split along the developed-emerging market line in the efficient markets hypothesis context. They concluded that the capital markets of Germany, the United Kingdom, Belgium, and France are among the relatively efficient markets, while the most efficient market is the Greek capital market.

Li et al (2014), drawing on daily data from the 2000:01 to 2012:12 period, investigated the efficiency of stocks of the 50 largest companies in various sectors on the Australian Stock Exchange. They employed correlation tests, runs test, and the filter rules test as their research methods. The researchers stated that the results indicated that there was no significant autocorrelation between returns for most stocks and that the ability to predict current short-term returns using historical return information was very limited. The prices of these stocks follow a random walk. The Australian market is a weak-form efficient market. Therefore, investors and fund managers will not be able to achieve abnormal returns in the Australian market from trading strategies based on past stock prices.

Gan et al. (2005) examined market efficiencies in the New Zealand Stock Exchange (NZSE) and Australian Stock Exchange (ASX) stock indices in the post-financial liberalization period (January 1990-January 2003) to test whether their findings, citing Groenewold's (1997), still hold. Similar to Groenewold's findings, they found evidence of weak-form efficiency for the NZSE and ASX stock indices using the ADF and PP unit root tests.

Simmons (2011) investigated the validity of the efficient market hypothesis in Australia using daily data from 2000 to 2008. In the study, they used a method called Data Envelopment Analysis (DEA), which differs from unit root tests, runs tests, or variance ratio tests. The findings indicate that while above-average returns are achieved in some periods, these returns are not sustained, and losses are experienced in other periods. This suggests that the market is not fully efficient and that partial/ periodic inefficiencies exist in some periods.

Neifar and Gharbi (2023) examined the efficiency of the Canadian stock market using linear and nonlinear unit root tests. Drawing on monthly data from the 2004:08 to 2018:04, the weak form of the efficient markets hypothesis was tested for two Canadian stock indices—the Islamic (DJICPI) and the conventional (CCSI). The linearity test results indicated that the Canadian Conventional Stock Index exhibited nonlinear behavior, while the Dow Jones Canadian Islamic Price Index followed a linear process path. The findings suggest that only the Canadian Islamic Price Index exhibits a random walk, suggesting that conventional stock markets are not the only efficient markets. The main conclusion is that fund managers and investors in Canada can achieve excess returns on their investments in conventional stock markets.

Shiller and Radikoko (2014) tested the validity of the efficient market hypothesis in the Canadian stock market using Canadian TSX index returns. Using autocorrelation tests, ADF, PP, and KPSS-type unit root tests, and the BDS test, the study concluded that the Canadian stock market is not weakly efficient.

Showalter and Gropp (2019) tested whether the United States stock market is weakly efficient by using the ADF unit root test, the variance ratio test, and five different machine learning algorithms. As a result of their research, they found that the American stock market is weakly efficient.

It is useful to categorize the literature reviewed thus far within the contextual framework of the markets under investigation. Six studies focusing on the German stock market utilized research methods such as unit root tests, runs tests, and variance ratio tests. Four of these studies concluded that the efficient market hypothesis was valid for the German stock market. Similar tests were also used to investigate the efficiency of the Canadian stock market, and three of the four studies summarized above yielded evidence that the Canadian stock market was inefficient. A study using the RALS-ADF unit root test (Küçükkaplan et al., 2023) found that the efficient market hypothesis was valid for the Toronto stock market. All studies investigating the efficiency of the American stock market using variance ratio tests, unit root tests, and various machine learning algorithms concluded that the American stock market is efficient. Three of the five studies focusing on the French stock market concluded that the stock market behaves in accordance with the efficient market hypothesis. These studies utilized unit root tests, runs tests, and variance ratio tests as research methods. Studies conducted in the United Kingdom have largely found that the efficient market hypothesis is valid in this stock market. Testing methods include the Ljung Box test, the Shannon entropy test, and other tests such as the unit root test, runs test, autocorrelation test, and variance ratio. Three studies have investigated the validity of the efficient market hypothesis in the Australian stock market. Two of these concluded that the efficient market hypothesis is valid in this stock market, while one concluded that it is partially valid but not fully effective.

In the literature review, no study analyzing financial markets with the RALS-LM method that takes into account the non-normal distribution assumption was found. Therefore, unlike other studies in the literature, this paper first examines the normal distribution for the financial markets of Canada, the United States, Australia, Germany, France, and the United Kingdom, and then uses the RALS-LM type test, which is known to be more powerful than other tests in the literature, in addition to unit root tests that take into account structural breaks and unit root tests that do not take into account structural breaks. Then, the divergence in the findings of the tests that do and do not rely on the assumption of a normal distribution is illustrated.

3. METHODOLOGY

Unit root tests are built on the assumption that there is no correlation between the error terms and that the series is first-order autoregressive. The first test developed in this context was the DF unit root test developed by Dickey-Fuller (1979). However, correlation between error terms is a common occurrence in series. Consequently, this assumption used in the model was subject to intense criticism, and this deficiency was addressed by the same authors with the Augmented Dickey-Fuller (ADF) test (1981) by adding lagged difference terms to the model. Another test developed by Phillips-Perron (PP) (1989) utilized nonparametric statistical methods and eliminated this shortcoming of the DF test by allowing for weak dependence and heterogeneous distribution among the error terms. The Ng-Perron (1996) procedure aimed to address the dimensionality distortions that arise in the PP test. The Kwiatkowski, Phillips, Schmidt, and Shin (KPSS) Unit Root Test (1992) suggested that standard unit root tests fail to reject the null hypothesis of a unit root. Unlike the other tests described above, this approach tests the stationarity of the series by interchanging the null and alternative hypotheses. In other words, the primary hypothesis indicates that the series is stationary, while the alternative hypothesis indicates that the series contains a unit root (Varoğlu, 2022). In summary, unit root tests test for the presence of a unit root in the series, while the KPSS test, also known as the stationarity test, tests the stationarity of the series. The tests described above are tclassical unit root tests and are based on the assumption that the series has a linear structure and contains no structural breaks. In this traditional approach, the properties of the series are assumed to remain constant throughout the sample. However, in real life, data sets are subject to shocks such as political tensions, economic crises, reforms, wars, and technological breakthroughs. Such situations, which occur particularly in developing economies, cause deterioration in the structure of the series and sudden or smooth breaks, and are called structural breaks. Classical tests ignore structural breaks. Therefore, if a series contains a structural break, using a test appropriate to the series' structure is important to achieve a more reliable conclusion. The first test developed in this context was introduced to the literature by Perron in 1989. In the Perron procedure, a single structural break is allowed, and the structural break point is determined by the user. In other words, it is determined exogenously. This led to criticism of the test, and subsequently, new tests were developed that determine the break point(s) endogenously. Zivot-Andrews (2002), Lumsdaine-Papell (1997), and Lee-Strazicich (2003) are among these tests that can endogenously detect one or more structural breaks in the series.

Nonlinear unit root tests can be considered as the third step in the development of unit root tests. In classical tests and tests with structural breaks, the linearity of the series is the fundamental assumption. This assumption, used in classical unit root tests, is not present in nonlinear unit root tests. If the series has a nonlinear structure and is attempted to be modeled with a classical unit root test, the results obtained will be weak. Therefore, this is the fundamental difference between classical tests and unit root tests based on nonlinear structure. The fundamental difference between tests based on nonlinear structure and tests with structural breaks stems from the structure of the series before and after the break point(s). Structural break unit root tests model sudden regime changes that occur at specific points in the series. In these tests, the linear structure of the series is preserved before and after the break points, changing only the slope coefficient and the constant term. In tests based on nonlinear structure, the series cannot be fitted into a linear model, and the existence of a linear structure before and after the break is not required. While linearity is observed at certain points between the breakpoints, other points are nonlinear, or the series follows a nonlinear process from beginning to end.

Nonlinear unit root tests are also classified among themselves. Briefly, nonlinear tests can be classified as SETAR-type unit root tests, which assume that the regime change is abrupt and sharp, MTAR-type unit root tests, and ESTAR-type unit root tests (Kapetanios-Shin-Snell unit root test (2006), Sollis unit root test (2009), Kruse unit root test (2011) etc.), which assume that the transition between regimes is smooth (Güriş, 2020). These types of tests are powerful alternatives to linear tests when there is a suspicion that the series exhibits nonlinear dynamics. However, the details of nonlinear tests will not be discussed here.

The development of statistical tests is constantly advancing. Another group of tests to be discussed is unit root tests based on the Fourier procedure. The most important difference between these tests and other tests is the inclusion of deterministic terms, called Fourier terms, in the modeling phase. The primary purpose of developing these tests is to provide researchers with a robust alternative when they are concerned about the number and nature of structural breaks. The difference between unit root tests based on Fourier functions and tests with structural breaks stems from the way the break points of the series are modeled. Unlike the Fourier approach, unit root tests with structural breaks utilize dummy variables to model break points. This requires predetermining the break points, either endogenously or exogenously. If the break dates are uncertain or the series contains numerous breaks, utilizing dummy variables creates confusion and complexity. Furthermore, due to the use of dummy variables in structural break unit root tests, only sharp and sudden breaks can be detected. This is precisely where the strength of tests based on Fourier functions over tests with structural breaks becomes apparent. In Fourier-based unit root tests, the location, number, and shape of structural breaks (hard or soft) can be easily modeled without requiring dummy variables, regardless of their nature (Becker et al., 2006). In other words, with the help of Fourier functions, model complexity is reduced and the effects of breaks can be captured more accurately.

All of the testing approaches discussed so far are based on certain common fundamental assumptions, such as the series having a stochastic structure, the error terms being white noise, a normal distribution, and constant variance. They also assume that stationarity-unit root tests can be determined based on the I(0)/I(1) dichotomy. Consequently, time series are not always normally distributed.

Big data strengthens the asymptotic assumptions of test statistics, allowing for more reliable and consistent results. According to the central limit theorem, as sample size increases, the series approaches a normal distribution. While this provides a significant advantage for econometric research involving large data sets, researchers cannot always work with large data sets. Especially in financial data sets, the series frequently deviates from the normal distribution. In this case, the power of the unit root tests used is weakened and can deprive the researcher of reliable results. Unit root tests, which can be used in cases where non-normally distributed data sets, such as financial data sets, are based on the extended least squares (RALS) method with residuals. These tests allow for more reliable results in non-normally distributed data sets. Unit root tests developed in this context include the tests such as Im, Lee, and Tieslau (RALS-ADF) unit root test (2014), Meng, Im, Lee, and Tieslau (RALS-LM) unit root test (2014), and Meng, Lee, and Payne (RALS-LM) unit root test (2017) etc.

In this section of the study, the theoretical framework of RALS-LM unit root tests developed by Meng et al. (2017) will be discussed. Unit root tests used in time series analysis are based on the assumption of normal distribution. As mentioned earlier, financial data may not conform to the normal distribution. The distribution of financial data tends to exhibit a leptokurtic pattern, with a more pronounced peak and heavier tails than the normal distribution. Im and Schmidt (2008) state that least squares are efficient under normality, but when errors are not normally distributed, there is information about the distribution of errors at higher moments. They state that tests that allow the use of information available at higher moments may yield more efficient results. Therefore, the use of unit root and cointegration tests based on the assumption of normal distribution in the analysis of non-normally distributed data has started to be discussed.

In the study, Im and Schmidt (2008) laid the foundation of the RALS approach by arguing that the least squares approach is efficient in the case of normal distribution, but the RALS estimator will give more efficient results when the normal distribution condition is not met. Hence, to increase the power of unit root tests, the RALS-DF approach was developed by extending the traditional Dickey-Fuller (1979) unit root test by adding the non-normal distribution of residuals to these tests.

Meng et al. (2014) developed unit root tests that do not take into account structural breaks, that take into account one break in the constant and two breaks in the constant. The RALS-LM type unit root test that does not consider structural breaks is obtained by extending the Schmidt and Phillips (1992) test with residuals. The RALS-LM type unit root test, which takes into account one and two breaks in the constant, was obtained by extending the LM type test with one and two breaks in the constant developed by Lee-Strazicich (2003, 2004) with residuals (Hepsağ, 2022).

In the study, Meng et al. (2014) compared the LM test and the RALS-DF test developed by Im et al. (2014) with their own RALS-LM test. As a result of their study, they concluded that the RALS-LM test is significantly stronger than the LM and RALS-DF tests when the errors are not normally distributed.

To begin, the following data generation process (DGP) is considered based on the unobserved component representation;

$$\Delta y_t = \delta' \Delta Z_t + e_t \qquad (1)$$
$$e_t = \beta e_{t-1} + \varepsilon_t$$

The unit root null hypothesis is $\beta = 1$. $Z_t = 1, t, D_{1t}, \ldots. D_{Rt}, DT_{1t}, \ldots\ldots., DT_{Rt}]'$ where D_{it} and DT_{it}(i=1,...,R) are dummy variables indicating the positions of the ith level and trend breaks, respectively.

$D_{it} = 1$ for $t \geq T_{Bi}+1$, and zero otherwise;
$DT_{it} = t - T_{Bi}$ for $t \geq T_{Bi}+1$, and zero otherwise;

R is the maximum number of structural breaks; and T_{Bi} is the location of the ith structural break.

According to the LM principle, the unit root test statistic is obtained from regression:

$$\Delta y_t = \delta' \Delta Z_t + \phi \tilde{S}_{t-1} + e_t \qquad (2)$$

$\widetilde{S}_t$ is the LM trend-free y_t series calculated using $\widetilde{S}_t = y_t - \widetilde{\psi} - \widetilde{\delta}' Z_t$. $\widetilde{\delta}'$ is the coefficient vector obtained in the regression of Δy_t on ΔZ_t where the first difference of y_t and Z_t are used.

The dependency of tests on the nuisance parameter in the trend-shift models has been an issue in the literature. As such, Meng et al. (2017) consider a simple transformation that can make the unit root test statistic free of the dependency on the break location. The following transformation can remove the dependency on the nuisance parameter:

$$\widetilde{S}_t^* = \begin{cases} \frac{T}{T_{B1}} fort \leq T_{B1} \\ \frac{T}{T_{B2} - T_{B1}} \widetilde{S}_t forT_{B1} \leq t \leq T_{B1} \\ \vdots \\ \frac{T}{T - T_{BR}} \widetilde{S}_t forT_{BR} \leq t \leq T \end{cases} \tag{3}$$

where $\widetilde{S}_t$ is the untransformed series and $\widetilde{S}_t^*$ is the transformed series. We then replace $\widetilde{S}_{t-1}$ in the testing regression (2) with $\widetilde{S}_{t-1}^*$ such that we have a new testing regression, $\Delta y_t = \delta' \Delta Z_t + \phi \widetilde{S}_{t-1}^* + e_t$, and denote τ_{LM}^* as the t-statistic testing the null hypothesis $\varphi = 0$. The RALS procedure augments the following term $\widehat{w}_t$ to testing regression (2).

$$\widehat{w}_t = h(\hat{e}_t) - \widehat{K} - \hat{e}_t \widehat{D}_2 \tag{4}$$

where $\hat{e}_t$ is the OLS residual from regression (2), $\widehat{K} = \frac{1}{T} \sum_{t=1}^{T} h(\hat{e}_t)$ and $\widehat{D}_2 = \frac{1}{T} \sum_{t=1}^{T} h'(\hat{e}_t)$.

To capture the information of non-normal errors, $h(\hat{e}_t) = [\hat{e}_t^2, \hat{e}_t^3]$, which involves the second and third moments of $\hat{e}_t$. The moments can be calculated as follows; $\widehat{m}_j = T^{-1} \sum_{t=1}^{T} \hat{e}_i^j$. After the moments are calculated, $\widehat{w}_t$ variables expanded with residues are obtained, $\widehat{w}_t = [\hat{e}_t^2 - \widehat{m}_2, \hat{e}_t^3 - \widehat{m}_3 - 3\widehat{m}_2 \hat{e}_t]'$.

By adding the residuals expanded variables ($\widehat{w}_t$) to the test regression, the following transformed RALS-LM test statistic is obtained;

$$\Delta y_t = \delta' \Delta Z_t + \phi \widetilde{S}_{t-1}^* \gamma' \widehat{w}_t + u_t \tag{5}$$

After the (5) regression is estimated by the least squares method, the following test statistic is used to test the existence of unit root.

$$\tau_{RALS-LM} - \rho \tau_{LM} \sqrt{1 - \rho^2 Z} \tag{6}$$

ρ reflects the relative ratio of the variance of two error terms.

4. DATA

This study selects Canada, the United States, Australia, Germany, France, and the United Kingdom. These countries are chosen because they possess a well-developed financial market structure that drives the global economy, as well as being recognized for their reliability of financial data, market transparency, and regulatory effectiveness. These economies, with their deep capital markets, high transaction volumes, and the potential to attract international investor interest, constitute a suitable sample for testing the efficient markets hypothesis.

A more detailed analysis reveals that the German stock exchange is not only Europe's largest financial market but also plays a key role in global capital flows. Like the other selected stock exchanges, it boasts high liquidity and depth. The German economy boasts a strong export volume and potential. The stock exchange boasts a robust institutional structure, along with high regulatory standards. It boasts a significant volume of domestic and international players. Therefore, the German stock exchange, centrally located in Europe, is in a critical position in transmitting positive and negative financial shocks to other financial markets. It was selected as a suitable stock exchange for our research due to its high participation volume, the dominance of the industrial and service sectors, extensive algorithmic trading volume, transparency, strict EU financial regulations, and its status as a central European economy.

Similarly, the UK stock exchange is a major financial player in Europe. The London Stock Exchange, along with the New York Stock Exchange, is the oldest and largest global financial player. The London Stock Exchange is highly sensitive to global economic and geopolitical developments. Its algorithmic and high-frequency trading can influence market efficiency. Its deep, established institutional standards and sensitivity to international dynamism make it a suitable platform for testing the efficient markets hypothesis. Therefore, it is particularly important for this type of research. The Toronto Stock Exchange was chosen because it boasts advanced technological infrastructure, a strong regulatory framework, and is open to global capital flows. The Toronto Stock Exchange is closely integrated with the New York Stock Exchange. This exchange, with its stable financial structure, offers researchers a practical framework for investigating the efficient markets hypothesis due to factors such as regulatory quality and information transparency.

The Australian Stock Exchange is one of the most liquid and largest exchanges in the Asia-Pacific region. Its prominent features include advanced technological infrastructure, high institutional transparency, high global investor participation, robust financial infrastructure, high-frequency trading volume, and advanced regulatory compliance. Furthermore, its strong relationship with both Asian and Western financial centers was crucial in selecting this exchange for this study. The study of Australia provides geographical and economic diversity.

Finally, the New York Stock Exchange is the largest financial player globally. The negative and positive shocks that may occur in this exchange react to global stock markets extremely quickly and strongly. Data transparency, high trading volume, broad global investor volume, and the concentration of globally leading companies make the New York Stock Exchange valuable for such financial research. In brief, each of the selected country exchanges in this study serves as a critical laboratory for testing the efficient markets hypothesis.

The data used in the study is obtained daily from the Yahoo Finance database. The observation period is 02.01.2018-31.05.2024. In order to help the correct determination and interpretation of the methods to be used in the analysis, the variables and descriptive statistics of the variables are presented in Table 1.

Table 1. Variables and Descriptive Statistics

	S&P/TSX(Com.) Canada	**S&P500 USA**	**S&P/ASX200 Australia**	**DAX30 Germany**	**CAC40 France**	**FTSE100 UK**
Observations	1609	1612	1620	1627	1641	1616
Min	-0,1234	-0,1198	-0,0970	-0,1224	-0,1228	-0,1087
Max	0,1196	0,0938	0,0700	0,1098	0,0839	0,0905
Mean	0,0002	0,0005	0,0002	0,0003	0,0003	0,0001
S.Devia.	0,0106	0,0127	0,0102	0,0124	0,0120	0,0104
Median	0,0008	0,0008	0,0009	0,0008	0,0009	0,0006
Skewness	-1,1259	-0,5129	-1,0485	-0,4135	-0,7502	-0,8561
Kurtosis	41,7080	16,1293	16,3617	15,9976	15,8141	17,3891
JB Stats	100789	11649	12348	11499	11381	14139
JB Prob	0,0000	0,0000	0,0000	0,0000	0,0000	0,0000

When the information in Table 1 is examined, it is understood that the data used in the analysis do not conform to the normal distribution as a result of a priori tests. The histograms of these series are presented in Figure 1.

Fig. 1. Histogram Graphs of the Series

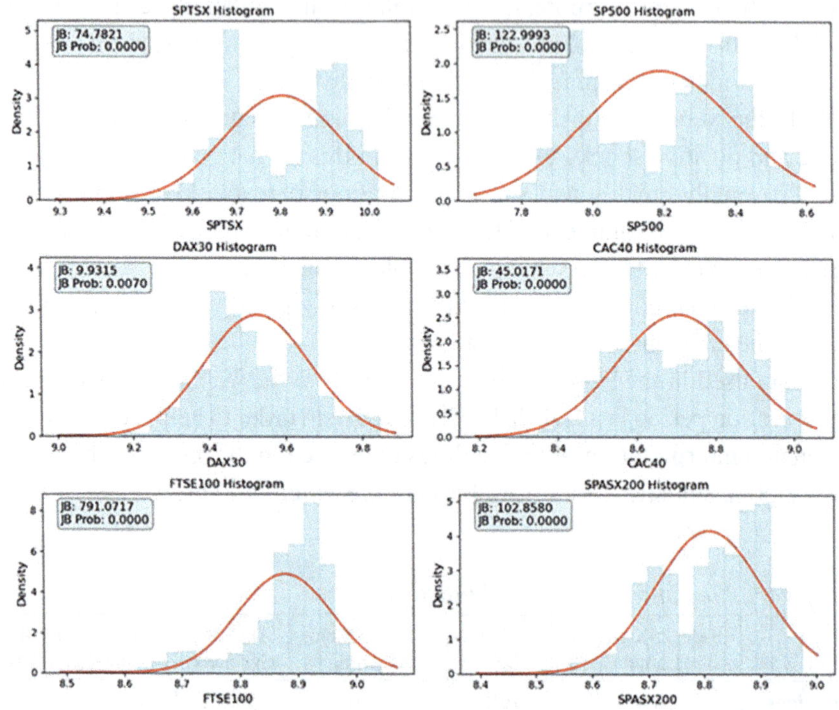

Histogram plots also indicate that the series do not conform to a normal distribution. Figure 2 presents the graph of logarithmic price indices. The natural logarithms of the price indices were used to investigate the efficiency of the markets in these countries. Upon examining Figure 2, it can be observed that the price indices of the stock markets in these countries exhibit asymmetric behavior.

Fig. 2. Graphs of the Series

In the study, various tests were utilized to test weak-form efficiency. ADF (1979) unit root test, one of the classical unit root tests, was applied to the series respectively. If a series contains structural breaks, conventional unit root tests may give biased results in the direction of non-stationarity (Perron, 1989). Therefore, in addition to the ADF unit root test, the single-break Lee-Strazicich (2004) unit root test, which is one of the unit root tests that take into account structural breaks, is applied to the series.

Finally, all of the above-mentioned unit root tests are based on the assumption that the series are normally distributed. However, most of the financial series are not normally distributed. In this study, we apply the RALS-LM unit root test developed by Meng et al. (2017), which is one of the unit root tests that allows for stronger findings in the case of non-normal distribution of the series. The series analyzed in the paper are not normally distributed, as evidenced by the results of the Jarque-Bera normality test in Table 1. Therefore, RALS-LM unit root tests are of significant importance in testing the efficiency of these markets.

5. EMPIRICAL RESULTS

The results of the ADF unit root test applied to the series are presented in Table 2. For the Canada stock index series, the null hypothesis (the existence of a unit root) is rejected at the 5% significance level. For the United States, Australia, Germany, France, France, and the United Kingdom, the null hypothesis could not be rejected at the 5% significance level and the series were found to have unit roots. Therefore, it is concluded that the Canadian market is not weakly efficient, but the United States, Australia, Germany, France, and the United Kingdom markets are weakly efficient markets. In other words, in these markets, excess returns are possible by using information that is not available in the market.

Table 2 and Table 3 present the results of two different tests developed with the assumption that the series is normally distributed. RALS-LM unit root test results, which give stronger results for non-normally distributed series, are reported in Table 4.

Table 2. Results of ADF Unit Root Tests

	S&P/TSX(Com.) Canada	**S&P500 The U.S.**	**S&P/ASX200 Australia**	**DAX30 Germany**	**CAC40 France**	**FTSE100 The U.K.**
Test Stats	-3,5512**	-2,9049	-3,3285	-2,9417	-2.8372	-2,7231
Max Lags	7	9	3	0	0	0

* indicates that the coefficients are significant at 1%; ** indicates that the coefficients are significant at 5%.

The results of the Lee-Strazicich (2004) unit root test taking into account structural breaks are provided in Table 3. The test results indicate that the null hypothesis (the existence of a unit root) is rejected at the 1% significance level for the Canada stock index series and at the 5% significance level for the United Kingdom stock index series. These series are stationary, indicating that these markets are not weakly efficient. For Australia, Germany, France, and the United States, the null hypothesis cannot be rejected at the 5% significance level and the series are found to have unit roots. Therefore, it is concluded that the United States, Australia, Germany, and France are weakly efficient markets.

Table 3. Lee-Strazicich Unit Root Tests Analysis Results

	S&P/TSX(Com.) Canada	**S&P500 The U.S.**	**S&P/ASX200 Australia**	**DAX30 Germany**	**CAC40 France**	**FTSE100 The U.K.**
Test Stats	-4,9755*	-3,2455	-3,9027	-3,2212	-3,3789	-4,0906**
Max Lags	7	9	11	18	17	15
Break Date	18/10/2023	02/02/2022	05/03/2020	26/09/2018	10/06/2020	10/06/2020

* indicates that the coefficients are significant at 1%; ** indicates that the coefficients are significant at 5%.

All studies reviewed in the literature have identified the Toronto Stock Exchange as a weakly inefficient stock exchange. The results of the ADF and LS unit root tests conducted in this section are consistent with studies reviewed in the literature for the Toronto Stock Exchange. Regarding the UK Stock Exchange, some studies in the literature (Küçükkaplan et al., 2023; Kaushik, 2020; Paskaleva and Stoykova,

2021) have demonstrated that this stock exchange is weakly efficient, while others (Hepsağ and Akçalı, 2015) have indicated that it is not weakly efficient. As a result of this study, the LS unit root test revealed that the London Stock Exchange is weakly inefficient, consistent with the study by Hepsağ and Akçalı (2015). However, in this study, the results of the RALS-LM unit root test, which does not assume a normal distribution, are more relevant than the classical unit root tests and structural break tests. The RALS-LM unit root test results are presented in Table 4.

Table 4. RALS-LM (2017) Unit Root Tests Analysis Results

	S&P/TSX(Com.) Canada	**S&P500 The U.S.**	**S&P/ASX200 Australia**	**DAX30 Germany**	**CAC40 France**	**FTSE100 The U.K.**
Test Stats	-1,1891	-0,8227	-4,0392**	-0,4385	-2,4438	-2,6181
Max Lags	23	23	22	23	24	24
Ro2	0,8599	0,8387	0,7799	0,8059	0,8333	0,8242
Break Date	16/11/2023	11/03/2020	27/10/2023	11/03/2020	28/09/2022	11/03/2020

* indicates that the coefficients are significant at 1%; ** indicates that the coefficients are significant at 5%.

The test results indicate that the null hypothesis (the existence of a unit root) is rejected at the 5% significance level for the Australia stock index series. The series is stationary and the Australian market is not weakly efficient. On the other hand, for Canada, the United States, Germany, France, and the United Kingdom, the null hypothesis cannot be rejected at 5% significance level and the series are found to have unit root. Therefore, it is concluded that Canada, the United States, Germany, France, and the United Kingdom are weakly efficient markets. The results of the RALS-LM unit root test indicate that investors can earn excessive profits by using non-market information in the financial markets of Canada, the United States, Germany, France, and the United Kingdom. The findings are exactly the same as the results obtained from the Rals-ADF unit root test by Küçükkaplan et al (2023).

CONCLUSION

The efficient market hypothesis states that financial asset prices contain all available information in the market, preventing investors from profiting excessively. While increased market efficiency leads to increased price stability, the potential for excessive profits decreases as the degree of efficiency increases for investors.

In markets with low efficiency, transparency controls and policies that strengthen market integrity are necessary to increase efficiency.

Several factors influence the validity of the efficient market hypothesis. The first of these is the level of development of the current market. The efficiency of a market can be influenced by its level of development. Efficiency in emerging markets is lower than in developed countries. This is due to factors such as information asymmetry, low liquidity, insufficient foreign investors, limited financial depth, and high volatility, which exist in emerging markets. Therefore, the flow of information in emerging markets is relatively slower than in developed markets. While lower efficiency is expected in these markets than in developed markets, the efficiency level in developed markets is expected to be higher due to reasons such as a robust institutional structure, rapid information flow, and advanced regulation. Another factor affecting efficiency is factors such as pandemics, economic/social crises, and natural disasters, which create sudden structural changes at a national, regional, or global level. Due to market disruptions during the Covid-19 pandemic, investors were unable to access information quickly and completely, creating uncertainty in investor behavior, which was then reflected in the market as uncertainty. During times of emergency, the degree of access to information varies, and market reactions vary accordingly. When the full flow of information in the markets is disrupted, financial asset prices fail to reflect information quickly and effectively. All of this disrupts market efficiency and presents excessive profit opportunities to those with faster access to information.

Sudden and disruptive events such as pandemics and natural disasters not only disrupt market behavior but also create structural breaks. These structural breaks lead to reforms and create new regulatory frameworks. During the pandemic, global digital transformation, regulations in financial markets, and changes in central bank attitudes led to changes. Therefore, excessive market intervention during the pandemic led to a decrease in transparency, while post-pandemic digitalization, fintech developments, financial integration, the acceleration of global capital movements, and regulations increased market transparency.

Generally speaking, the 2018-2024 period covered in this study also encompasses this period of uncertainty in the markets. Therefore, the findings should be evaluated not only statistically but also within the economic, structural, and regulatory framework of the period under consideration.

In other words, weakening efficiency does not imply rejection of the efficient market hypothesis. The finding that the efficient market hypothesis was invalid in Australia should be considered in this context, and its possible relevance to the structural conditions of the period should not be overlooked.

On the other side, examining non-normally distributed series with tests that assume normal distribution will lead to controversial results. Therefore, using non-normally distributed series with tests that do not assume normality will lead the researcher to more accurate findings. Financial series are generally non-normally distributed series. In this study, financial series of Canada, the United States, Australia, Germany, France, and the United Kingdom are used to investigate whether the RALS-LM unit root test without the assumption of normality is a more powerful test than the tests based on the assumption of normality. The findings of the RALS-LM unit root test are surprisingly different from those of the ADF and Lee-Strazicich (2004) tests.

In the study, according to the ADF unit root test, which is based on the assumption of normality and does not take into account structural breaks, the Canadian market is weakly inefficient; according to the Lee-Strazicich (2004) test, which is based on the assumption of normality and takes into account structural breaks, the Canada and the United Kingdom financial markets are weakly inefficient. On the contrary, according to the RALS-LM unit root test, which is not based on the normality assumption, all markets except Australia are identified as weak form efficient markets.

Based on the results, the Australian market is sensitive to historical information. Policies should be adopted to increase information transparency in Australian markets. In this context, financial reporting strategies should be strengthened. Policies should be adopted to reduce transaction costs and improve electronic transaction systems to improve market infrastructure. Financial literacy programs should be developed, and insider information should not be allowed to leak. In short, policies should be adopted to develop financial markets.

Weak forms of efficiency have been observed in other economies outside of Australia. Historical information in the markets is incorporated into financial asset prices. Therefore, these exchanges should implement macroprudential policies that maintain current efficiency, transparency, and regulatory levels, while also fostering financial stability. Maintaining a fair and competitive environment for investors and strengthening the independence of regulatory institutions are crucial for maintaining and developing current market efficiency. In short, investors in countries outside of Australia can benefit from short-term price fluctuations, but these opportunities are risky and volatile.

Future research could utilize advanced analytical techniques, such as machine learning algorithms, to uncover nonlinear relationships and complex interactions. Furthermore, repeating such analyses across different countries/sectors and time periods would provide a more comprehensive understanding of how market efficiency evolves with institutional and structural conditions.

DECLARATIONS

Conflict of Interest

The author has no conflict of interest to declare regarding the content of this article. The author declares that she has no financial interest.

REFERENCES

Bachelier, L. (1900). Théorie de la spéculation. *Annales Scientifiques de l'Ecole Normale Supérieure*, *17*, 21–86.

Becker, R., Enders, W., & Lee, J. (2006). A Stationarity Test In The Presence Of An Unknown Number Of Smooth Breaks. *Journal of Time Series Analysis*, *27*(3), 381–409.

Black, F., & Scholes, M. (1973). The Pricing of Options and Corporate Liabilities. *Journal of Political Economy*, *81*(3), 637–654.

Borges, M. R. (2010). Efficient Market Hypothesis in European Stock Markets. *European Journal of Finance*, *16*(7), 711–726. DOI: 10.1080/1351847X.2010.495477

Brock, W. A., Dechert, W., & Scheinkman, J. (1987). A Test For Independence Based On The Correlation Dimension. University of Wisconsin-Madison. Working Paper.

Cowles, A.III. (1933). Can Stock Market Forecasters Forecast? *Econometrica*, 309–324.

Cowles, A.III, & Jones, H. E. (1937). Some A Posteriori Probabilities In Stock Market Action. *Econometrica*, 280–294.

Dickey, D. A., & Fuller, W. A. (1979). Distribution Of The Estimators For Autoregressive Time Series With A Unit Root. *Journal of the American Statistical Association*, *74*(366a), 427–431.

Dickey, D. A., & Fuller, W. A. (1981). Likelihood Ratio Statistics For Autoregressive Time Series With A Unit Root. *Econometrica*, 1057–1072.

Dimand, R. W., & Veloce, W. (2010). Alfred Cowles And Robert Rhea on the Predictability Of Stock Prices. *Journal of Business Inquiry*, *9*(1), 56–64.

Fama, E. F. (1960). Efficient Capital Markets: A Review Of Theory And Empirical Work. *The Journal of Finance*, *25*(2), 383–417.

Fattahi, S. (2010). Weak-Form Efficiency In The German Stock Market. *Iranian Economic Review*, *15*(27), 77–94. DOI: 10.22059/ier.2010.32707

Gan, C., Lee, M., Yong, H. H. A., & Zhang, J. (2005). Revisiting Share Market Efficiency: Evidence From The New Zealand, Australia, US and Japan Stock Indices. *American Journal of Applied Sciences*, *2*(5), 996–1002.

Görgel, B., & Ege, İ. (2024). E7 Ülkeleri Pay Piyasaları Üzerine Etkin Piyasa Hipotezi ile Adaptif Piyasa Hipotezinin Araştırılması. *Fiscaoeconomia*, *8*(3), 1392–1423. DOI: 10.25295/fsecon.1448436

Groenewold, N. (1997). Share Market Efficiency Tests Using Daily Data For Australia and New Zealand. *Applied Financial Economics*, *7*, 645–657.

Güriş, B. (2020). *R Uygulamalı Doğrusal Olmayan Zaman Serileri Analizi*. Der Yayınevi.

Hatipoğlu, M. (2024). Rassal Yürüyüş Hipotezinin Varyans Rasyo Testi ile ESG Endeksleri İçin Araştırılması Üzerine Bir Deneme. *Uluslararası Finansal Ekonomi ve Bankacılık Uygulamaları Dergisi*, *5*(2), 235–252. DOI: 10.57085/ufebud.1551550

Hepsağ, A. (2022). *Ekonometrik Zaman Serileri Analizlerinde Güncel Yöntemler (Winrats Uygulamalı)*. Der Yayınevi.

Hepsağ, A., & Akçalı, B. Y. (2015). The Analysis Of Weak Form Efficiency With Asymmetric Nonlinear Unit Root Test: The Case Of G-7 And E-7 Countries. *Journal of BRSA Banking and Financial Markets*, *9*, 73–90.

Im, K. S., Lee, J., & Tieslau, M. A. (2014). More Powerful Unit Root Tests With Non-Normal Errors. In *Festschrift in honor of Peter Schmidt: Econometric Methods And Applications* (pp. 315–342). Springer New York.

Im, K. S., & Schmidt, P. (2008). More Efficient Estimation Under Non-Normality When Higher Moments Do Not Depend On The Regressors, Using Residual Augmented Least Squares. *Journal of Econometrics*, *144*, 219–233.

Kapetanios, G., Shin, Y., & Snell, A. (2006). Testing For Cointegration In Nonlinear Smooth Transition Error Correction Models. *Econometric Theory*, *22*(2), 279–303.

Kaushik, M. (2020). Testing The Degree Of Efficiency Of Ireland Capital Market With Efficient Market Hypothesis (EMH): A Comparative Analysis Of Ireland Capital Market Efficiency With Its Neighboring Capital Markets Of The UK, Belgium And The Netherlands (Master's thesis, National College of Ireland, Dublin).

Kendall, M. G. (1953). The Analysis of Economic Time-Series —*Part I: Prices. Journal of the Royal Statistical Society. Series A (General), 116*(1), 11–25. DOI: 10.2307/2980947

Kruse, R. (2011). A New Unit Root Test Against ESTAR Based On A Class Of Modified Statistics. *Statistische Hefte*, *52*(1), 71–85.

Küçükkaplan, İ., Kılıç, E., Pazarcı, Ş., & Kar, A. (2023). Testing The Efficient Market Hypothesis In G8 Countries: New Evidence From Unit Root Tests With Fourier Shifts. [Istanbul University, Faculty of Economics.]. *Journal of Economic Policy Researches*, *10*(1), 1–18.

Kwiatkowski, D., Phillips, P. C., Schmidt, P., & Shin, Y. (1992). Testing The Null Hypothesis Of Stationarity Against The Alternative Of A Unit Root: How Sure Are We That Economic Time Series Have A Unit Root? *Journal of Econometrics*, *54*(1-3), 159–178.

Lee, J., & Strazicich, M. C. (2003). Minimum Lagrange Multiplier Unit Root Test With Two Structural Breaks. *The Review of Economics and Statistics*, *85*, 1082–1089.

Lee, J., & Strazicich, M. C. (2004). Minimum LM Unit Root Test With One Structural Break (Manuscript, Department of Economics, Appalachian State University), 1–16.

Li, B., Tong, T., & Benkato, O. (2014). *Revisiting the weak-form efficiency of the Australian stock market.* Corporate Ownership & Control.

Lo, A. W., & MacKinlay, A. C. (1988). Stock Market Prices Do Not Follow Random Walks: Evidence From A Simple Specification Test. *Review of Financial Studies*, *1*(1), 41–66.

Lo, A. W. (2018). Efficient Markets Hypothesis. In The New Palgrave Dictionary of Economics (pp. 3543-3560). Palgrave Macmillan, London.

Lumsdaine, R. L., & Papell, D. H. (1997). Multiple Trend Breaks And The Unit-Root Hypothesis. *The Review of Economics and Statistics*, *79*(2), 212–218.

Meng, M., Im, K. S., Lee, J., & Tieslau, M. A. (2014). More Powerful LM Unit Root Tests With Non-Normal Errors. In *Festschrift In Honor Of Peter Schmidt*. Springer.

Meng, M., Lee, J., & Payne, J. E. (2017). RALS-LM Unit Root Test With Trend Breaks and Non-Normal Errors: Application to the Prebisch-Singer Hypothesis. *Studies in Nonlinear Dynamics and Econometrics*, *21*(1), 31–45.

Merton, R. C. (1973). An Intertemporal Capital Asset Pricing Model. *Econometrica*, 867–887.

Neifar, M., & Gharbi, L. (2023). Weak Efficient Markets Hypothesis And Canadian Stock Markets: Evidence From Linear And Nonlinear Unit Root Tests. *Journal of Islamic Accounting and Business Research, 14*(4), 629–651.

Paskaleva, M., & Stoykova, A. (2021). The Influence Of Uncertainty On Market Efficiency: Evidence From Selected European Financial Markets. [Bulgarian Academy of Sciences–Economic Research Institute.]. *Economic Studies Journal, 8*, 175–198.

Perron, P. (1989). The Great Crash, The Oil Price Shock, And The Unit Root Hypothesis. *Econometrica, 57*, 1391–1401. DOI: 10.2307/1913712

Perron, P., & Ng, S. (1996). Useful Modifications To Some Unit Root Tests With Dependent Errors And Their Local Asymptotic Properties. *The Review of Economic Studies, 63*(3), 435–463.

Phillips, P. C., & Perron, P. (1988). Testing for a Unit Root in Time Series Regression. *Biometrika, 75*(2), 335–346.

Samuelson, P. A. (1965). Proof that Properly Anticipated Prices Fluctuate Randomly In. *Industrial Management Review, 1965*(6), 41–49.

Schmidt, P., & Phillips, P. C. (1992). LM Tests For A Unit Root In The Presence Of Deterministic Trends. *Oxford Bulletin of Economics and Statistics, 54*, 257–287.

Shiller, I., & Radikoko, I. (2014). Testing weak-form market efficiency on the TSX. *Journal of Applied Business Research, 30*(3), 647.

Showalter, S., & Gropp, J. (2019). Validating Weak-form Market Efficiency in United States Stock Markets with Trend Deterministic Price Data and Machine Learning. arXiv preprint arXiv:1909.05151.

Simmons, P. (2011). How efficient was the Australian share market between 2000 and 2008? Business, Economics and Public Policy Working Papers, (2011-1), 2-15.

Sollis, R. (2009). A Simple Unit Root Test Against Asymmetric Star Nonlinearity With An Application To Real Exchange Rates In Nordic Countries. *Economic Modelling, 26*(1), 118–125.

Szylar, C. (2014). *Handbook Of Market Risk*. John Wiley & Sons.

Varoglu, N. (2022). Karbon Vergisinin CO_2 Emisyonu Üzerine Etkisi: Çevresel Kuznets Eğrisi Hipotezi Işığında Bir Uygulama (Yüksek lisans tezi, İstanbul Üniversitesi).

Wald, A., & Wolfowitz, J. (1940). On A Test Whether Two Samples Are From The Same Population. *Annals of Mathematical Statistics*, *11*(2), 147–162. DOI: 10.1214/aoms/1177731909

Yahoo Finance. (2024). Yahoo Finance. https://finance.yahoo.com/

Zivot, E., & Andrews, D. W. K. (2002). Further Evidence on the Great Crash, the Oil-Price Shock, and the Unit-Root Hypothesis. *Journal of Business & Economic Statistics*, *20*(1), 25–44. DOI: 10.1198/073500102753410372

Compilation of References

Abdulla, A., & Baryannis, G. (2024). A hybrid multi-criteria decision-making and machine learning approach for explainable supplier selection. *Supply Chain Analytics*, *7*, 100074. DOI: 10.1016/j.sca.2024.100074

Abraham, S., Noriega, B. R., & Shin, J. Y. (2018). College students eating habits and knowledge of nutritional requirements. *Journal of Nutrition and Human Health*, *2*(1), 13–17. DOI: 10.35841/nutrition-human-health.2.1.13-17

Acciarini, C., Cappa, F., Boccardelli, P., & Oriani, R. (2023). How Can Organizations Leverage Big Data to Innovate Their Business Models? A Systematic Literature Review. *Technovation*, *123*, 102713. DOI: 10.1016/j.technovation.2023.102713

Adejo, O. W., & Connolly, T. (2018). Predicting student academic performance using multi-model heterogeneous ensemble approach. *Journal of Applied Research in Higher Education*, *10*(1), 61–75.

Adekitan, A. I., & Salau, O. (2019). The impact of engineering students' performance in the first three years on their graduation result using educational data mining. *Heliyon*, *5*(2), 1–20.

Adenubi, A. O., Oduroye, A., & Akanni, A. (2024, February). Data Security in Big Data: Challenges, Strategies, and Future Trends. *International Journal of Research in Education Humanities and Commerce*, *05*(02). Advance online publication. DOI: 10.37602/IJREHC.2024.5201

Adeyemo, A. B., & Kuyoro, S. O. (2013). Investigating the Effect of Students Socio-Economic/ Family Background on Students Academic Performance in Teritary Institutions Using Decision Tree Algorithms. *Acta Phytotaxonomica et Geobotanica*, *4*(2), 61–78.

Adiba, M., Castrejon-Castillo, J.-C., Espinosa Oviedo, J. A., Vargas-Solar, G., & Zechinelli-Martini, J.-L. (2016). Big data management challenges, approaches, tools and their limitations. In S. Yu, X. Lin, J. Misic, & X. S. Shen (Eds.), Networking for big data (pp. 1–25). Chapman and Hall/CRC. https://hal.science/hal-01270335

Agarwal, M., & Agarwal, B. B. (2023). Methodological Implementation of Data Miming Classifiers and ANN for Prediction of Accomplishment of Student Education. *International Journal of Performability Engineering*, *19*(9), 587–597.

Ahmadi, A., Keshavarz, M., & Ejlali, F. (2025). Resilience to climate change in agricultural water-scarce areas: The major obstacles and adaptive strategies. *Water Resources Management*, *39*, 1195–1214. DOI: 10.1007/s11269-024-04019-z

Ahmad, Z., & Shahzadi, E. (2018). Prediction of students' academic performance using artificial neural netwok. *Bulletin of Education and Research*, *40*(3), 157–164.

Akdemirel, E. (2024). OECD Ülkelerinde Sağlığın Sosyal Belirleyicileri Üzerine Bir Panel Veri Analizi. *Hacettepe Sağlık İdaresi Dergisi*, *27*(1), 65–90.

Akdemirel, S. (2024). Health outcomes and environmental risk factors in OECD countries: A panel data analysis. *Journal of Health Economics and Policy*, *12*(2), 45–63.

Aladejare, S. A. (2020). Are public spending determinants significant in per capita budget spending decisions in Nigeria? *International Journal of Finance & Economics*, •••, 1–15. DOI: 10.1002/ijfe.2146

Alarcón Garavito, G. A., Moniz, T., Mansilla, C., Iqbal, S., Dobrogowska, R., Bennin, F., Talwar, S., Khalid, A. F., & Vindrola-Padros, C. (2024). Activities used by evidence networks to promote evidence-informed decision-making in the health sector: A rapid evidence review. *BMC Health Services Research*, *24*, 261. DOI: 10.1186/s12913-024-10744-3

Alation. (2025). AI data quality explained: Tools, challenges, fixes. Alation Blog. https://atlan.com/know/ai-data-quality/

Alber, M., Buganza Tepole, A., Cannon, W. R., De, S., Dura-Bernal, S., Garikipati, K., Karniadakis, G. E., Lytton, W. W., Perdikaris, P., Petzold, L., & Kuhl, E. (2019). Integrating machine learning and multiscale modeling—Perspectives, challenges, and opportunities in the biological, biomedical, and behavioral sciences. *NPJ Digital Medicine*, *2*(1), 115. DOI: 10.1038/s41746-019-0193-y PMID: 31799423

Albreiki, B., Zaki, N., & Alashwal, H. A. (2021). A Systematic Literature Review of Student' Performance Prediction Using Machine Learning Techniques. *Education Sciences*, *11*(552), 1–27.

Alfiani, A. P., & Wulandari, F. A. (2015). Mapping Student's Performance Based on Data Mining Approach. *Agriculture and Agricultural Science Procedia, 3*, 173–177.

Algarni, A. (2016). Data mining in Education. *International Journal of Advanced Computer Science and Applications, 7*(6), 456–461.

Al-Hebaish, S. M. (2012). The correlation between general self-confidence and academic achievement in the oral presentation course. *Theory and Practice in Language Studies, 2*(1).

Alhumaidi, N. H., Dermawan, D., Kamaruzaman, H. F., & Alotaiq, N. (2025). The use of machine learning for analyzing real-world data in disease prediction and management: Systematic review. *JMIR Medical Informatics, 13*, e68898. DOI: 10.2196/68898 PMID: 40537090

Alibaba Cloud. (2024). *Safeguarding digital frontiers: A deep dive into Alibaba Cloud fraud detection*. Alibaba Cloud Blog https://www.alibabacloud.com/blog/safeguarding-digital-frontiers-a-deep-dive-into-alibaba-cloud-fraud-detection_602116

Alkaya, A., & Alkaş, C. (2021). *OECD ülkelerinin sağlık göstergelerine göre kümeleme analizi sınıflaması*. Sosyal Güvence., DOI: 10.21441/sosyalguvence.1050881

Almutairi, K. M., Alonazi, W. B., Vinluan, J. M., Almigbal, T. H., Batais, M. A., Alodhayani, A. A., Alsadhan, N., Tumala, R. B., Moussa, M., & Aboshaiqah, A. E. (2018). Health promoting lifestyle of university students in Saudi Arabia : A cross-sectional assessment. *BMC Public Health, 18*, 1–10. DOI: 10.1186/s12889-018-5999-z

Al-Naimi, S., Sooud, M., Alyahyaee, M., Al-Saedi, Z., & Feroz, H. (2025). Relationship Between Academic Stress and Eating Habits Among University of Sharjah Medical Students. *Clinical Psychology Review, 119*, 103890. DOI: 10.70706/j.cpr.2024.103890

Alnassar, F., Blackwell, T., Homayounvala, E., & Yee-king, M. (2021). How Well a Student Performed? A Machine Learning Approach to Classify Students' Performance on Virtual Learning Environment, 2nd *IEEE International Conference on Intelligent Engineering and Management (ICIEM),* 1-6.

Al-Qahtani, A. M. (2022). Lifestyle habits among Najran University students, Najran, Saudi Arabia. *Frontiers in Public Health, 10*, 938062. DOI: 10.3389/fpubh.2022.938062

Alshboul, O., Shehadeh, A., Almasabha, G., & Almuflih, A. S. (2022). Extreme gradient boosting-based machine learning approach for green building cost prediction. *Sustainability*, *14*(11), 6651. DOI: 10.3390/su14116651

Alzahib, R. H., Migdadi, H. M., Al Ghamdi, A. A., Alwahibi, M. S., Ibrahim, A. A., & Al-Selwey, W. A. (2021). Assessment of Morpho-Physiological, Biochemical and Antioxidant Responses of Tomato Landraces to Salinity Stress. *Plants*, *10*(4), 696. DOI: 10.3390/plants10040696

Alzamil, H. A., Alhakbany, M. A., Alfadda, N. A., Almusallam, S. M., & Al-Hazzaa, H. M. (2019). A profile of physical activity, sedentary behaviors, sleep, and dietary habits of Saudi college female students. *Journal of Family & Community Medicine*, *26*(1), 1–8. DOI: 10.4103/jfcm.JFCM_58_18

Amin, A., Zaman, W., & Park, S. (2025). Harnessing multi-omics and predictive modeling for climate-resilient crop breeding: From genomes to fields. *Genes*, *16*(7), 809. DOI: 10.3390/genes16070809 PMID: 40725465

Amrieh, E. A., Hamtini, T., & Aljarah, I. (2016). Mining Educational Data to predict Students' academic performance using Ensemble methods. *International Journal of Database Theory and Application*, *9*(8), 119–136.

Anderegg, L. D. L. (2023). Why can't we predict traits from the environment? *The New Phytologist*, *237*(6), 1998–2004. DOI: 10.1111/nph.18586 PMID: 36308517

Anderson, D.F., (1974), Multivariate Statistical Methods.

Andrae, S. (2024). Explainable Artificial Intelligence in risk management: A framework. In Corazza, M., Shah Khan, F., Torre, D. La, & Masri, H. (eds.). *Artificial intelligence and beyond for finance*. World Scientific Publishing, 149–206. https://doi.org/DOI: 10.1142/9781800615212_0004

Anfal, A.-D., Batarseh, N., Al-Bashabsheh, Z., Batarseh, D., & Karablieh, A. (2024). Stress levels and eating behavior among university students in Jordan : A cross-sectional study. *Journal of Education and Health Promotion*, *13*(1), 285. https://doi.org/DOI. DOI: 10.4103/jehp.jehp_168_24

Angrist, J. D., & Pischke, J.-S. (2009). *Mostly harmless econometrics: An empiricist's companion*. Princeton University Press.

Aouifi, H. E., Hajji, M. E., Es-Saady, Y., & Douzi, H. (2021). Predicting learner's performance through video sequences viewing behavior analysis using educational data-mining. *Education and Information Technologies*, *26*(5), 5799–5814.

Aras, Y. (2024). Sosyoekonomik ve Yaşam Tarzı Faktörlerinin Ruhsal ve Davranışsal Bozukluklara Bağlı Ölümler Üzerindeki Etkisi: Bir Panel Veri Analizi. *Hacettepe Sağlık İdaresi Dergisi*, *27*(3), 445–458.

Ashraf, M., Zaman, M., & Ahmed, M. (2018). Using Ensemble StackingC Method and Base Classifiers to Ameliorate Prediction Accuracy of Pedagogical Data. *Procedia Computer Science*, *132*, 1021–1040.

Ashraf, M., Zaman, M., & Ahmed, M. (2020). An Intelligent Prediction System for Educational Data Mining Based on Ensemble and Filtering approaches. *Procedia Computer Science*, *167*, 1471–1483.

Asif, R., Merceron, A., Ali, S. A., & Haider, N. G. (2017). Analyzing undergraduate students' performance using educational data mining. *Computers & Education*, *113*, 177–194.

Asselman, A., Khaldi, M., & Aammou, S. (2021). Enhancing the prediction of student performance based on the machine learning XGBoost algorithm. *Interactive Learning Environments*, *31*(3), 1–20.

Ayindi, A. Q., Odeniyi, O. A., & Sarumi, O. A. (2013). Mining Parent Socio-Economic Factors to predict Students' Academic Performance in Osum State College of Technology, Esa Oke. *International Journal of Engineering Research & Technology (Ahmedabad)*, *2*(12), 1677–1683.

Bachelier, L. (1900). Théorie de la spéculation. *Annales Scientifiques de l'Ecole Normale Supérieure*, *17*, 21–86.

Badr, G., Algobail, A., Almutairi, H., & Almutery, M. (2016). Predicting Student's Performance in University Courses: A Case Study and Tool in KSU Mathematics Department, Symposium on Data Mining Applications. *Procedia Computer Science*, *82*, 80–89.

Badshah, A., Daud, A., Alharbey, R., Banjar, A., Bukhari, A., & Alshemaimri, B. (2024). Big Data Applications: Overview, Challenges, and Future. *Artificial Intelligence Review*, *57*(290). Advance online publication. DOI: 10.1007/s10462-024-102713

Bagheri, R., Fordoei, A. R., Mousavi, H., & Tahmasebi, P. (2021). Climate-driven abrupt changes in plant communities of desert and semi-desert region. *Theoretical and Applied Climatology*, *146*, 331–348. DOI: 10.1007/s00704-021-03720-y

Bakhshinategh, B., Zaiane, O. R., ElAtia, S., & Ipperciel, D. (2018). Educational data mining applications and tasks: A survey of the last 10 years. *Education and Information Technologies*, *23*, 537–553.

Barredo Arrieta, A., Díaz-Rodríguez, N., Del Ser, J., Bennetot, A., Tabik, S., Barbado, A., García, S., Gil-López, S., Molina, D., Benjamins, R., Chatila, R., & Herrera, F. (2020). Explainable artificial intelligence (XAI): Concepts, taxonomies, opportunities and challenges toward responsible AI. *Information Fusion*, *58*, 82–115. DOI: 10.1016/j.inffus.2019.12.012

Barrett, R. C., Carter, B. R., Fassbender, A. J., Tilbrook, B., Woosley, R. J., Azetsu-Scott, K., Feely, R. A., Goyet, C., Ishii, M., Murata, A., & Pérez, F. F. (2025). Biological responses to ocean acidification are changing the global ocean carbon cycle. Global Biogeochemical Cycles, 39(3), e2024GB008358. https://doi.org/DOI: 10.1029/2024GB008358

Barrington-Trimis, J. L., Urman, R., Leventhal, A. M., ... (2019). *Trends in electronic cigarette use among US youth, 2011–2018*. Nature Medicine. https://www.nature.com/articles/s41533-022-00277-9

Bates, L.S., waldren, R.P., Teare, I.D. (1973). Rapid determination of free proline for water studies. *Plant and Soil*, *39*, 205–208.

Bayram, N., & Bilgel, N. (2008). The prevalence and socio-demographic correlations of depression, anxiety and stress among a group of university students. *Social Psychiatry and Psychiatric Epidemiology*, *43*, 667–672. DOI: 10.1007/s00127-008-0345-x

Beceren, E., Balcı İzgi, B., & Fırat, E. (2021). *Sağlık ve kalkınma ilişkisi OECD ülkelerinde sağlık harcamalarını belirleyen faktörler*. Aksaray Üniversitesi İİBF Dergisi., DOI: 10.52791/aksarayiibd.915196

Becker, R., Enders, W., & Lee, J. (2006). A Stationarity Test In The Presence Of An Unknown Number Of Smooth Breaks. *Journal of Time Series Analysis*, *27*(3), 381–409.

Bedoya, D., Manolakos, E. S., & Novotny, V. (2011). Characterization of biological responses under different environmental conditions: A hierarchical modeling approach. *Ecological Modelling*, *222*(3), 532–545. DOI: 10.1016/j.ecolmodel.2010.10.007

Begum, S., & Padmannavar, S. S. (2022). Genetically optimized ensemble classifiers for multiclass student performance prediction. *Int. J. Intell. Eng. Syst*, *15*(2), 316–328.

Beiter, R., Nash, R., McCrady, M., Rhoades, D., Linscomb, M., Clarahan, M., & Sammut, S. (2015). The prevalence and correlates of depression, anxiety, and stress in a sample of college students. *Journal of Affective Disorders*, *173*, 90–96. DOI: 10.1016/j.jad.2014.10.054

Belli, L., Davoli, L., Oddi, G., Preite, L., Galaverni, M., Ganino, T., & Ferrari, G. (2024). IoT-based dataset of a tomato cultivation under different irrigation regimes. *Mendeley Data*, 2. https://doi.org/DOI: 10.17632/35WH56287Y.2

Belton, V., & Stewart, T. J. (2002). *Multiple criteria decision analysis: An integrated approach.* Springer.

Benlemlih, N., Auajjar, N., & Aammouri, EL, S., Nizar, Y., & IBRIZ, M. (2023). Physicochemical characterization of the sludge generated by the wastewater treatment plant of Kenitra. *Research Journal of Pharmacy and Technology*, *16*(11), 5366–5371. DOI: 10.52711/0974-360X.2023.00869

Benlemlih, N., Khiyati, M. E., El Aammouri, S., & Ibriz, M. (2024). A review on wastewater and its use in agriculture in Morocco: Situation, case study and recommendations. *Research Journal of Pharmacy and Technology*, *17*(10), 5132–5140. DOI: 10.52711/0974-360X.2024.00787

Bennett, E. A., & Grabs, J. (2025). How can sustainable business models distribute value more equitably in global value chains? Introducing “value chain profit sharing” as an emerging alternative to fair trade, direct trade, or solidarity trade. *Business Ethics, the Environment & Responsibility*, *34*(2), 581–601. DOI: 10.1111/beer.12666

Ben-Tal, A., El Ghaoui, L., & Nemirovski, A. (2009). *Robust optimization.* Princeton University Press.

Bere, A., Sithuba, G. H., Mabvuu, C., Mashabela, R., Sigauke, C., & Kyei, K. (2021). A COMPARISON OF LASSO AND GRADIENT BOOSTING. *South African Statistical Journal*, *55*(1), 29–44. DOI: 10.37920/sasj.2021.55.1.3

Bhardwaj, B. K., & Pal, S. (2011). Data Mining: A prediction for performance improvement using classification. *International Journal of Computer Science and Information Security*, *9*(4), 136–140.

Bindhia, K. F., & Suvanam, S. B. (2019). Predicting Academic Performance of Students Using a Hybrid Data Mining Approach. *Journal of Medical Systems*, *43*(162), 1–15.

BioMed Central. (2022). *Global e-cigarette prevalence: Lifetime and current use.* https://archpublichealth.biomedcentral.com/articles/10.1186/s13690-022-00998-w

Black, F., & Scholes, M. (1973). The Pricing of Options and Corporate Liabilities. *Journal of Political Economy*, *81*(3), 637–654.

Blanchet, J., & Murthy, K. (2017). Quantifying distributional model risk via optimal transport. *arXiv*. https://doi.org//arXiv.1604.01446DOI: 10.48550

Blogs, S. A. S. (2024). 6 unique ethical challenges for generative AI in 2025. SAS Blogs. https://blogs.sas.com/content/sascom/?p=70982

Borger, J. G., Ng, A. P., Anderton, H., Ashdown, G. W., Auld, M., Blewitt, M. E., Brown, D. V., Call, M. J., & Collins, P. (2023). Artificial intelligence takes center stage: Exploring the capabilities and implications of ChatGPT and other AI-assisted technologies in scientific research and education. *Immunology and Cell Biology*, *101*(10), 923–935. DOI: 10.1111/imcb.12689

Borges, M. R. (2010). Efficient Market Hypothesis in European Stock Markets. *European Journal of Finance*, *16*(7), 711–726. DOI: 10.1080/1351847X.2010.495477

Borup, D., Christensen, B. J., Mühlbach, N. S., & Nielsen, M. S. (2023). Targeting predictors in random forest regression. *International Journal of Forecasting*, *39*(2), 841–868. DOI: 10.1016/j.ijforecast.2022.02.010

Bradley, R. A., & Terry, M. E. (1952). Rank analysis of incomplete block designs: The method of paired comparisons. *Biometrika*, *39*(3–4), 324–345. DOI: 10.1093/biomet/39.3-4.324

Breiman, L. (2001). Random forests. *Machine Learning*, *45*(1), 5–32. DOI: 10.1023/A:1010933404324

Brewis, A., SturtzSreetharan, C., & Wutich, A. (2018). Obesity stigma as a globalizing health challenge. *Globalization and Health*, *14*, 1–6. DOI: 10.1186/s12992-018-0337-x

Brock, W. A., Dechert, W., & Scheinkman, J. (1987). A Test For Independence Based On The Correlation Dimension. University of Wisconsin-Madison. Working Paper.

Brownstein, N. C., Louis, T. A., O'Hagan, A., & Pendergast, J. (2019). The Role of Expert Judgment in Statistical Inference and Evidence-Based Decision-Making. *The American Statistician, 73*(sup1), 56–68. https://doi.org/DOI: 10.1080/00031305.2018.1529623

Bufacchi, V. (2020). Justice as non-maleficence. *Theoria*, *67*(162), 1–27.

Buldu, A., & Üçgün, K. (2010). Data mining application on students' data. *Procedia: Social and Behavioral Sciences*, *2*, 5251–5259.

Byrne, B. M. (2001). *Structural equation modeling with AMOS: Basic concepts, applications, and programming* (3rd ed.). Routledge.

Bzdok, D., Altman, N., & Krzywinski, M. (2018). Statistics versus machine learning. *Nature Methods*, *15*(4), 233–234. DOI: 10.1038/nmeth.4642 PMID: 30100822

Cai, Y., Cao, X., Liu, B., Lin, H., Luo, H., Liu, F., Su, D., Lv, S., Lin, Z., & Lin, D. (2025). Saline–alkali tolerance evaluation of giant reed (Arundo donax) genotypes under saline–alkali stress at seedling stage. *Agronomy (Basel), 15*(2), 463. DOI: 10.3390/agronomy15020463

Castro García, L., Wang, J., Wang, S., Qiao, X., Kuhl, S., Zhang, H., Rangel de la Tejera, R., & Reed, D. (2025). I-Canopy: A resilient IoT platform for adaptive environmental monitoring at the edge. *Smart Agricultural Technology, 12*, 101611. DOI: 10.1016/j.atech.2025.101611

CDC. (2023). *Youth and tobacco use: Electronic cigarettes.* Centers for Disease Control and Prevention. https://www.cdc.gov/tobacco/e-cigarettes/youth.html

Çevik, N. K. (2021). OECD ülkeleri sağlık sistemi göstergelerine çok boyutlu bir yaklaşım. Iğdır Üniversitesi Sosyal Bilimler Dergisi, 213-245.

Chen, G., Zheng, Y.-D., Wang, J., Xu, J., Huang, Y., Pan, J., Wang, Y., ... "Videollm: Modeling video sequence with large language models." *arXiv preprint arXiv:2305.13292* (2023).

Chen, J. (2025). Data quality quantized framework: Ensuring large-scale data integration in gig economy platforms. *Data quality quantized framework: Ensuring large-scale data integration in gig economy platforms.* https://www.researchgate .net/publication/388769132

Chen, H., Yu, S., Yu, Z., Ma, M., Liu, M., & Pei, H. (2024). Phycoremediation potential of salt-tolerant microalgal species: Motion, metabolic characteristics, and their application for saline–alkali soil improvement in eco-farms. *Microorganisms, 12*, 676. DOI: 10.3390/microorganisms12040676

Chen, J., Hu, H., Wu, H., Jiang, Y., & Wang, C. "Learning the best pooling strategy for visual semantic embedding." In *Proceedings of the IEEE/CVF conference on computer vision and pattern recognition*, pp. 15789-15798. 2021.

Chen, M., Mao, S., & Liu, Y. (2014). Big data: A survey. *Mobile Networks and Applications, 19*(2), 171–209. DOI: 10.1007/s11036-013-0489-0

Chen, X., Singh, M. M., & Geyer, P. (2024). Utilizing domain knowledge: Robust machine learning for building energy performance prediction with small, inconsistent datasets. *Knowledge-Based Systems, 294*, 111774. DOI: 10.1016/j.knosys.2024.111774

Choi, J. W., Hidayat, M. S., Cho, S. B., Hwang, W.-H., Lee, H., Cho, B.-K., Kim, M. S., Baek, I., & Kim, G. (2025). Recent trends in machine learning, deep learning, ensemble learning, and explainable artificial intelligence techniques for evaluating crop yields under abnormal climate conditions. *Plants*, *14*(18), 2841. DOI: 10.3390/plants14182841 PMID: 41011993

ChungHsuanKao. (n.d.). movie_metadata.csv [Data set]. GitHub. https://raw.githubusercontent.com/ChungHsuanKao/1stCapstoneProject_github/master/movie_metadata.csv

Chupryna, R., & Leleko, S. (2025). Predictive maintenance with machine learning: A complete guide. *SPD Technology*. https://spd.tech/machine-learning/predictive-maintenance

Chu, W., & Ghahramani, Z. (2005). Preference learning with Gaussian processes. In *Proceedings of the 22nd International Conference on Machine Learning*, 137–144. https://doi.org/DOI: 10.1145/1102351.1102369

Ciano, T., & Ferrara, M. (2025). *Explainable Multi-criteria Decision Making for tourism economics: integrating XAI with MCDM for a robust accommodation performance assessment*. Decisions in Economics and Finance., DOI: 10.1007/s10203-025-00553-6

Ciccone, D. K., Vian, T., Maurer, L., & Bradley, E. H. (2014). Linking governance mechanisms to health outcomes: A review of the literature in low- and middle-income countries. *Social Science & Medicine*, *117*, 86–95. DOI: 10.1016/j.socscimed.2014.07.010

Codewave. (2024). Machine Learning Pattern Recognition: A Beginner's Guide. https://codewave.com/insights/pattern-recognition-machine-learning/

Cohen, S. M., Iyer, G. C., Brown, M., Macknick, J., Wise, M., Binsted, M., Voisin, N., Rice, J., & Hejazi, M. (2021). How structural differences influence cross-model consistency: An electric sector case study. *Renewable & Sustainable Energy Reviews*, *144*, 111009.

Collado-Soler, R., Alférez-Pastor, M., Torres, F. L., Trigueros, R., Aguilar-Parra, J. M., & Navarro, N. (2023). A systematic review of healthy nutrition intervention programs in kindergarten and primary education. *Nutrients*, *15*(3), 541. DOI: 10.3390/nu15030541

Corso, G., Stark, H., Jegelka, S., Jaakkola, T., & Barzilay, R. (2024). Graph neural networks. *Nature Reviews. Methods Primers*, *4*(1), 17.

Cortés-Ibáñez, J. A., González, S., Valle-Alonso, J. J., Luengo, J., García, S., & Herrera, F. (2020). Preprocessing methodology for time series: An industrial world application case study. *Information Sciences*, *514*, 385–401.

Costa, C. S., Carlos, C., Oliveira, A. A., & Barros, A. N. (2025). Reuse of treated wastewater to address water scarcity in viticulture: A comprehensive review. *Agronomy (Basel)*, *15*(4), 941. DOI: 10.3390/agronomy15040941

Costa, E. B., Fonseca, B., Santana, M. A., Araujo, F. F. D., & Rego, J. (2017). Evaluating the effectiveness of educational data mining techniques for early prediction of students' academic failure in introductory programming courses. *Computers in Human Behavior*, *73*, 247–256.

Cote, C. (2021). 4 types of data analytics to improve decision-making. Harvard Business School Online. https://online.hbs.edu/blog/post/types-of-data-analysis

Coursera. (2024). 7 statistical analysis methods beginners should know. https://www.coursera.org/articles/statistical-analysis-methods

Cowles, A.III. (1933). Can Stock Market Forecasters Forecast? *Econometrica*, 309–324.

Cowles, A.III, & Jones, H. E. (1937). Some A Posteriori Probabilities In Stock Market Action. *Econometrica*, 280–294.

Danach, K., Fouad Aly, W. H., Tarhini, A., & Laouadi, S. (2025). Toward transparent optimization: A systematic review of Explainable AI in decision-making systems. *European Journal of Pure and Applied Mathematics*, *18*(4), 6707. DOI: 10.29020/nybg.ejpam.v18i4.6707

Darmon, N., & Drewnowski, A. (2008). Does social class predict diet quality? *The American Journal of Clinical Nutrition*, *87*(5), 1107–1117. DOI: 10.1093/ajcn/87.5.1107

Datta, S. D., Islam, M., Sobuza, M. H. R., Ahmed, S., & Kar, M. (2024). Artificial intelligence and machine learning applications in the project lifecycle of the construction industry: A comprehensive review. *Advanced Engineering Informatics*, *10*(5), e26888.

David, M. E., David, F. R., & David, F. R. (2009). The quantitative strategic planning matrix (QSPM) applied to a retail computer store. Coastal Business Journal, 8(1), Article 4. https://digitalcommons.coastal.edu/cbj/vol8/iss1/4

David, F. R. (1986). The strategic planning matrix—A quantitative approach. *Long Range Planning*, *19*(5), 102–107. DOI: 10.1016/0024-6301(86)90015-4

David, M. E., David, F. R., & David, F. R. (2017). The quantitative strategic planning matrix: A new marketing tool. *Journal of Strategic Marketing*, *25*(4), 342–352. DOI: 10.1080/0965254X.2016.1148763

de Miranda, R. C., Rauber, F., & Levy, R. B. (2021). Impact of ultra-processed food consumption on metabolic health. *Current Opinion in Lipidology*, *32*(1), 24–37. https://doi.org/DOI. DOI: 10.1097/MOL.0000000000000728

Dean, J. W., & Sharfman, M. P. (1996). Does decision process matter? A study of strategic decision-making effectiveness. *Academy of Management Journal*, *39*(2), 368–392. DOI: 10.2307/256784

Dear, K. (2019). Artificial intelligence and decision-making. *RUSI Journal*, *164*(5–6), 18–25.

Dehghanian, Z., Ahmadabadi, M., Asgari Lajayer, B., Gougerdchi, V., Hamedpour-Darabi, M., Bagheri, N., Sharma, R., Vetukuri, R. R., Astatkie, T., & Dell, B. (2024). Quinoa: A Promising Crop for Resolving the Bottleneck of Cultivation in Soils Affected by Multiple Environmental Abiotic Stresses. *Plants*, *13*(11), 2117. DOI: 10.3390/plants13152117

Del Vecchio, M., Kharlamov, A., Parry, G., & Pogrebna, G. (2021). Improving productivity in Hollywood with data science: Using emotional arcs of movies to drive product and service innovation in entertainment industries. *The Journal of the Operational Research Society*, *72*(5), 1110–1137. DOI: 10.1080/01605682.2019.1705194

Delen, D., Topuz, K., & Eryarsoy, E. (2020). Development of a Bayesian Belief Network-based DSS for predicting and understanding freshmen student attrition. *European Journal of Operational Research*, *281*(3), 575–587.

Deliens, T., Clarys, P., De Bourdeaudhuij, I., & Deforche, B. (2014). Determinants of eating behaviour in university students : A qualitative study using focus group discussions. *BMC Public Health*, *14*, 1–12. DOI: 10.1186/1471-2458-14-53

Demiral, G. A., Şahan, C., Ozgur, E. A., Vatansever, M., & Demiral, Y. (2023). Bir Tekstil Fabrikasında Psikososyal Riskler ve Mental Sağlik Durumu. *Eskişehir Türk Dünyası Uygulama ve Araştırma Merkezi Halk Sağlığı Dergisi*, *8*(1), 54–70.

Demir, S., & Sahin, E. K. (2023). An investigation of feature selection methods for soil liquefaction prediction based on tree-based ensemble algorithms using AdaBoost, gradient boosting, and XGBoost. *Neural Computing & Applications*, *35*, 3173–3190. DOI: 10.1007/s00521-022-07856-4

Deng, D. "DBSCAN clustering algorithm based on density." In *2020 7th international forum on electrical engineering and automation (IFEEA)*, pp. 949-953. IEEE, 2020.

Derici, S., & Dogan, N. O. (2025). A decision model based on artificial intelligence for disease prediction and patient treatment process planning. *Asia Pacific Journal of Health Management*, *20*(2), 1–12.

Desiani, A., Yahdin, S., & Kartikasari, A., & Irmeilyana. (2021). Handling the imbalanced data with missing value elimination SMOTE in the classification of the relevance education background with graduates employment. *International Journal of Artificial Intelligence*, *10*(2), 346–354.

dev-ezzy. (n.d.). tn.movie_budgets.csv [Data set]. GitHub. https://raw.githubusercontent.com/dev-ezzy/phase-1-project/main/tn.movie_budgets.csv

Dev, K. L., Kumar, D. R., & Wipulanusat, W. (2024). Machine learning prediction of the unconfined compressive strength of controlled low strength material using fly ash and pond ash. *Scientific Reports*, *14*, 27540. DOI: 10.1038/s41598-024-78050-w

Dhanekula, A. (2025). AI-driven business intelligence framework for predictive decision-making and strategic resource optimization. *International Journal of Business and Economics Insights*, *5*(3), 1238–1270. DOI: 10.63125/dkt2w457

Dicken, S. J., & Batterham, R. L. (2024). Ultra-processed Food and Obesity : What Is the Evidence? *Current Nutrition Reports*, *13*(1), 23–38. DOI: 10.1007/s13668-024-00517-z

Dickey, D. A., & Fuller, W. A. (1979). Distribution Of The Estimators For Autoregressive Time Series With A Unit Root. *Journal of the American Statistical Association*, *74*(366a), 427–431.

Dickey, D. A., & Fuller, W. A. (1981). Likelihood Ratio Statistics For Autoregressive Time Series With A Unit Root. *Econometrica*, 1057–1072.

Dilmegani, C. (2025). Top 10 applications of deep learning in manufacturing. AI-Multiple. https://research.aimultiple.com/deep-learning-in-manufacturing/

Dimand, R. W., & Veloce, W. (2010). Alfred Cowles And Robert Rhea on the Predictability Of Stock Prices. *Journal of Business Inquiry*, *9*(1), 56–64.

Dissertation India. (2024, February 20). What is the difference between SPSS and Amos? https://www.dissertationindia.com/article/what-is-the-difference-between-spss-and-amos-60

Domingos, P. (2012). A few useful things to know about machine learning. *Communications of the ACM*, *55*(10), 78–87. DOI: 10.1145/2347736.2347755

Dovetail. (2024). 10 statistical analysis methods for research. https://dovetail.com/research/key-statistical-analysis-methods-explained/

Drăgulin, B., tefan, V., Tăbîrcă, A.-I., cheau, M.-C., Radu, V., & Munteanu, V. (2025). AI-driven models for forecasting public expenditures in the digital era. *Electronics (Basel)*, *14*(20), 4047. DOI: 10.3390/electronics14204047

Du, M., Manjunatha, V., Jain, R., Deshpande, R., Dernoncourt, F., Gu, J., Sun, T., & Hu, X. "Towards interpreting and mitigating shortcut learning behavior of NLU models." *arXiv preprint arXiv:2103.06922* (2021).

Duan, Y., Edwards, J. S., & Dwivedi, Y. K. (2019). Artificial intelligence for decision making in the era of Big Data: Evolution, challenges and research agenda. *International Journal of Information Management*, *48*, 63–71.

Dubois, M. K., Gils, J. K., Hanniton, P. A., & Smith, F. (1956). Use of phenol reagent for the determination of total sugar. *Analytical Chemistry*, *28*, 350–356.

Efron, B., & Tibshirani, R. (1993). *An introduction to the bootstrap*. Chapman & Hall.

Eisenhardt, K. M., & Zbaracki, M. J. (1992). Strategic decision making. *Strategic Management Journal*, *13*(S2), 17–37. DOI: 10.1002/smj.4250130904

El Aammouri, S., Benlemlih, N., & Ibriz, M. (2023).. . *Research Journal of Pharmacy and Technology*, *16*(12), 5722–5727. DOI: 10.52711/0974-360X.2023.00925

El Aammouri, S., Benlemlih, N., & Ibriz, M. (2024). Sludge recovery from wastewater treatment plants: A review of current practices and tips. *Research Journal of Pharmacy and Technology*, *17*(8), 4117–4124. DOI: 10.52711/0974-360X.2024.00638

El Ansari, W., Stock, C., & Mikolajczyk, R. T. (2012). Relationships between food consumption and living arrangements among university students in four European countries-a cross-sectional study. *Nutrition Journal*, *11*, 1–7. DOI: 10.1186/1475-2891-11-28

El Arab, R. A., & Al Moosa, O. A. (2025). Systematic review of cost effectiveness and budget impact of artificial intelligence in healthcare. npj. *Digital Medicine*, *8*, 548. DOI: 10.1038/s41746-025-01722-y

Elastic. (2024). Deep learning vs. machine learning: Understanding the differences. Elastic Blog. https://www.elastic.co/blog/deep-learning-vs-machine-learning

El-Khadir, I., Mouniane, Y., Chriqui, A., Kouighat, M., & Hmouni, D. (2025). Effects of Salt Stress on *Origanum majorana* and Improvement via Association with Halophytes Plants: *Spergularia salina* and *Aptenia cordifolia. Acta Universitatis Agriculturae et Silviculturae Mendelianae Brunensis*, *73*(1), 27–35. DOI: 10.11118/actaun.2025.003

El-Khadir, I., Mouniane, Y., Chriqui, A., Oudghiri, M., El Bakkali, M., & Hmouni, D. (2026). Synergistic effects of halophyte intercropping on the salt stress tolerance of Rosmarinus officinalis L.: A sustainable approach for climate-resilient agriculture. In Shah, I., & Jhanjhi, N. (Eds.), *Advances in computational intelligence for climate change security and sustainability* (pp. 187–222). IGI Global Scientific Publishing., DOI: 10.4018/979-8-3693-9132-7.ch008

Elsaraiti, M., & Merabet, A. (2021). A comparative analysis of the arima and lstm predictive models and their effectiveness for predicting wind speed. *Energies*, *14*(20), 6782.

Enders, C. K., Du, H., & Keller, B. T. (2020). A model-based imputation procedure for multilevel regression models with random coefficients, interaction effects, and nonlinear terms. *Psychological Methods*, *25*(1), 88. PMID: 31259566

Ersöz, F. (2008). Türkiye ile OECD ülkelerinin sağlık düzeyleri ve sağlık harcamalarının analizi.

Erwig, M., & Kumar, P. (2022). Explaining results of multi-criteria decision making. *Journal of Multi-Criteria Decision Analysis*, *32*(1), e70011. DOI: 10.1002/mcda.70011

Esteva, A., Kuprel, B., Novoa, R. A., Ko, J., Swetter, S. M., Blau, H. M., & Thrun, S. (2017). Dermatologist-level classification of skin cancer with deep neural networks. *Nature*, *542*(7639), 115–118. DOI: 10.1038/nature21056 PMID: 28117445

Estrera, P. J. M., Natan, P. E., Rivera, B. G. T., & Colarte, F. B. (2017). Student Performance Analysis for Academic Ranking Using Decision Tree Approach in University of Science and Technology of Southern Philippines Senior High School. *International Journal of Engineering and Techniques*, *3*(5), 147–154.

Fama, E. F. (1960). Efficient Capital Markets: A Review Of Theory And Empirical Work. *The Journal of Finance*, *25*(2), 383–417.

Fattahi, S. (2010). Weak-Form Efficiency In The German Stock Market. *Iranian Economic Review*, *15*(27), 77–94. DOI: 10.22059/ier.2010.32707

Feng, G., Fan, M., & Chen, Y. (2022). Analysis and prediction of students' academic performance based on educational data mining. *IEEE Access : Practical Innovations, Open Solutions*, *10*, 19558–19571.

Fernandes, E., Holanda, M., Victorino, M., Borges, V., Carvalho, R., & Van Erven, G. (2019). Educational data mining: Predictive analysis of academic performance of public school students in the capital of Brazil. *Journal of Business Research*, *94*, 335–343.

Fernández-Triana, I., Rubilar, O., Parada, J., Fincheira, P., Benavides-Mendoza, A., Durán, P., Fernández-Baldo, M., Seabra, A. B., & Tortella, G. R. (2024). Metal nanoparticles and pesticides under global climate change: Assessing the combined effects of multiple abiotic stressors on soil microbial ecosystems. *The Science of the Total Environment*, *942*, 173494. DOI: 10.1016/j.scitotenv.2024.173494

Field, A. (2018). *Discovering statistics using IBM SPSS statistics*. Sage publications limited.

Filiz, E., Ertaş, H., & Fidan, Y. (2022). Sağlık Çalışanlarında Psiko-sosyal Risk Faktörlerinin Değerlendirilmesi. Generıc Page, 81.

Främling, K. (2020). Decision theory meets Explainable AI. In Calvaresi, D., Najjar, A., Winikoff, M., & Främling, K. (Eds.), Lecture Notes in Computer Science: Vol. 12175. *Explainable, transparent autonomous agents and multi-agent systems. EXTRAAMAS 2020* (pp. 57–74)., DOI: 10.1007/978-3-030-51924-7_4

Friedman, J. H. (2001). Greedy function approximation: A gradient boosting machine. *Annals of Statistics*, *29*(5), 1189–1232. DOI: 10.1214/aos/1013203451

Gallego-García, S., & García-García, M. (2021). Predictive sales and operations planning based on a statistical treatment of demand to increase efficiency: A supply chain simulation case study. *Applied Sciences (Basel, Switzerland)*, *11*(1), 233. DOI: 10.3390/app11010233

Gan, C., Lee, M., Yong, H. H. A., & Zhang, J. (2005). Revisiting Share Market Efficiency: Evidence From The New Zealand, Australia, US and Japan Stock Indices. *American Journal of Applied Sciences*, *2*(5), 996–1002.

Gandomi, A., & Haider, M. (2015). Beyond the hype: Big data concepts, methods, and analytics. *International Journal of Information Management*, *35*(2), 137–144. DOI: 10.1016/j.ijinfomgt.2014.10.007

Garbanzo, G., Cameira, M. R., & Paredes, P. (2024). The mangrove swamp rice production system of Guinea Bissau: Identification of the main constraints associated with soil salinity and rainfall variability. *Agronomy (Basel)*, *14*(3), 468. DOI: 10.3390/agronomy14030468

GeeksforGeeks. (2017). Decision Tree in Machine Learning. https://www.geeksforgeeks.org/machine-learning/decision-tree-introduction-example/

GeeksforGeeks. (2024). Supervised vs Unsupervised vs Reinforcement Learning. https://www.geeksforgeeks.org/machine-learning/supervised-vs-reinforcement-vs-unsupervised/

Gelman, A., Carlin, J. B., Stern, H. S., Dunson, D. B., Vehtari, A., & Rubin, D. B. (2013). *Bayesian data analysis* (3rd ed.). CRC Press.

Gentsch, P. (2019). *AI in marketing, sales and service: How marketers without a data science degree can use AI, big data and bots*. Springer.

Germano, R. P., Melito, S., Cacini, S., Carmassi, G., Leoni, F., Maggini, R., Montesano, F. F., Pardossi, A., & Massa, D. (2022). Sweet basil can be grown hydroponically at low phosphorus and high sodium chloride concentration: Effect on plant and nutrient solution management. *Scientia Horticulturae*, *304*, 111324. DOI: 10.1016/j.scienta.2022.111324

Ghorbani, R., & Ghousi, R. (2020). Comparing Different Resampling Methods in Predicting Student's Performance Using Machine Learning Techniques. *IEEE Access : Practical Innovations, Open Solutions*, *8*, 7899–67911.

Gil-Garcia, J. R., Helbig, N., & Ojo, A. (2014). Being smart: Emerging technologies and innovation in the public sector. *Government Information Quarterly*, *31*(Suppl. 1), I1–I8. DOI: 10.1016/j.giq.2014.09.001

Gil, P. D., da Cruz Martins, S., Moro, S., & Costa, J. M. (2021). A data-driven approach to predict first-year students' academic success in higher education institutions. *Education and Information Technologies*, *26*(2), 2165–2190.

Gizir, C. A. (2016). *Psikolojik Sağlamlık*. Risk Faktörleri Ve Koruyucu Faktörler Üzerine Bir Derleme Çalışması.

Glantz, S. A., & Bareham, D. W. (2018). E-cigarettes: Use, effects on smoking, risks, and policy implications. *Annual Review of Public Health*, *39*, 215–235.

Global Burden of Disease. (2023). *Global Burden of Disease Study 2023*. Institute for Health Metrics and Evaluation. https://www.healthdata.org/gbd

Google Cloud. (2025). *What is big data?* Google Cloud. Retrieved September 2, 2025, from https://cloud.google.com/learn/what-is-big-data

Görgel, B., & Ege, İ. (2024). E7 Ülkeleri Pay Piyasaları Üzerine Etkin Piyasa Hipotezi ile Adaptif Piyasa Hipotezinin Araştırılması. *Fiscaoeconomia*, *8*(3), 1392–1423. DOI: 10.25295/fsecon.1448436

Gosmann, L., Geitner, C., & Wieler, N. (2022). Data-driven forward osmosis model development using multiple linear regression and artificial neural networks. *Computers & Chemical Engineering*, *165*, 107933. DOI: 10.1016/j.compchemeng.2022.107933

Gray, C. C., & Perkins, D. (2019). Utilizing early engagement and machine learning to predict student outcomes. *Computers & Education*, *131*, 22–32.

Gray, G., McGuinness, C., & Owende, P. (2014). An application of classification models to predict learner progression in tertiary education, *IEEE International Advance Computing Conference (IACC)*, 549–554.

Grobler-Dębska, K., Mularczyk, R., Gawęda, B., & Kucharska, E. (2025). Time series methods and business intelligent tools for budget planning—Case study. *Applied Sciences (Basel, Switzerland)*, *15*(1), 287. DOI: 10.3390/app15010287

Groenewold, N. (1997). Share Market Efficiency Tests Using Daily Data For Australia and New Zealand. *Applied Financial Economics*, *7*, 645–657.

Guan, Z., & Zhao, Y. (2024). Optimizing stock market volatility predictions based on the SMVF-ANP approach. *International Review of Economics & Finance*, *95*, 103502. DOI: 10.1016/j.iref.2024.103502

Guardian. (2025). *Tobacco exposure killed more than 7 million people worldwide: 2023 study*. https://www.theguardian.com/society/2025/jun/24/tobacco-exposure-killed-more-than-7m-people-worldwide-2023-study

Güriş, B. (2020). *R Uygulamalı Doğrusal Olmayan Zaman Serileri Analizi*. Der Yayınevi.

Habibi, N., Terada, N., Sanada, A., & Koshio, K. (2024). Alleviating salt stress in tomatoes through seed priming with polyethylene glycol and sodium chloride combination. *Stresses*, *4*(2), 210–224. DOI: 10.3390/stresses4020012

Ha, D. T., Giap, C. N., Loan, P. T. T., & Huong, N. T. L. (2020). An Empirical Study for Student Academic Performance Prediction Using Machine Learning Techniques. *International Journal of Computer Science and Information Security*, *18*(3), 21–28.

Hair, J. F. (2009). *Multivariate data analysis*. Pearson.

Hair, J. F., Anderson, R. E., Tahtam, R. L., & Black, W. C. (1998). *Multivariate Data Analysis*. Prentice-Hall International.

Hai, T., Basem, A., Alizadeh, A., Singh, P. K., Rajab, H., Maatki, C., Becheikh, N., Kolsi, L., Sawaran Singh, N., & Maleki, H. (2025). Optimizing ternary hybrid nanofluids using neural networks, gene expression programming, and multi-objective particle swarm optimization: A computational intelligence strategy. *Scientific Reports*, *15*, 1986. DOI: 10.1038/s41598-025-85236-3

Haj-Amor, Z., Araya, T., Kim, D.-G., Bouri, S., Lee, J., Ghiloufi, W., Yang, Y., Kang, H., Jhariya, M. K., Banerjee, A., & Lal, R. (2022). Soil salinity and its associated effects on soil microorganisms, greenhouse gas emissions, crop yield, biodiversity and desertification: A review. *The Science of the Total Environment*, *843*, 156946. DOI: 10.1016/j.scitotenv.2022.156946

Halder, R. K., Uddin, M. N., Uddin, M. A., Aryal, S., & Khraisat, A. (2024). Enhancing K-nearest neighbor algorithm: A comprehensive review and performance analysis of modifications. *Journal of Big Data*, *11*(1), 113.

Hamsa, H., Indiradevi, S., & Kizhakkethottam, J. J. (2016). Student academic performance prediction model using decision tree and fuzzy genetic algorithm. *Procedia Technology*, *25*, 326–332.

Hardesty, L. (2017). Explained: Neural networks. *MIT News*. https://news.mit.edu/2017/explained-neural-networks-deep-learning-0414

Harrington, E., Parsons, K., Neptune, L., Winn, C., Watras, E., Raymond, K., Zhou, W., Barr, M., & McNamara, J. (2023). Cooking Self-efficacy Predicts Diet Quality in College Students. *Journal of the Academy of Nutrition and Dietetics*, *123*(9), A59.

Harvard Business School Online. (2019). The advantages of data-driven decision-making. https://online.hbs.edu/blog/post/data-driven-decision-making

Hasan, R., Palaniappan, S., Mahmood, S., Abbas, A., Sarker, K. U., & Sattar, M. U. (2020). Predicting Student Performance in Higher Educational Institutions using Video Learning Analytics and Data Mining Techniques. *Applied Sciences (Basel, Switzerland)*, *10*(3894), 1–20.

Hashim, A. S., Hamoud, A. K., & Awadh, W. A. (2018). Analyzing students' answers using association rule mining based on feature selection. *Journal of Southwest Jiaotong University*, *53*(5).

Hastie, T., Tibshirani, R., & Friedman, J. (2009). *The elements of statistical learning: Data mining, inference, and prediction* (2nd ed.). Springer.

Hatipoğlu, M. (2024). Rassal Yürüyüş Hipotezinin Varyans Rasyo Testi ile ESG Endeksleri İçin Araştırılması Üzerine Bir Deneme. *Uluslararası Finansal Ekonomi ve Bankacılık Uygulamaları Dergisi*, *5*(2), 235–252. DOI: 10.57085/ufebud.1551550

Heiba, Y., Ibrahim, M. G., Mohamed, A. E., Fujii, M., & Nasr, M. (2023). Developing smart sustainable irrigation matrix (SIM)-based model for selection of best irrigation techniques: A framework to achieve SDGs. *Journal of Cleaner Production*, *420*, 138404. DOI: 10.1016/j.jclepro.2023.138404

Helms, M. M., & Nixon, J. (2010). Exploring SWOT analysis – where are we now? A review of academic research from the last decade. *Journal of Strategy and Management*, *3*(3), 215–251. DOI: 10.1108/17554251011064837

He, P., Cai, T., & Gu, J. (2024). A survey of Internet of Medical Things: Technology, applications, and challenges. *Sensors (Basel)*, *24*(2), 36. DOI: 10.1016/j.dcan.2024.11.013

Hepsağ, A. (2022). *Ekonometrik Zaman Serileri Analizlerinde Güncel Yöntemler (Winrats Uygulamalı)*. Der Yayınevi.

Hepsağ, A., & Akçalı, B. Y. (2015). The Analysis Of Weak Form Efficiency With Asymmetric Nonlinear Unit Root Test: The Case Of G-7 And E-7 Countries. *Journal of BRSA Banking and Financial Markets*, *9*, 73–90.

Hernández Hernández, G. C., Gómez Gómez, J., & Jiménez-Cabas, J. (2025). Predictive models based on artificial intelligence to estimate crop yield: A literature review. *Agriculture*, *15*(23), 2438. DOI: 10.3390/agriculture15232438

Hoban, S., Archer, F. I., Bertola, L. D., Bragg, J. G., Breed, M. F., Bruford, M. W., Coleman, M. A., Ekblom, R., & Funk, W. C. & al. (2022). Global genetic diversity status and trends: Towards a suite of Essential Biodiversity Variables (EBVs) for genetic composition. *Biological Reviews of the Cambridge Philosophical Society*, *98*(4), 994–1016. DOI: 10.1111/brv.12852 PMID: 35415952

Hong, L., Luo, M., Wang, R., Lu, P., Lu, W., & Lu, L. (2018, December 1). Big Data in Health Care: Applications and Challenges. *Data and Information Management*, *2*(3), 175–197. DOI: 10.2478/dim-2018-0014

Hort, M., Chen, Z., Zhang, J. M., Harman, M., & Sarro, F. (2024). Bias mitigation for machine learning classifiers: A comprehensive survey. *ACM Journal on Responsible Computing*, *1*(2), 1–52.

Hosseini Nasab, H., & Milani, A. S. (2012). An improvement of quantitative strategic planning matrix using multiple criteria decision making and fuzzy numbers. *Applied Soft Computing*, *12*(8), 2246–2253. DOI: 10.1016/j.asoc.2012.03.010

Huang, L., Qin, J., Zhou, Y., Zhu, F., Liu, L., & Shao, L. (2023). Normalization techniques in training dnns: Methodology, analysis and application. *IEEE Transactions on Pattern Analysis and Machine Intelligence*, *45*(8), 10173–10196. PMID: 37027763

Huang, S., & Fang, N. (2013). Predicting student academic performance in an engineering dynamics course: A comparison of four types of predictive mathematical models. *Computers & Education*, *61*, 133–145.

Hu, K., Gong, S., Zhang, Q., Seng, C., Xia, M., & Jiang, S. (2024). An overview of implementing security and privacy in federated learning. *Artificial Intelligence Review, 57*(8), 204. DOI: 10.1007/s10462-024-10846-8

Hussain, O. A. I., Moehler, R. C., Walsh, S. D. C., & Ahiaga-Dagbui, D. D. (2023). Minimizing cost overrun in rail projects through 5D-BIM: A systematic literature review. *Infrastructures, 8*(5), 93. DOI: 10.3390/infrastructures8050093

Hussain, S., Dahan, N. A., Ba-Alwi, F. M., & Ribata, N. (2018). Educational Data Mining and Analysis of Students' Academic Performance Using WEKA. *Indonesian Journal of Electrical Engineering and Computer Science, 9*(2), 447–459.

Hussain, S., Gaftandzhieva, S., Maniruzzaman, M., Doneva, R., & Muhsin, Z. F. (2021). Regression analysis of student academic performance using deep learning. *Education and Information Technologies, 26*, 783–798.

Hyperstack. (2025). Exploring risk assessment with machine learning in finance. https://www.hyperstack.cloud/blog/case-study/exploring-risk-assessment-with-machine-learning-in-finance

IBM. (2021). What is a Decision Tree? https://www.ibm.com/think/topics/decision-trees

IBM. (2021a). What are Convolutional Neural Networks? IBM Cloud Learn Hub. https://www.ibm.com/think/topics/convolutional-neural-networks

IBM. (2021b). What is a Recurrent Neural Network (RNN)? IBM Research Blog. https://www.ibm.com/think/topics/recurrent-neural-networks

IBM. (2024). What is Big Data? IBM. https://www.ibm.com/think/topics/big-data

IBM. (2024). What is data-driven decision-making? https://www.ibm.com/think/topics/data-driven-decision-making

IBM. (2025). What is a Transformer Model? IBM Cloud Learn Hub. https://www.ibm.com/think/topics/transformer-model

IBM. (2025). What is Machine Learning (ML)? https://www.ibm.com/think/topics/machine-learning

Ibrahim, I. A., Jabbour, A. A., Abdulmajeed, A. M., Elhady, M. E., Almaroai, Y. A., & Hashim, A. M. (2022). Adaptive Responses of Four Medicinal Plants to High Altitude Oxidative Stresses through the Regulation of Antioxidants and Secondary Metabolites. *Agronomy (Basel), 12*, 3032. DOI: 10.3390/agronomy12123032

Idris, M., Hussain, S., & Ahmad, N. (2020). Relationship between parents' education and their children's academic achievement [JASS]. *Journal of Arts & Social Sciences*, *7*(2), 82–92.

Ijiga, O. M., Balogun, S. A., Okika, N., Agbo, O. J., & Enyejo, L. A. (2025). An In-Depth Review of Blockchain-Integrated Logging Mechanisms for Ensuring Integrity and Auditability in Relational Database Transactions. *International Journal of Social Science and Human Research*, *13*(3), 245–259.

Im, K. S., Lee, J., & Tieslau, M. A. (2014). More Powerful Unit Root Tests With Non-Normal Errors. In *Festschrift in honor of Peter Schmidt: Econometric Methods And Applications* (pp. 315–342). Springer New York.

Im, K. S., & Schmidt, P. (2008). More Efficient Estimation Under Non-Normality When Higher Moments Do Not Depend On The Regressors, Using Residual Augmented Least Squares. *Journal of Econometrics*, *144*, 219–233.

Indeed Editorial Team. (2025). 7 types of statistical analysis techniques (and process steps). Indeed. https://www.indeed.com/career-advice/career-development/types-of-statistical-analysis

Injadat, M., Moubayed, A., Nassif, A. B., & Shami, A. (2020). Systematic ensemble model selection approach for educational data mining. *Knowledge-Based Systems*, *200*, 1–16.

Ioannis, B., & Maria, K. (2019). Gender and student course preferences and course performance in computer science departments: A case study. *Education and Information Technologies*, *24*(2), 1269–1291.

IONOS. (2025). What are recurrent neural networks? IONOS Digital Guide. https://www.ionos.co.uk/digitalguide/websites/web-development/recurrent-neural-network/

İpekçi Çetin, E. (2003). Çok Değişkenli Analizlerin Pazarlama ile İlgili Araştırmalarda Kullanımı: 1995-2002 Arası Yazın Taraması, Akdeniz Üniversitesi. İ.İ.B.F. *Dergisi*, *5*, 32–47.

Ishizaka, A., & Labib, A. (2011). Review of the main developments in the analytic hierarchy process. *Expert Systems with Applications*, *38*(11), 14336–14345. DOI: 10.1016/j.eswa.2011.03.063

Ishizaka, A., & Nemery, P. (2013). *Multi-criteria decision analysis: Methods and software*. Wiley.

Jakupi, S., Statovci, B., & Hajrizi, B. L. (2017). Breakeven analysis as a powerful tool in decision-making. *International Journal of Management Excellence*, *9*(3), 1169–1171. DOI: 10.17722/ijme.v9i3.940

James, G., Witten, D., Hastie, T., Tibshirani, R., & Taylor, J. (2023). Tree-based methods. In *An introduction to statistical learning: With applications in Python* (pp. 331–366). Springer International Publishing.

Jang, H. (2019). A decision support framework for robust R&D budget allocation using machine learning and optimization. *Decision Support Systems*, *121*, 1–12. DOI: 10.1016/j.dss.2019.03.010

Jaokar, A. (2025). Statistical inference vs machine learning inference vs deep learning inference. LinkedIn Pulse. https://www.linkedin.com/pulse/statistical-inference-vs-machine-learning-deep-ajit-jaokar-khfue

Javaid, M., Haleem, A., Khan, I. H., & Suman, R. (2023). Understanding the potential applications of artificial intelligence in the agriculture sector. *Advanced Agrochem*, *2*(1), 15–30. DOI: 10.1016/j.aac.2022.10.001

Jia, W., Ma, M., Chen, J., & Wu, S. (2021). Plant morphological, physiological, and anatomical adaptation to flooding stress and the underlying molecular mechanisms. *International Journal of Molecular Sciences*, *22*(3), 1088. DOI: 10.3390/ijms22031088

Jobson, J. D. (1992). *Applied Multivariate Data Analysis*. Springer-Verlag.

Johnson, R. A., & Wichern, D. A. (1992). *Applied Multivariate Statistical Analysis*. Prentice-Hall.

José, M. E. R., Ramos, I. E. C., de Sousa, T. M., & Canella, D. S. (2025). Food consumption associated with depression, anxiety and stress in students entering a public university. *Journal of Nutritional Science*, *14*, e3. https://doi.org/doi:10.1017/jns.2024.90

Kabakchieva, D. (2013). Predicting Student Performance by Using Data Mining Methods for Classification. *Cybernetics and Information Technologies*, *13*(1), 61–72.

Kalaycı, Ş. (2006). *SPSS uygulamalı çok değişkenli istatistik teknikleri, 2.baskı*. Asil Yayın Dağıtım Ltd.Şti.

Kamal, P., & Ahuja, S. (2019). An ensemble-based model for prediction of academic performance of students in undergrad professional course, *Journal of Engineering. Defence Technology*, *17*(4), 769–781.

Kamilaris, A., & Prenafeta-Boldú, F. X. (2018). Deep learning in agriculture: A survey. *Computers and Electronics in Agriculture*, *147*, 70–90. DOI: 10.1016/j.compag.2018.02.016

Kanerika. (2025). *How to address key AI ethical concerns in 2025*. Kanerika Blog. https://kanerika.com/blogs/ai-ethical-concerns/

Kapetanios, G., Shin, Y., & Snell, A. (2006). Testing For Cointegration In Nonlinear Smooth Transition Error Correction Models. *Econometric Theory*, *22*(2), 279–303.

Karthikeyan, V. G., Thangaraj, P., & Karthik, S. (2020). Towards developing hybrid educational data mining model (HEDM) for efficient and accurate student performance evaluation. *Soft Computing*, *24*(24), 18477–18487.

Kassout, J., Hmimsa, Y., Fatehi, S. E., Kadaoui, K., Houssni, M., Chakkour, S., Sahli, A., El Chami, M. A., Ariza-Mateos, D., & Palacios-Rodríguez, G.. (2023). Aridity Gradients Shape Intraspecific Variability of Morphological Traits in Native *Ceratonia siliqua* L. of Morocco. *Plants*, *12*(19), 3447. DOI: 10.3390/plants12193447 PMID: 37836187

Kaushik, M. (2020). Testing The Degree Of Efficiency Of Ireland Capital Market With Efficient Market Hypothesis (EMH): A Comparative Analysis Of Ireland Capital Market Efficiency With Its Neighboring Capital Markets Of The UK, Belgium And The Netherlands (Master's thesis, National College of Ireland, Dublin).

Keeney, R. L., & Raiffa, H. (1993). *Decisions with multiple objectives: Preferences and value tradeoffs* (2nd ed.). Cambridge University Press.

Kendall, M. G. (1953). The Analysis of Economic Time-Series —*Part I: Prices. Journal of the Royal Statistical Society. Series A (General)*, *116*(1), 11–25. DOI: 10.2307/2980947

Kennedy, M. C., Ford, E. D., Singleton, P., Finney, M., & Agee, J. K. (2008). Informed multi-objective decision-making in environmental management using Pareto optimality. *Journal of Applied Ecology*, *45*(1), 181–192. DOI: 10.1111/j.1365-2664.2007.01367.x

Khan, A., & Ghosh, S. K. (2020). Student performance analysis and prediction in classroom learning: A review of educational data mining studies. *Education and Information Technologies*, *26*(1), 205–240.

Khan, S. K. S. M. H., & Rehman, K. (2017). Comparison Between the Academic Performance of Public and Private Schools' Male Students on the Basis of Regularity, Discipline, Punctuality and Environment at Secondary School Level in District Lakki Marwat. *Humanity & Social Sciences Journal*, *7*(7), 53–57.

Khoee, A. G., Yu, Y., & Feldt, R. (2024). Domain generalization through meta-learning: A survey. *Artificial Intelligence Review*, *57*, 285. DOI: 10.1007/s10462-024-10922-z

Khosravi, M., Zare, Z., Mojtabaeian, S. M., & Izadi, R. (2024). Artificial intelligence and decision-making in healthcare: A thematic analysis of a systematic review of reviews. *Health Services Research and Managerial Epidemiology*, *11*. Advance online publication. DOI: 10.1177/23333928241234863 PMID: 38449840

Kijne, J. W. (2006). Abiotic stress and water scarcity: Identifying and resolving conflicts from plant level to global level. *Field Crops Research*, *97*(1), 3–18. DOI: 10.1016/j.fcr.2005.08.011

Kim, H., Sclar, M., Zhou, X., Bras, R., Kim, G., Choi, Y., & Sap, M. "FANToM: A benchmark for stress-testing machine theory of mind in interactions." In *Proceedings of the 2023 Conference on Empirical Methods in Natural Language Processing*, pp. 14397-14413. 2023.

Kitchenham, B., Pfleeger, S. L., McColl, B., & Eagan, S. (2002). An empirical study of maintenance and development estimation accuracy. *Journal of Systems and Software*, *64*(1), 57–77. DOI: 10.1016/S0164-1212(02)00021-3

Kiu, C.-C. (2018). Data Mining Analysis on Student's Academic Performance through Exploration of Student's Background and Social Activities, *IEEE Conference*, 1-5.

Koç, I., & Saraç, M. (2020). Socio-economic, demographic and health determinants of the coronavirus pandemic: Analysis of data from OECD countries.

Koh, E., Sunil, R. S., Lam, H. Y. I., & Mutwil, M. (2024). Confronting the data deluge: How artificial intelligence can be used in the study of plant stress. *Computational and Structural Biotechnology Journal*, *23*, 3454–3466. DOI: 10.1016/j.csbj.2024.09.010

Kotha, K. (2024). "Demystifying Quantum Databases." *Medium*. https://medium.com

Kousar, S., Ahmed, F., Afzal, M., & Trinidad Segovia, J. E. (2023). Is government spending in the education and health sector necessary for human capital development? *Humanities & Social Sciences Communications*, *10*, 62. DOI: 10.1057/s41599-023-01514-3

Kruse, R. (2011). A New Unit Root Test Against ESTAR Based On A Class Of Modified Statistics. *Statistische Hefte*, *52*(1), 71–85.

Küçükkaplan, İ., Kılıç, E., Pazarcı, Ş., & Kar, A. (2023). Testing The Efficient Market Hypothesis In G8 Countries: New Evidence From Unit Root Tests With Fourier Shifts. [Istanbul University, Faculty of Economics.]. *Journal of Economic Policy Researches, 10*(1), 1–18.

Kumar, M., Quteishat, A., & Qtaishat, A. (2022). Performance Comparison of the Optimized Ensemble Model with Existing Classifier Models. *International Journal of Modern Education & Computer Science, 14*(3).

Kumar, S., Datta, S., Singh, V., Datta, D., Singh, S. K., & Sharma, R. (2024). Applications, challenges, and future directions of human-in-the-loop learning. *IEEE Access : Practical Innovations, Open Solutions, 12*, 75735–75760.

Kurttila, M., Pesonen, M., Kangas, J., & Kajanus, M. (2000). Utilizing the analytic hierarchy process (AHP) in SWOT analysis: A hybrid method and its application to a forest-certification case. *Forest Policy and Economics, 1*(1), 41–52. DOI: 10.1016/S1389-9341(99)00004-0

Kwiatkowski, D., Phillips, P. C., Schmidt, P., & Shin, Y. (1992). Testing The Null Hypothesis Of Stationarity Against The Alternative Of A Unit Root: How Sure Are We That Economic Time Series Have A Unit Root? *Journal of Econometrics, 54*(1-3), 159–178.

Lacerda, B., Marcondes, F. S., Lima, H., Durães, D., & Novais, P. (2023). Prediction of Students' Grades Based on Non-academic Data. *International Conference in Methodologies and intelligent Systems for Technology Enhanced Learning,* 87-95.

Larson, N., Neumark-Sztainer, D., Laska, M. N., & Story, M. (2011). Young adults and eating away from home : Associations with dietary intake patterns and weight status differ by choice of restaurant. *Journal of the American Dietetic Association, 111*(11), 1696–1703. DOI: 10.1016/j.jada.2011.08.007

LeCun, Y., Bengio, Y., & Hinton, G. (2015). Deep learning. *Nature, 521*(7553), 436–444. DOI: 10.1038/nature14539 PMID: 26017442

Lee, J., & Strazicich, M. C. (2004). Minimum LM Unit Root Test With One Structural Break (Manuscript, Department of Economics, Appalachian State University), 1–16.

Lee, J., & Strazicich, M. C. (2003). Minimum Lagrange Multiplier Unit Root Test With Two Structural Breaks. *The Review of Economics and Statistics, 85*, 1082–1089.

Lee, Y. K., Oh, H., Lee, B.-M., & Hur, J. (2025). Deciphering and predicting algal bloom variability using size-fractionated organic matter and machine learning in a complex watershed. *Ecological Indicators, 178*, 114067. DOI: 10.1016/j.ecolind.2025.114067

Le, T. T. T., Williams, B., & Mundree, S. G. (2018). An osmotin from the resurrection plant Tripogon loliiformis (TlOsm) confers tolerance to multiple abiotic stresses in transgenic rice. *Physiologia Plantarum, 162*(1), 13–34. DOI: 10.1111/ppl.12585

Li, B., Tong, T., & Benkato, O. (2014). *Revisiting the weak-form efficiency of the Australian stock market.* Corporate Ownership & Control.

Liu, H.-W., Lin, C.-F., & Chen, Y.-J. (2019). Beyond State v Loomis: Artificial intelligence, government algorithmization and accountability. *International Journal of Law and Information Technology, 27*(2), 122–141.

Lo, A. W. (2018). Efficient Markets Hypothesis. In The New Palgrave Dictionary of Economics (pp. 3543-3560). Palgrave Macmillan, London.

Lo, A. W., & MacKinlay, A. C. (1988). Stock Market Prices Do Not Follow Random Walks: Evidence From A Simple Specification Test. *Review of Financial Studies, 1*(1), 41–66.

Lucas, L. V., Brown, C. J., Robertson, D. M., Baker, N. T., Johnson, Z. C., Green, C. T., Cho, S. J., Erickson, M. L., Gellis, A. C., & Jasmann, J. R.. (2025). Gaps in water quality modeling of hydrologic systems. *Water (Basel), 17*(8), 1200. DOI: 10.3390/w17081200

Luce, R. D. (1959). *Individual choice behavior: A theoretical analysis.* Wiley.

Lumsdaine, R. L., & Papell, D. H. (1997). Multiple Trend Breaks And The Unit-Root Hypothesis. *The Review of Economics and Statistics, 79*(2), 212–218.

Lu, T., Luo, P., Wang, J., Lu, Y., Huo, A., & Liu, L. (2025). Soil salinity accumulation and groundwater degradation due to overexploitation over recent 40-year period in Yaoba Oasis, China. *Soil & Tillage Research, 248*, 106398. DOI: 10.1016/j.still.2024.106398

Macieira, A., Barbosa, J., & Teixeira, P. (2021). Food safety in local farming of fruits and vegetables. *International Journal of Environmental Research and Public Health, 18*(18), 9733. DOI: 10.3390/ijerph18189733 PMID: 34574658

Magalhães, V., Severo, M., Correia, D., Torres, D., de Miranda, R. C., Rauber, F., Levy, R., Rodrigues, S., & Lopes, C. (2021). Associated factors to the consumption of ultra-processed foods and its relation with dietary sources in Portugal. *Journal of Nutritional Science, 10*, e89. https://doi.org/doi:10.1017/jns.2021.61

Mahajan, A., Heydari, K., & Powell, D. (2025). Wearable AI to enhance patient safety and clinical decision-making. *npj. Digital Medicine, 8*, 176. DOI: 10.1038/s41746-025-01554-w PMID: 40121336

Malik, V. S., Popkin, B. M., Bray, G. A., Després, J.-P., & Hu, F. B. (2010). Sugar-sweetened beverages, obesity, type 2 diabetes mellitus, and cardiovascular disease risk. *Circulation, 121*(11), 1356–1364. DOI: 10.1161/CIRCULATIONAHA.109.8761

Mardani, A., Jusoh, A., Nor, K. M., Khalifah, Z., Zakwan, N., & Valipour, A. (2015). Multiple criteria decision-making techniques and their applications – A review of the literature from 2000 to 2014. *Economic Research-Ekonomska Istraživanja, 28*(1), 516–571. DOI: 10.1080/1331677X.2015.1075139

Márquez-Vera, C., Cano, A., Romero, C., & Ventura, S. (2012). Predicting student failure at school using genetic programming and different data mining approaches with high dimensional and imbalanced data. *Applied Intelligence, 38*(3), 315–330.

Márquez-Vera, C., Morales, C. R., & Soto, S. V. (2013). Predicting school failure and dropout by using data mining techniques. *IEEE Journal of Latin-American Learning Technologies, 8*(1), 7–14.

Marr, B. (2021). *What are the 4 Vs of Big Data?* BernardMarr.com. https://bernardmarr.com/what-are-the-4-vs-of-big-data/

Marr, B. (2016). *Big data in practice: how 45 successful companies used big data analytics to deliver extraordinary results.* John Wiley & Sons.

Masud, S., Mufarrih, S. H., Qureshi, N. Q., Khan, F., Khan, S., & Khan, M. N. (2019). Academic performance in adolescent students: The role of parenting styles and socio-demographic factors–a cross sectional study from eshawar, Pakistan. *Frontiers in Psychology, 10*, 2497.

Mat, U. B., Buniyamin, N., Arsad, P. M., & Kassim, R. (2013). An overview of using academic analytics to predict and improve students' achievement: A proposed proactive intelligent intervention, in Engineering Education, *2013 IEEE 5th Conference*, 126-130.

Matejová, M., & Paralič, J. (2025). A multi-criteria decision-making approach for the selection of Explainable AI methods. *Machine Learning & Knowledge Extraction, 7*, 158. DOI: 10.3390/make7040158

Matheri, A. N., Mohamed, B., Ntuli, F., Nabadda, E., & Ngila, J. C. (2022). Sustainable circularity and intelligent data-driven operations and control of the wastewater treatment plant. *Physics and Chemistry of the Earth Parts A/B/C, 126*, 103152. DOI: 10.1016/j.pce.2022.103152

Mathew, D. E., Ebem, D. U., Ikegwu, A. C., Ukeoma, P. E., & Dibiaezue, N. F. (2025). Recent emerging techniques in explainable artificial intelligence to enhance the interpretable and understanding of AI models for human. *Neural Processing Letters*, *57*, 16. DOI: 10.1007/s11063-025-11732-2

Mathotaarachchi, K. V., Hasan, R., & Mahmood, S. (2024). Advanced machine learning techniques for predictive modeling of property prices. *Information (Basel)*, *15*(6), 295. DOI: 10.3390/info15060295

Matthew, P., Mchale, S., Deng, X., Nakhla, G., Trovati, M., Nnamoko, N., Pereira, E., Zhang, H., & Raza, M. (2025). A Review of the State of the Art for the Internet of Medical Things. *Sci*, *7*(2), 36. DOI: 10.3390/sci7020036

McLaren, L., & Dutton, D. J. (2020). The social determinants of pandemic impact: An opportunity to rethink what we mean by "public health spending". *Canadian Journal of Public Health = Revue Canadienne de Santé Publique*, *111*(4), 451–453.

Mendez, S., Kubota, J., Widaman, A. M., & Gieng, J. (2021). Advance quantity meal preparation pilot program improves home-cooked meal consumption, cooking attitudes, and self-efficacy. *Journal of Nutrition Education and Behavior*, *53*(7), 608–613. DOI: 10.1016/j.jneb.2020.12.014

Menezes, N. (2024, January). The role of artificial intelligence in modern database management. SQLOPS Blogs. https://www.sqlops.com/tag/artificial-intelligence

Meng, M., Im, K. S., Lee, J., & Tieslau, M. A. (2014). More Powerful LM Unit Root Tests With Non-Normal Errors. In *Festschrift In Honor Of Peter Schmidt*. Springer.

Meng, M., Lee, J., & Payne, J. E. (2017). RALS-LM Unit Root Test With Trend Breaks and Non-Normal Errors: Application to the Prebisch-Singer Hypothesis. *Studies in Nonlinear Dynamics and Econometrics*, *21*(1), 31–45.

Merton, R. C. (1973). An Intertemporal Capital Asset Pricing Model. *Econometrica*, 867–887.

Mikolajczyk, R. T., El Ansari, W., & Maxwell, A. E. (2009). Food consumption frequency and perceived stress and depressive symptoms among students in three European countries. *Nutrition Journal*, *8*, 1–8. DOI: 10.1186/1475-2891-8-31

Milačič Ščančar, R., Kozlica, K., Marković, S., Leban, P., Vidmar, J., Heath, E., Kacjan Maršić, N., Železnikar, Š., & Ščančar, J. (2025). The uptake and translocation of lead, chromium, cadmium, and zinc by tomato plants grown in nutrient and contaminated nutrient solutions: Implications for food safety. *Toxics*, *13*(9), 738. DOI: 10.3390/toxics13090738 PMID: 41012359

Milvus. (2025). How does reinforcement learning differ from supervised and unsupervised learning? Milvus AI Quick Reference. https://milvus.io/ai-quick-reference/how-does-rl-differ-from-supervised-and-unsupervised-learning

Mishra, T., Kumar, D., & Gupta, S. (2014). Mining students' data for prediction performance, *Advanced Computing & Communication Technologies (ACCT), Fourth International Conference IEEE,* 255-262.

MJ, G. M. (2013). Food habits and nutritional assessment in a university population. *Nutrición Hospitalaria, 28*(2), 438–446. DOI: 10.3305/nh.2013.28.2.6303

Mohamad, S. K., & Tasir, Z. (2013). Educational data mining: A review. *Procedia: Social and Behavioral Sciences, 97*, 320–324.

Mokrani, S., Nabti, E.-h., & Cruz, C. (2020). Current advances in plant growth promoting bacteria alleviating salt stress for sustainable agriculture. *Applied Sciences (Basel, Switzerland), 10*(20), 7025. DOI: 10.3390/app10207025

Molaei, A., Aminimehr, A., Tavakoli, A., Kazerouni, A., Azad, B., Azad, R., & Merhof, D. "Implicit neural representation in medical imaging: A comparative survey." In *Proceedings of the IEEE/CVF International Conference on Computer Vision*, pp. 2381-2391. 2023.

Molnar, C. (2022). *Interpretable machine learning*. 2nd ed. Lulu.com. https://christophm.github.io/interpretable-ml-book/

Mouniane, Y., Chriqui, A., El-Khadir, I., Bensaid, A., Arabi, R., Zoubir, M. H., Mabrouki, J., & Hmouni, D. (2024). Investigation on the Phytosociology and Soil Characteristics of Carob Tree (*Ceratonia siliqua* L.) in Northeast Morocco: Implications for Biodiversity Conservation and Ecosystem Restoration. In Mabrouki, J., & Azrour, M. (Eds.), *Integrated Solutions for Smart and Sustainable Environmental Conservation* (Vol. 527, pp. 133–142). Springer., DOI: 10.1007/978-3-031-55787-3_10

Mouniane, Y., Chriqui, A., El-Khadir, I., Hbyaj, K., El-Moustaqim, K., Mezouara, M., Zirari, M., Mabrouki, J., & Hmouni, D. (2024). Impact of remote sensing on irrigation management in precision agriculture. In Mabrouki, J., & Mourade, A. (Eds.), *Technical and technological solutions towards a sustainable society and circular economy* (pp. 1–X). World Sustainability Series. Springer., DOI: 10.1007/978-3-031-56292-1_19

Mouniane, Y., Kadmi, Y., Doubi, M., Tabi, S., El-Khadir, I., Chriqui, A., Soufiani, A., El Ouardi, B., Rachmoune, K., Souissi, A., Sharma, A., Joshi, S., Belkadi, M. C., & Hmouni, D. (2025). Secondary Metabolites of Carob Tree (*Ceratonia siliqua* L.): Potential Applications. In Alexiou, A. (Eds.), *Secondary Metabolites and Their Applications in Various Diseases* (pp. 381–408). IGI Global., DOI: 10.4018/979-8-3693-9112-9.ch011

Mouniane, Y., Vibhute, A. D., Aitouhanni, I., Souissi, A., Kadmi, Y., & Hmouni, D. (2025). Exploring water and salt stress impact on the germination of carob (*Ceratonia siliqua*, Fabaceae) germplasm ecotypes from Morocco, using machine learning for sustainable cultivation. *Journal of Crop Health*, *77*, 159. DOI: 10.1007/s10343-025-01224-7

Mqolombeni, N., Tewari, D. D., & Ilesanmi, K. D. (2023). Exploring the role of high government debt on economic growth: A nonlinearity and threshold analysis for Africa's developing countries. *Economies*, *11*(2), 51. DOI: 10.3390/economies11020051

Mucsi, M., Borsodi, A. K., Megyes, M., & Szili-Kovács, T. (2024). Response of the metabolic activity and taxonomic composition of bacterial communities to mosaically varying soil salinity and alkalinity. *Scientific Reports*, *14*, 7460. DOI: 10.1038/s41598-024-57430-2

Mueen, A., Zafar, B., & Manzoor, U. (2016). Modeling and Predicting Students' Academic Performance Using Data Mining Techniques, *I. J. Modern Education and Computer Science*, *11*, 36–42.

Muller, C. (2018). Parent involvement and academic achievement: An analysis of family resources available to the child, Parents, their children, and schools, *Routledge*, 77–114.

Musso, M. F., Hernandez, C. F. R., & Cascallar, E. C. (2020). Predicting key educational outcomes in academic trajectories: A machine learning approach. *Higher Education*, *80*(5), 875–894.

Mutanov, G. (2015). *Mathematical methods and models in economic planning, management and budgeting*. Springer Nature.

Nahar, K., Shova, B. I., Ria, T., Rashid, H. B., & Islam, A. S. (2021). Mining educational data to predict students performance: A comparative study of data mining techniques. *Education and Information Technologies*, *26*(5), 6051–6067.

Nakhla, G. (in press). Structural equation modelling. In Kaur, B., & Atkins, L. (Eds.), *Elgar encyclopaedia of educational research*. Edward Elgar Publishing., https://research.edgehill.ac.uk/en/publications/structural-equation-modelling

Nandeshwar, A., Menzies, T., & Nelson, A. (2011). Learning patterns of university student retention. *Expert Systems with Applications*, *38*(12), 14984–14996.

Naorem, A., Jayaraman, S., Dang, Y. P., Dalal, R. C., Sinha, N. K., Rao, C. S., & Patra, A. K. (2023). Soil Constraints in an Arid Environment—Challenges, Prospects, and Implications. *Agronomy (Basel)*, *13*(1), 220. DOI: 10.3390/agronomy13010220

Nasiri, M., & Minaei, B. (2012). Predicting GPA and academic dismissal in LMS using educational data mining: A case mining, In E-Learning and E-Teaching (ICE-LET), *Third International Conference, IEEE*, 53-58.

Navarro-Prado, S., Schmidt-RioValle, J., Fernández-Aparicio, Á., Montero-Alonso, M. Á., Perona, J. S., & González-Jiménez, E. (2024). Assessing the Impact of Religion and College Life on Consumption Patterns of Ultra-Processed Foods by Young Adults : A Cross-Sectional Study. *Nutrients*, *16*(11). Advance online publication. DOI: 10.3390/nu16111619

Ndlovu, H. S., Odindi, J., Sibanda, M., Mutanga, O., Clulow, A., Chimonyo, V. G. P., & Mabhaudhi, T. (2021). A comparative estimation of maize leaf water content using machine learning techniques and unmanned aerial vehicle (UAV)-based proximal and remotely sensed data. *Remote Sensing*, *13*(20), 4091. DOI: 10.3390/rs13204091

Neifar, M., & Gharbi, L. (2023). Weak Efficient Markets Hypothesis And Canadian Stock Markets: Evidence From Linear And Nonlinear Unit Root Tests. *Journal of Islamic Accounting and Business Research*, *14*(4), 629–651.

Neural Concept. (2024). Applications of machine learning in mechanical engineering. https://www.neuralconcept.com/post/applications-of-machine-learning-in-mechanical-engineering

Newell, A., & Simon, H. A. (1972). *Human problem solving*. Prentice-Hall.

Nguyen, T., Zhou, L., Spiegler, V., Ieromonachou, P., & Lin, Y. (2018). Big data analytics in supply chain management: A state-of-the-art literature review. *Computers & Operations Research*, *98*, 254–264. DOI: 10.1016/j.cor.2017.07.004

Nile, B. K., Al-Saadi, R. J. M., Abdulameer, L., Al Maimuri, N. M. L., & Al-Dujaili, A. N. (2025). Climate change impacts on river hydraulics: A global synthesis of hydrological shifts, ecological consequences, and adaptive strategies. *Water Conservation Science and Engineering*, *10*, 48. DOI: 10.1007/s41101-025-00375-y

Nkomo, L. M., & Nat, M. (2021). Student Engagement Patterns in a Blended Learning Environment: An Educational Data Mining Approach. *TechTrends*, *65*(5), 808–817.

Noh, Y., Jafarinejad, S., & Anand, P. (2024). A Review on Harnessing Renewable Energy Synergies for Achieving Urban Net-Zero Energy Buildings: Technologies, Performance Evaluation, Policies, Challenges, and Future Direction. *Sustainability, 16*, 3444. DOI: 10.3390/su16083444

Nutt, P. C. (1993). Identifying solution ideas during organizational decision-making. *Management Science, 39*(9), 1071–1085. DOI: 10.1287/mnsc.39.9.1071

Nutt, P. C. (2001). A taxonomy of strategic decisions and tactics for uncovering alternatives. *European Journal of Operational Research, 132*(3), 505–527. DOI: 10.1016/S0377-2217(00)00141-7

Nutt, P. C. (2005). Search during decision making. *European Journal of Operational Research, 160*(3), 851–876. DOI: 10.1016/j.ejor.2003.07.009

Nutt, P. C. (2008). Investigating the success of decision-making processes. *Journal of Management Studies, 45*(2), 425–455. DOI: 10.1111/j.1467-6486.2007.00756.x

O'neil, A., Quirk, S. E., Housden, S., Brennan, S. L., Williams, L. J., Pasco, J. A., Berk, M., & Jacka, F. N. (2014). Relationship between diet and mental health in children and adolescents : A systematic review. *American Journal of Public Health, 104*(10), e31–e42. DOI: 10.2105/AJPH.2014.302110

Oberlender, G. D., & Trost, S. M. (2001). Predicting accuracy of early cost estimates based on estimate quality. *Journal of Construction Engineering and Management, 127*(3), 173–182.

Obesity and overweight. (s. d.). Consulté 18 janvier 2025, à l'adresse https://www.who.int/news-room/fact-sheets/detail/obesity-and-overweight

Obi, J. C., & Jecinta, I. C. (2023). A review of techniques for regularization. *International Journal of Research in Engineering and Science, 11*(1), 360–367.

Obijianya, C. C., Yakamercan, E., Karimi, M., Veluru, S., Simko, I., Eshkabilov, S., & Simsek, H. (2025). Agricultural irrigation using treated wastewater: Challenges and opportunities. *Water (Basel), 17*(14), 2083. DOI: 10.3390/w17142083

OECD. (2023a). [*OECD Indicators*. OECD Publishing. https://www.oecd.org/health/health-at-a-glance]. *Health at a Glance, 2023*, •••.

OECD. (2023b). *Daily smokers and vaping prevalence, 15+ population.* OECD Health Statistics. https://www.oecd.org/data-portal

Oeda, S., & Hashimoto, G. (2017). Log-data clustering analysis for dropout prediction in beginner programming classes. *Procedia Computer Science, 112*, 614–621.

Oğuzlar, A., (1995) "Çok Boyutlu Ölçekleme ve Kümeleme Analizi Arasındaki İlişkiler", Uludağ Üniversitesi İ.İ.B.F. Dergisi. 18(2).

Omotunde, H., & Ahmed, M. (2023). A comprehensive review of security measures in database systems: Assessing authentication, access control, and beyond. *Mesopotamian Journal of CyberSecurity*, *2023*, 115–133.

Ondrasek, G., Rathod, S., Manohara, K. K., Gireesh, C., Anantha, M. S., Sakhare, A. S., Parmar, B., Yadav, B. K., Bandumula, N., Raihan, F., Zielińska-Chmielewska, A., Meriño-Gergichevich, C., Reyes-Díaz, M., Khan, A., Panfilova, O., Seguel Fuentealba, A., Meier Romero, S., Nabil, B., Wan, C., & Shepherd, J. (2022). Salt Stress in Plants and Mitigation Approaches. *Plants*, *11*(6), 717. DOI: 10.3390/plants11060717

Onlinemanipal. (2025, June 2). 10 Popular Regression Algorithms in Machine Learning. https://www.onlinemanipal.com/blogs/popular-regression-algorithms-in-machine-learning

Oracle. (2024, July 16). What is Unsupervised Learning? https://www.oracle.com/uk/artificial-intelligence/machine-learning/unsupervised-learning/

Osmanbegorvic, E., & Suljic, M. (2012). Data Mining Approach for Predicting Student Performance, Economic Review-. *Journal of Economics and Business*, *10*(1), 3–12.

Oudghiri, M., Mouniane, Y., Bouhassoun, S., El Aammouri, S., Yamani, B., Benlemlih, N., Abid, N., Elwahab, F. A., Zoubir, M. H., El-Khadir, I., Chriqui, A., Aitouhanni, I., El Bakkali, M., El Ouardi, B., & Ibriz, M. (2025). Abiotic stress and climate change: Technological innovations and adaptation strategies for sustainable resilience. In Ahmadi, A., & Mna, A. (Eds.), *Nexus of environmental quality and technology innovation* (pp. 1–16). IGI Global., DOI: 10.4018/979-8-3373-1822-6.ch001

Oussous, A., Benjelloun, F.-Z., Lahcen, A. A., & Belfkih, S. (2018, October). Big Data Technologies: A Survey. *Journal of King Saud University. Computer and Information Sciences*, *30*(4), 431–448. DOI: 10.1016/j.jksuci.2017.06.001

Ozcelik, M. M., Kok, I., & Ozdemir, S. (2025). A Survey on Internet of Medical Things (IoMT): Enabling Technologies, Security and Explainability Issues, Challenges, and Future Directions. *Expert Systems: International Journal of Knowledge Engineering and Neural Networks*, *42*(5), e70010. DOI: 10.1111/exsy.70010

Pal, A. K., & Pal, S. (2013). Analysis and Mining of Educational Data for Predicting the Performance of Students. *International Journal of Electronics Communication and Computer Engineering*, *4*(5), 1560–1565.

Palipudi, K. M.. (2016). Age and gender differences in tobacco use: Findings from the Global Adult Tobacco Survey. *Nicotine & Tobacco Research: Official Journal of the Society for Research on Nicotine and Tobacco, 18*(5), 1230–1241.

Pande, S. M. (2023). Machine Learning Models for Student Performance Prediction. *International Conference on Innovative Data Communication Technologies and Application,* 27-32.

Pandey, M., & Taruna, S. (2016). Towards the integration of multiple classifier pertaining to the Student's performance prediction. *Perspectives in Science, 8,* 364–366.

Pantic, I., Paunovic, J., Pejic, S., Drakulic, D., Todorovic, A., Stankovic, S., Vucevic, D., Cumic, J., & Radosavljevic, T. (2022). Artificial intelligence approaches to the biochemistry of oxidative stress: Current state of the art. *Chemico-Biological Interactions, 358,* 109888. DOI: 10.1016/j.cbi.2022.109888

Parajuli, M., & Thapa, A. (2017). Gender differences in the academic performance of students. *Journal of Development and Social Engineering, 3*(1), 39–47.

Parra-López, C., Ben Abdallah, S., Garcia-Garcia, G., Hassoun, A., Sánchez-Zamora, P., Trollman, H., Jagtap, S., & Carmona-Torres, C. (2024). Integrating digital technologies in agriculture for climate change adaptation and mitigation: State of the art and future perspectives. *Computers and Electronics in Agriculture, 226,* 109412. DOI: 10.1016/j.compag.2024.109412

Pasipanodya, J. T., Horn, L. N., Achigan-Dako, E. G., Musango, R., & Sibiya, J. (2022). Utilization of Plant Genetic Resources of Bambara Groundnut Conserved Ex Situ and Genetic Diversification of Its Primary Genepool for Semi-Arid Production. *Agriculture, 12,* 492. DOI: 10.3390/agriculture12040492

Paskaleva, M., & Stoykova, A. (2021). The Influence Of Uncertainty On Market Efficiency: Evidence From Selected European Financial Markets. [Bulgarian Academy of Sciences–Economic Research Institute.]. *Economic Studies Journal, 8,* 175–198.

Paubox. (2025). Deep learning in healthcare. Paubox Blog. https://www.paubox.com/blog/deep-learning-in-healthcare

Pecan, A. I. (2025). Three Types of Machine Learning You Should Know. https://www.pecan.ai/blog/3-types-of-machine-learning/

Perron, P. (1989). The Great Crash, The Oil Price Shock, And The Unit Root Hypothesis. *Econometrica, 57,* 1391–1401. DOI: 10.2307/1913712

Perron, P., & Ng, S. (1996). Useful Modifications To Some Unit Root Tests With Dependent Errors And Their Local Asymptotic Properties. *The Review of Economic Studies*, *63*(3), 435–463.

Phadermrod, B., Crowder, R. M., & Wills, G. B. (2019). Importance–performance analysis based SWOT analysis. *International Journal of Information Management*, *44*, 194–203. DOI: 10.1016/j.ijinfomgt.2016.03.009

Phillips, P. C., & Perron, P. (1988). Testing for a Unit Root in Time Series Regression. *Biometrika*, *75*(2), 335–346.

Pichler, M., & Hartig, F. (2023). Machine learning and deep learning—A review for ecologists. *Methods in Ecology and Evolution*, *14*(4), 994–1016. DOI: 10.1111/2041-210X.14061

Pickton, D. W., & Wright, S. (1998). What's SWOT in strategic analysis? *Strategic Change*, *7*(2), 101–109. DOI: 10.1002/(SICI)1099-1697(199803/04)7:2

Pierson, K., Hand, M. L., & Thompson, F. (2015). The Government Finance Database: A common resource for quantitative research in public financial analysis. *PLoS One*, *10*(6), e0130119. DOI: 10.1371/journal.pone.0130119

Pikhart, H., & Pikhartova, J. (2015b). The relationship between psychosocial risk factors and health outcomes of chronic diseases: a review of the evidence for cancer and cardiovascular diseases.

Pikhart, H., & Pikhartova, J. (2015a). Psychosocial factors and health in Europe: Evidence from multi-country studies. *Social Science & Medicine*, *138*, 73–80.

Pinek, L., Mansour, I., Lakovic, M., Ryo, M., & Rillig, M. C. (2020). Rate of environmental change across scales in ecology. *Biological Reviews of the Cambridge Philosophical Society*, *95*(6), 1798–1811. DOI: 10.1111/brv.12639 PMID: 32761787

Pisal, P., Reddy, K. K., Kishore, J., Jonnalagadda, R. R., Kumar, M., & Band, G. (2025). An integrated TOPSIS and ARAS method multi-criteria decision-making approach for optimizing investment portfolios using goal programming and genetic algorithm model. *Scientific Reports*, *15*, 34450. DOI: 10.1038/s41598-025-17604-y PMID: 41039038

Polireddi, N. S. A. (2024). An effective role of artificial intelligence and machine learning in banking sector. *Measurement. Sensors*, *33*, 101135. DOI: 10.1016/j.measen.2024.101135

Popkin, B. M., Corvalan, C., & Grummer-Strawn, L. M. (2020). Dynamics of the double burden of malnutrition and the changing nutrition reality. *Lancet, 395*(10217), 65–74. DOI: 10.1016/S0140-6736(19)32497-3

Poturić, V. Č., Čandrlić, S., & Dražić, I. (2023). Predicting Students' Academic Success Based on Various Course Activities: A Case Study. In *The Learning Ideas Conference,* 665-675.

Puhl, R. M., & Heuer, C. A. (2010). Obesity stigma : Important considerations for public health. *American Journal of Public Health, 100*(6), 1019–1028.

Puhl, R. M., Himmelstein, M. S., & Pearl, R. L. (2020). Weight stigma as a psychosocial contributor to obesity. *The American Psychologist, 75*(2), 274. DOI: 10.1037/amp0000538

Pure Search. (2025). Quantitative analytics in the age of AI and big data. https://www.puresearch.com/blog/quantitative-analytics-in-the-age-of-ai-and-big-data

Qadir, M. (2024). Transition from traditional machine learning to neural networks. LinkedIn Pulse. https://www.linkedin.com/pulse/transition-from-traditional-machine-learning-neural-networks-qadir-ujowc

Qadir, M., Drechsel, P., Jiménez Cisneros, B., Kim, Y., Pramanik, A., Mehta, P., & Olaniyan, O. (2020). Global and regional potential of wastewater as a water, nutrient and energy source. *Natural Resources Forum, 44*(1), 40–51. DOI: 10.1111/1477-8947.12187

Rabie, A. B., Elhag, M., & Subyani, A. (2025). Remote sensing, GIS, and machine learning in water resources management for arid agricultural regions: A review. *Water (Basel), 17*(21), 3125. DOI: 10.3390/w17213125

Rahman, S. R., Islam, M. A., Akash, P. P., Parvin, M., Moon, N. N., & Nur, F. N. (2021). Effects of co-curricular activities on student's academic performance by machine learning. *Current Research in Behavioral Sciences, 2*(100057), 1–7.

Rahul, K., Banyal, R. K., & Arora, N. (2023). A Systematic Review on Big Data Applications and Scope for Industrial Processing and Healthcare Sectors. *Journal of Big Data, 10,* 133. DOI: 10.1186/s40537-023-00808-2

Rai, A., & Dong, Y. (2025). A scoping review of irrigation scheduling methods in potato (*Solanum tuberosum* L.) production. *American Journal of Potato Research, 102,* 348–371. DOI: 10.1007/s12230-025-09997-3

Ramaswami, G., Susnjak, T., & Mathrani, A. (2022). On developing generic models for predicting student outcomes in educational data mining. *Big Data and Cognitive Computing*, *6*(1), 1–16.

Rane, J., Mallick, S. K., Kaya, O., & Rane, N. L. (2024). Scalable and adaptive deep learning algorithms for large-scale machine learning systems. In *Future Research Opportunities for Artificial Intelligence in Industry 4.0 and 5.0* (pp. 39–92). Deep Science Publishing., DOI: 10.70593/978-81-981271-0-5_2

Redman, B. K. (2023). Rebalancing commercial and public interests in prioritizing biomedical, social and environmental aspects of health through defining and managing conflicts of interest. *Frontiers in Medicine*, *10*, 1247258. DOI: 10.3389/fmed.2023.1247258 PMID: 37809337

Rentsen, E., Chuluundorj, B., & Natsagdorj, T. (2023). Optimization approach to constrained breakeven points with respect to price. iBusiness, 15(3), 183–191. https://doi.org/DOI: 10.4236/ib.2023.153014

Riahi, Y. (2018, November). Big Data and Big Data Analytics: Concepts, Types, and Technologies. *International Journal of Research and Engineering*, *5*(9), 524–528. DOI: 10.21276/ijre.2018.5.9.5

Ribeiro, M. T., Singh, S., & Guestrin, C. (2016). "Why should I trust you?" Explaining the predictions of any classifier. Proceedings of the 22nd ACM SIGKDD International Conference on Knowledge Discovery and Data Mining, 1135–1144. https://doi.org/DOI: 10.1145/2939672.2939778

Ribeiro, N., Malafaia, C., Neves, T., & Menezes, I. (2024). The impact of extra-curricular activities on university students' academic success and employability. *European Journal of Higher Education*, *14*(3), 389–409.

Rico-Chávez, A. K., Franco, J. A., Fernandez-Jaramillo, A. A., Contreras-Medina, L. M., Guevara-González, R. G., & Hernandez-Escobedo, Q. (2022). Machine learning for plant stress modeling: A perspective towards hormesis management. *Plants*, *11*(7), 970. DOI: 10.3390/plants11070970

Rivera, P., Moya, C., & O'Brien, J. A. (2022). Low salt treatment results in plant growth enhancement in tomato seedlings. *Plants*, *11*(6), 807. DOI: 10.3390/plants11060807

Rodgers, R. F., & DuBois, R. H. (2016). Cognitive biases to appearance-related stimuli in body dissatisfaction : A systematic review. *Clinical Psychology Review*, *46*, 1–11. DOI: 10.1016/j.cpr.2016.04.006

Rodrigues, M. W., Zárate, L. E., & Isotani, S. (2018). Educational Data Mining: A review of evaluation process in the e-learning. *Telematics and Informatics*, •••, 1–25.

Rokach, L., Maimon, O., & Shmueli, E. (2023). *Machine learning for data science handbook*. Springer., DOI: 10.1007/978-3-031-24628-9

Roslan, M. H. B., & Chen, C. J. (2023). Predicting students' performance in English and Mathematics using data mining techniques. *Education and Information Technologies*, *28*(2), 1427–1453.

Ross, S. M. (2014). *Introduction to probability models* (11th ed.). Academic Press.

Rouyendegh, B. D., & Savalan, Ş. (2022). An integrated fuzzy MCDM hybrid methodology to analyze agricultural production. *Sustainability*, *14*(8), 4835. DOI: 10.3390/su14084835

Roy, B. (1996). *Multicriteria methodology for decision aiding*. Springer.

Russell, S., & Norvig, P. (2002). *Artificial intelligence: A modern approach*. Prentice Hall.

Saa, A. A., Al-Emran, M., & Shaalan, K. (2019). Factors affecting students' performance in higher education: A systematic review of predictive data mining techniques, *Technology. Knowledge and Learning*, *24*(4), 567–598.

Saaty, T. L. (1980). *The analytic hierarchy process: Planning, priority setting, resource allocation* (2nd ed.). McGraw-Hill., https://archive.org/details/analytichierarch0000saat

Sabates-Wheeler, R., & Devereux, S. (2008). Transformative social protection: The currency of social justice. In Barrientos, A., & Hulme, D. (Eds.), *Social protection for the poor and poorest (Palgrave Studies in Development)*. Palgrave Macmillan.

Saha, B. (2025). AI-driven predictive analytics for financial risk management and financial risk prevention with ranking cost scheduling. [IJARI]. *International Journal of Advance Research and Innovation*, *13*(1), 12–22. DOI: 10.69996/ijari.2025002

Şahman, O., & Gün, İ. (2024). Sağlığın sosyal belirleyicilerinin MOORA yöntemi ile analizi: OECD ülkeleri örneği. *Sosyal Güvence*, *26*, 1376–1399. DOI: 10.21441/sosyalguvence.1507454

Salcedo, F., Pérez Cutillas, P., Alarcón Cabañero, J. J., & Vivaldi, A. G. (2022). Use of remote sensing to evaluate the effects of environmental factors on soil salinity in a semi-arid area. *The Science of the Total Environment*, *815*, 152524. DOI: 10.1016/j.scitotenv.2021.152524

Salimi, A., Rostami, J., Moormann, C., & Delisio, A. (2016). Application of non-linear regression analysis and artificial intelligence algorithms for performance prediction of hard rock TBMs. *Tunnelling and Underground Space Technology, 58*, 236–246. DOI: 10.1016/j.tust.2016.05.009

Samuelson, P. A. (1965). Proof that Properly Anticipated Prices Fluctuate Randomly In. *Industrial Management Review, 1965*(6), 41–49.

San Cristóbal Mateo, J. R. (2015). *Management science, operations research and project management: Modelling, evaluation, scheduling, monitoring* (1st ed.). Routledge., DOI: 10.4324/9781315593449

Santiago-Rosario, L. Y., Harms, K. E., Elderd, B. D., Hart, P. B., & Dassanayake, M. (2021). No escape: The influence of substrate sodium on plant growth and tissue sodium responses. *Ecology and Evolution, 11*(20), 14231–14249. DOI: 10.1002/ece3.8138

Saravi, B., Nejadhashemi, A. P., Jha, P., & Tang, B. (2021). Reducing deep learning network structure through variable reduction methods in crop modeling. *Artificial Intelligence in Agriculture, 5*, 196–207. DOI: 10.1016/j.aiia.2021.09.001

Sarker, B. K., Sarker, D. K., Shaha, S. R., Saha, D., & Sarker, S. (2024). Why apply SPSS, SmartPLS and AMOS: An essential quantitative data analysis tool for business and social science research investigations. [IJRISS]. *International Journal of Research and Innovation in Social Science, 8*(9), 2688–2699. DOI: 10.47772/IJRISS.2024.8090224

Sarker, I. H. (2021). Machine learning: Algorithms, real-world applications and research directions. *SN Computer Science, 2*, 160. DOI: 10.1007/s42979-021-00592-x PMID: 33778771

Sarker, S., Paul, M. K., Thasin, S. T. H., & Hasan, M. A. M. (2024). Analyzing students' academic performance using educational data mining. *Computers and Education: Artificial Intelligence, 7*, 100263.

SAS. (2023). A guide to the types of machine learning algorithms. https://www.sas.com/en_gb/insights/articles/analytics/machine-learning-algorithms.html

Sayılı, U., Sayman, Ö.A., Vehid, S., Köksal, S.S., & Erginöz, E. (2017). Türkiye ve OECD Ülkelerinin Sağlık Göstergeleri ve Sağlık Harcamalarının Karşılaştırılması.

Schmidhuber, J. (2015). Deep learning in neural networks: An overview. *Neural Networks, 61*, 85–117. DOI: 10.1016/j.neunet.2014.09.003 PMID: 25462637

Schmidt, P., & Phillips, P. C. (1992). LM Tests For A Unit Root In The Presence Of Deterministic Trends. *Oxford Bulletin of Economics and Statistics*, *54*, 257–287.

Seldon. (2021). *What is machine learning?* Seldon. Retrieved September 10, 2025, from https://www.seldon.io/what-is-machine-learning/

Select Statistics. (2024). Evidence-based decision making. https://select-statistics.co.uk/videos/evidence-based-decision-making/

Selva86. (n.d.). BostonHousing.csv [Data set]. GitHub. https://raw.githubusercontent.com/selva86/datasets/master/BostonHousing.csv

Sembiring, S., Zarlis, M., Hartama, D., Ramliana, S., & Wani, E. (2011). Prediction of student academic performance by an application of data mining techniques, *International Conference on Management and Artificial Intelligence IPEDR*, 6, 110-114.

Senanayake, I. P., Pathira Arachchilage, K. R. L., Yeo, I.-Y., Khaki, M., Han, S.-C., & Dahlhaus, P. G. (2024). Spatial downscaling of satellite-based soil moisture products using machine learning techniques: A review. *Remote Sensing*, *16*(12), 2067. DOI: 10.3390/rs16122067

Shafiee, A., Nakhaee, Z., Bahri, R. A., Amini, M. J., Salehi, A., Jafarabady, K., Seighali, N., Rashidian, P., Fathi, H., & Esmaeilpur Abianeh, F. (2024). Global prevalence of obesity and overweight among medical students : A systematic review and meta-analysis. *BMC Public Health*, *24*(1), 1673. DOI: 10.1186/s12889-024-19184-4

Shahid, M. A., Sarkhosh, A., Khan, N., Balal, R. M., Ali, S., Rossi, L., Gómez, C., Mattson, N., Nasim, W., & Garcia-Sanchez, F. (2020). Insights into the physiological and biochemical impacts of salt stress on plant growth and development. *Agronomy (Basel)*, *10*(7), 938. DOI: 10.3390/agronomy10070938

Shahiri, A. M., Husain, W., & Rashid, N. A. (2015). A Review on Predicting Student's Performance using Data Mining Techniques. *Procedia Computer Science*, *72*, 414–422.

Shapiro, A., Dentcheva, D., & Ruszczyński, A. (2014). *Lectures on stochastic programming: Modeling and theory* (2nd ed.). SIAM.

Sharabiani, A., Karim, F., Sharabiani, A., Atanasov, M., & Darabi, H. (2014). An Enhanced Bayesian Network Model for Prediction of Students' Academic Performance in Engineering Programs, *IEEE Global Engineering Education Conference*, 832-837.

Sharma, S. (1996). *Applied Multivariate Techniques*. John Wiley&Sons, Inc.

Shen, T., Xu, F., Chen, D., Yan, R., Wang, Q., Li, K., Zhang, G., Ni, L., & Jiang, M. (2024). A B-box transcription factor OsBBX17 regulates saline-alkaline tolerance through the MAPK cascade pathway in rice. *The New Phytologist*, *241*(5), 2158–2175. DOI: 10.1111/nph.19480

Shethiya, A. S. (2023). Learning to Learn: Advancements and Challenges in Modern Machine Learning Systems. *Annals of Applied Sciences*, *4*(1).

Shiller, I., & Radikoko, I. (2014). Testing weak-form market efficiency on the TSX. *Journal of Applied Business Research*, *30*(3), 647.

Shi, Y., & Yao, J. (2025). Explainable multi-criteria decision-making: A three-way decision perspective. *International Journal of Approximate Reasoning*, *187*, 109528. DOI: 10.1016/j.ijar.2025.109528

Showalter, S., & Gropp, J. (2019). Validating Weak-form Market Efficiency in United States Stock Markets with Trend Deterministic Price Data and Machine Learning. arXiv preprint arXiv:1909.05151.

Simmons, P. (2011). How efficient was the Australian share market between 2000 and 2008? Business, Economics and Public Policy Working Papers, (2011-1), 2-15.

Singh, A., Rajput, V. D., Sharma, R., Ghazaryan, K., & Minkina, T. (2023). Salinity stress and nanoparticles: Insights into antioxidative enzymatic resistance, signaling, and defense mechanisms. *Environmental Research*, *235*, 116585. DOI: 10.1016/j.envres.2023.116585

Singh, D., Chaudhary, P., Taunk, J., Singh, C. K., Singh, D., Tomar, R. S. S., Aski, M., Konjengbam, N. S., Raje, R. S., Singh, S., Sengar, R. S., Yadav, R. K., & Pal, M. (2021). Fab advances in Fabaceae for abiotic stress resilience: From 'omics' to artificial intelligence. *International Journal of Molecular Sciences*, *22*(19), 10535. DOI: 10.3390/ijms221910535

Singh, K. M., Baksi, S., Rani, S., Jha, A. B., Dubey, R. S., & Sharma, P. (2024). NanoBoost: Maximizing crop resilience and yield via nanopriming under salt stress. *Environmental and Experimental Botany*, *226*, 105937. DOI: 10.1016/j.envexpbot.2024.105937

Singh, K., Yadav, M., Barak, D., Bansal, S., & Moreira, F. (2025). Machine-learning-based frameworks for reliable and sustainable crop forecasting. *Sustainability*, *17*(10), 4711. DOI: 10.3390/su17104711

Sinval, J., Oliveira, P., Novais, F., Almeida, C. M., & Telles-Correia, D. (2025). Exploring the impact of depression, anxiety, stress, academic engagement, and dropout intention on medical students' academic performance: A prospective study. *Journal of Affective Disorders*, *368*, 665–673.

Smith, R., Kelly, B., Yeatman, H., & Boyland, E. (2019). Food marketing influences children's attitudes, preferences and consumption : A systematic critical review. *Nutrients*, *11*(4), 875. DOI: 10.3390/nu11040875

Sogari, G., Velez-Argumedo, C., Gómez, M. I., & Mora, C. (2018). College students and eating habits : A study using an ecological model for healthy behavior. *Nutrients*, *10*(12), 1823. DOI: 10.3390/nu10121823

Sollis, R. (2009). A Simple Unit Root Test Against Asymmetric Star Nonlinearity With An Application To Real Exchange Rates In Nordic Countries. *Economic Modelling*, *26*(1), 118–125.

Soltabayeva, A., Ongaltay, A., Omondi, J. O., & Srivastava, S. (2021). Morphological, Physiological and Molecular Markers for Salt-Stressed Plants. *Plants*, *10*(2), 243. DOI: 10.3390/plants10020243

Sprake, E., Russell, J., Cecil, J., Cooper, R., Grabowski, P., Pourshahidi, L. K., & Barker, M. (2018). Dietary patterns of university students in the UK: A cross-sectional study. *Nutrition Journal*, *17*, 1–17. DOI: 10.1186/s12937-018-0398-y

Spronk, I., Kullen, C., Burdon, C., & O'Connor, H. (2014). Relationship between nutrition knowledge and dietary intake. *British Journal of Nutrition*, *111*(10), 1713–1726. https://doi.org/doi:10.1017/S0007114514000087

Stack, A. I. (2025). The 7 biggest AI adoption challenges for 2025. Stack AI Blog. https://www.stack-ai.com/blog/the-biggest-ai-adoption-challenges

Statistics Authority. (2024). Analytical leadership: Achieving better outcomes for citizens. https://osr.statisticsauthority.gov.uk/publication/analytical-leadership-achieving-better-outcomes-for-citizens/pages/3/

Statistics Solutions. (2025). An introduction to the AMOS software. https://www.statisticssolutions.com/academic-research-consulting/dissertation-consulting-services/spss-statistics-help/amos/

Strielkowski, W., Vlasov, A., Selivanov, K., Muraviev, K., & Shakhnov, V. (2023). Prospects and challenges of the machine learning and data-driven methods for the predictive analysis of power systems: A review. *Energies*, *16*(10), 4025. DOI: 10.3390/en16104025

Sun, J., & Hagedorn, L. S. (2016). Homesickness at college: Its impact on academic performance and retention. *Journal of College Student Development, 57*(8), 943–957.

Sun, T. Q., & Medaglia, R. (2019). Mapping the challenges of artificial intelligence in the public sector: Evidence from public healthcare. *Government Information Quarterly, 36*(2), 368–383. DOI: 10.1016/j.giq.2018.09.008

Sun, Y., Ma, L., Fu, Y., Dai, H., & Zhang, Y. (2021). The improvement of gel and physicochemical properties of porcine myosin under low salt concentrations by pulsed ultrasound treatment and its mechanism. *Food Research International, 141*, 110056. DOI: 10.1016/j.foodres.2020.110056

Sutton, R. S., & Barto, A. G. (2018). *Reinforcement learning: An introduction* (2nd ed.). MIT Press., https://mitpress.mit.edu/9780262039246/reinforcement-learning/

Szylar, C. (2014). *Handbook Of Market Risk*. John Wiley & Sons.

Tadayonrad, Y., & Balle Ndiaye, A. (2023). A new key performance indicator model for demand forecasting in inventory management considering supply chain reliability and seasonality. *Supply Chain Analytics, 3*, 100026. DOI: 10.1016/j.sca.2023.100026

Tama, E. H., & Mark, D. (2024, September). The Future of Database Management in the Era of Big Data and Cloud Computing. *International Journal of Information Technology and Computer Engineering, 4*(45), 48–60. DOI: 10.55529/ijitc.45.48.60

Tang, Y. (2020). Government spending on local higher education institutions (LHEIs) in China: Analysing the determinants of general appropriations and their contributions. *Studies in Higher Education*, •••, 1–14.

Tanner, T., & Toivonen, H. (2010). Predicting and preventing student failure – using the k-nearest neighbour method to predict student performance in an online course environment. *International Journal of Learning Technology, 5*(4), 356–377.

Tao, L., Feng, L., Wei, H., Yi, J., Huang, S.-J., & Chen, S. (2022). Can adversarial training be manipulated by non-robust features? *Advances in Neural Information Processing Systems, 35*, 26504–26518.

Tatlıdil, H. (2002). *Uygulamalı Çok Değişkenli İstatistiksel Analiz*. Ziraat Matbaacılık A.Ş. Ankara.

Technology Innovators. (2024). Data analytics for evidence-based decision-making. https://www.technology-innovators.com/data-analytics-for-evidence-based-policy-making-and-decision-making/

TechTarget. (2023). What are the 5 V's of big data? TechTarget. https://www.techtarget.com/searchdatamanagement/definition/5-Vs-of-big-data

Tekin Epik, M., & Öztürk, M. (2020). Sağlık hizmetlerinde psikososyal riskler. *Süleyman Demirel Üniversitesi İktisadi ve İdari Bilimler Fakültesi Dergisi*, *25*(4), 451–467.

Teoh, C. W., Ho, S. B., Dollmat, K. S., & Tan, C. H. (2022). Ensemble-Learning techniques for predicting student performance on video-based learning. *International Journal of Information and Education Technology (IJIET)*, *12*(8), 741–745.

Teradata. (2022). What are the 5 V's of Big Data? Teradata. https://www.teradata.com/glossary/what-are-the-5-v-s-of-big-data

Thammasiri, D., Delen, D., Meesad, P., & Kasap, N. (2014). A critical assessment of imbalanced class distribution problem: The case of predicting freshmen student attrition. *Expert Systems with Applications*, *41*(2), 321–330.

Thelin, R. (2020). "What Is Big Data? Characteristics, Types, and Technologies." Last modified December 15, 2020.

Tomasevic, N., Gvozdenovic, N., & Vranes, S. (2020). An overview and comparison of supervised data mining techniques for student exam performance prediction. *Computers & Education*, *143*, 1–18.

Tomaškinová, J., Tomaškin, J., Drimal, M., & Bellido, J. (2025). The impact of abiotic environmental stressors on fluorescence and chlorophyll content in Glycine max (L.) Merrill. *Agronomy (Basel)*, *15*, 263. DOI: 10.3390/agronomy15020263

Tomiyama, A. J. (2019). Stress and obesity. *Annual Review of Psychology*, *70*(1), 703–718. DOI: 10.1146/annurev-psych-010418-102936

Tomiyama, A. J., Dallman, M. F., & Epel, E. S. (2011). Comfort food is comforting to those most stressed : Evidence of the chronic stress response network in high stress women. *Psychoneuroendocrinology*, *36*(10), 1513–1519. DOI: 10.1016/j.psyneuen.2011.04.005

Torres-Quezada, E., & Gandini-Taveras, R. J. (2023). Tomato plant data [Dataset]. Virginia Tech Data Repository. https://data.lib.vt.edu/articles/dataset/Tomato_plant_data/24164079

Torres, S. J., & Nowson, C. A. (2007). Relationship between stress, eating behavior, and obesity. *Nutrition (Burbank, Los Angeles County, Calif.)*, *23*(11-12), 887–894. DOI: 10.1016/j.nut.2007.08.008

Toumi, I., Ghrab, M., Zarrouk, O., & Nagaz, K. (2024). Impact of deficit irrigation strategies using saline water on soil and peach tree yield in an arid region of Tunisia. *Agriculture*, *14*(3), 377. DOI: 10.3390/agriculture14030377

Triantaphyllou, E. (2000). Multi-criteria decision making methods: A comparative study. Kluwer Academic Publishers. https://archive.org/details/multicriteriadec0000tria

Tsai, C. F., Tsai, C. T., Hung, C. S., & Hwang, P. S. (2011). Data mining techniques for identifying students at risk of failing a computer proficiency test required for graduation. *Australasian Journal of Educational Technology*, *27*(3), 481–498.

Tyralis, H., & Papacharalampous, G. (2024). A review of predictive uncertainty estimation with machine learning. *Artificial Intelligence Review*, *57*, 94. DOI: 10.1007/s10462-023-10698-8

Umberson, D., & Montez, J. K. (2010). Social relationships and health: A flashpoint for health policy. *Journal of Health and Social Behavior*, *51*(1, suppl), S54–S66.

UNESCO. (2024). Ethics of artificial intelligence. UNESCO. https://www.unesco.org/en/artificial-intelligence/recommendation-ethics

Unser, M. (2021). A unifying representer theorem for inverse problems and machine learning. *Foundations of Computational Mathematics*, *21*, 941–960. DOI: 10.1007/s10208-020-09472-x

Unuriode, A., Yusuf, B., Durojaiye, O., & Okunade, L. (2023, October). The Integration of Artificial Intelligence into Database Systems (AI-DB Integration Review). *International Journal on Cybernetics & Informatics*, *12*(6). Advance online publication. DOI: 10.5121/ijci.2023.1206012

Vaidya, O. S., & Kumar, S. (2006). Analytic hierarchy process: An overview of applications. *European Journal of Operational Research*, *169*(1), 1–29. DOI: 10.1016/j.ejor.2004.04.028

Valle-Cruz, D., Fernandez-Cortez, V., & Gil-Garcia, J. R. (2022). From E-budgeting to smart budgeting: Exploring the potential of artificial intelligence in government decision-making for resource allocation. *Government Information Quarterly*, *39*(2), 101644. Advance online publication. DOI: 10.1016/j.giq.2021.101644

ValueCoders. (2023). The Role of Machine Learning in Predictive Analytics and Decision-making. https://www.valuecoders.com/blog/analytics/role-of-machine-learning-in-predictive-analytics-and-decision-making/

Van de Schoot, R., Depaoli, S., King, R., Kramer, B., Märtens, K., Tadesse, M. G., & Vannucci, M.. (2021). Bayesian statistics and modelling. *Nature Reviews. Methods Primers*, *1*(1), 1.

Varoglu, N. (2022). Karbon Vergisinin CO_2 Emisyonu Üzerine Etkisi: Çevresel Kuznets Eğrisi Hipotezi Işığında Bir Uygulama (Yüksek lisans tezi, İstanbul Üniversitesi).

Vartanian, L. R., Schwartz, M. B., & Brownell, K. D. (2007). Effects of soft drink consumption on nutrition and health : A systematic review and meta-analysis. *American Journal of Public Health*, *97*(4), 667–675.

Vasilchenko, A. (2024). Predictive maintenance in manufacturing using machine learning: Use cases, challenges and best practices. *MobiDev*. https://mobidev.biz/blog/predictive-maintenance-in-manufacturing-using-machine-learning

Vasuki, M., & Revathy, S. (2022). Analyzing performance of placement students record using different clustering algorithm [IJCSE]. *Indian Journal of Computer Science and Engineering*, *13*(2), 410–419.

Ventura, G., Coniglio, D., Losito, I., Cataldi, T. R. I., & Calvano, C. D. (2026). Impact of industrial processing on the phospholipid profile of canned tomatoes (Solanum lycopersicum). *Journal of Food Composition and Analysis*, *149*, 108692. DOI: 10.1016/j.jfca.2025.108692

Verma, S., Kholiya, K., & Bala, K. (2025). Leveraging Ensemble Model and Optimized Feature Selection to Boost Prediction Accuracy in Educational Data Mining. *SN Computer Science*, *6*(5), 487.

Verma, S., & Yadav, R. K. (2020). Effect of different attributes on the academic performance of engineering students, *IEEE International Conference on Advent Trends in Multidisciplinary Research and Innovation (ICATMRI),* 1-4.

Verma, S., Yadav, R. K., & Kholiya, K. (2022). A Scalable Machine Learning-based Ensemble Approach to Enhance the Prediction Accuracy for Identifying Students at-Risk. *International Journal of Advanced Computer Science and Applications*, *13*(8), 185–192.

Vilaro, M. J., Colby, S. E., Riggsbee, K., Zhou, W., Byrd-Bredbenner, C., Olfert, M. D., Barnett, T. E., Horacek, T., Sowers, M., & Mathews, A. E. (2018). Food choice priorities change over time and predict dietary intake at the end of the first year of college among students in the US. *Nutrients*, *10*(9), 1296. DOI: 10.3390/nu10091296

Vishnoi, S., & Goel, R. K. (2024). Climate smart agriculture for sustainable productivity and healthy landscapes. *Environmental Science & Policy*, *151*, 103600. DOI: 10.1016/j.envsci.2023.103600

Wahab, A., Abdi, G., Saleem, M. H., Ali, B., Ullah, S., Shah, W., Mumtaz, S., Yasin, G., Muresan, C. C., & Marc, R. A. (2022). Plants' physio-biochemical and phyto-hormonal responses to alleviate the adverse effects of drought stress: A comprehensive review. *Plants*, *11*(13), 1620. DOI: 10.3390/plants11131620

Wald, A., & Wolfowitz, J. (1940). On A Test Whether Two Samples Are From The Same Population. *Annals of Mathematical Statistics*, *11*(2), 147–162. DOI: 10.1214/aoms/1177731909

Wang, Z., Liu, Z., Yuan, M., Yin, W., Zhang, C., Zhang, Z., & Hu, X. (2025). A machine learning-based irrigation prediction model for cherry tomatoes in greenhouses: Leveraging optimal growth data for precision irrigation. Computers and Electronics in Agriculture, 237(A), 110558. https://doi.org/DOI: 10.1016/j.compag.2025.110558

Wang, D., Cao, W., Zhang, F., Li, Z., Xu, S., & Wu, X. (2022). A review of deep learning in multiscale agricultural sensing. *Remote Sensing*, *14*, 559. DOI: 10.3390/rs14030559

Wang, S., Yang, Y., Yin, H., Zhao, J., Wang, T., Yang, X., Ren, J., & Yin, C. (2025a). Towards digital transformation of agriculture for sustainable development in China: Experience and lessons learned. *Sustainability*, *17*(8), 3756. DOI: 10.3390/su17083756

Wang, X., Shi, J., Sun, S., Wang, D., Li, Z., Sun, Z., Sigrimis, N., & Li, T. (2025b). Understanding complex process sensor signal may be the source for deep learning: A smart IoT delving into a tomato weight sensor. *Computers and Electronics in Agriculture*, *231*, 110012. DOI: 10.1016/j.compag.2025.110012

Wang, X., & Xing, Y. (2017). Evaluation of the effects of irrigation and fertilization on tomato fruit yield and quality: A principal component analysis. *Scientific Reports*, *7*, 350. DOI: 10.1038/s41598-017-00373-8 PMID: 28336916

Wang, Y., Wei, W., Liu, Z., Liu, J., Lv, Y., & Li, X. (2025). Interpretable machine learning framework for corporate financialization prediction: A SHAP-based analysis of high-dimensional data. *Mathematics*, *13*(15), 2526. DOI: 10.3390/math13152526

Wang, Z., & Ren, F. (2025). Developing a decision support system for sustainable urban planning using machine learning-based scenario modeling. *Scientific Reports*, *15*, 13210. DOI: 10.1038/s41598-025-90057-5

Wankhade, M., Annavarapu, C. S. R., & Kulkarni, C. (2022). A survey on sentiment analysis methods, applications, and challenges. *Artificial Intelligence Review*, *55*(7), 5731–5780.

Warburton, D. E., Nicol, C. W., & Bredin, S. S. (2006). Health benefits of physical activity : The evidence. *Canadian Medical Association Journal, 174*(6), 801–809. https://doi.org/DOI. DOI: 10.1503/cmaj.051351

Weihrich, H. (1982). The TOWS matrix—A tool for situational analysis. *Long Range Planning, 15*(2), 54–66. DOI: 10.1016/0024-6301(82)90120-0

Weiner, E. B., Dankwa-Mullan, I., Nelson, W. A., & Hassanpour, S. (2025). Ethical challenges and evolving strategies in the integration of artificial intelligence into clinical practice. *PLOS Digital Health, 4*(4), e0000810. DOI: 10.1371/journal.pdig.0000810 PMID: 40198594

WHO. (2023). *Tobacco*. World Health Organization. https://www.who.int/news-room/fact-sheets/detail/tobacco

Widyahastutia, F., & Tjhin, V. U. (2018). Performance Prediction in Online Discussion Forum: State-of-the-art and comparative analysis. *Procedia Computer Science, 135*, 302–314.

Wirtz, B. W., Weyerer, J. C., & Sturm, B. J. (2020). The dark sides of artificial intelligence: An integrated AI governance framework for public administration. *International Journal of Public Administration, 43*(9), 818–829.

Wolfson, J. A., & Bleich, S. N. (2015). Is cooking at home associated with better diet quality or weight-loss intention? *Public Health Nutrition, 18*(8), 1397–1406. DOI: 10.1017/S1368980014001943

Wong, S. L. (2019). Relationship between interest and mathematics performance in a technology-enhanced learning context in Malaysia. *Research and Practice in Technology Enhanced Learning, 14*(1), 1–13.

Wooldridge, J. M. (2020). *Introductory econometrics: A modern approach* (7th ed.). Cengage Learning.

Worden, M. A., Bilir, T. E., Bloom, A. A., Fang, J., Klinek, L. P., Konings, A. G., Levine, P. A., & Milodowski, D. T.. (2025). Combining observations and models: A review of the CARDAMOM framework for data-constrained terrestrial ecosystem modeling. *Global Change Biology, 31*(8), e70462. DOI: 10.1111/gcb.70462 PMID: 40856273

World Health Organization (WHO). (2020). *WHO report on the global tobacco epidemic 2019: Offer help to quit tobacco use*. WHO.

Wunnasri, W., Musikawan, P., & So-In, C. (2023). A Two Phase Ensemble-Based Method for Predicting Learners' Grade in MOOCs. *Applied Sciences (Basel, Switzerland), 13*(1492), 1–15.

Xiang, F., Zhao, Y., Zhang, M., Zuo, Y., Zou, X., & Tao, F. (2024). Ensemble learning-based stability improvement method for feature selection towards performance prediction. *Journal of Manufacturing Systems*, *74*, 55–67.

Xiao, W., Ji, P., & Hu, J. (2021). A survey on educational data mining methods used for predicting students' performance. *Engineering Reports*, *4*(5), 1–23.

Yaakub, T. N. T., Ahmad, W. R. W., Husaini, Y., & Burham, N. (2018). Influence Factors in Academic Performance among Electronics Engineering Student: Geographic Background, Mathematics Grade and Psycographic Characteristics, *IEEE 10th International Conference on Engineering Education Kuala Lumpur, Malaysia*, 30-33.

Yagci, M. (2022). Educational data mining: Prediction of students' academic performance using machine learning algorithms. *Smart Learning Environments*, *9*(1), 1–19.

Yahoo Finance. (2024). Yahoo Finance. https://finance.yahoo.com/

Yang, C., Huang, Q., Li, Z., Liu, K., & Hu, F. (2016). Big Data and Cloud Computing: Innovation Opportunities and Challenges. *International Journal of Digital Earth*, *10*(1), 13–53. DOI: 10.1080/17538947.2016.1239771

Yanglet, X.-Y. L., Cao, Y., & Deng, L. "Multimodal financial foundation models (mffms): Progress, prospects, and challenges." *arXiv preprint arXiv:2506.01973* (2025).

Ye, L., & Zhang, X. (2018). Nonlinear Granger causality between health care expenditure and economic growth in the OECD and major developing countries. *International Journal of Environmental Research and Public Health*, *15*(9), 1953. DOI: 10.3390/ijerph15091953

Yigit, A., Karahan, F., & Demir, S. (2021). Noncommunicable disease risk factors and health inequalities across OECD countries. *European Journal of Public Health Studies*, *9*(3), 55–72.

Yiğit, P., Yılmaz, E., Gedikli, E., & Hayran, O. E. (2021). *The Relationship Between Noncommunicable Diseases Risk Factors, GINI Coefficient and Health Status in Organization of Economic Development and Cooperation Countries*. Turkiye Klinikleri Journal of Health Sciences.

Yuliansyah, H., Imaniati, R. A. P., Anggit Wirasto, A., & Wibowo, M. (2021). Predicting Students Graduate on Time Using C4.5 Algorithm. *Journal of Information Systems Engineering and Business Intelligence*, *7*(1), 67–73.

Zalai, M., Bujtás, O., Sárospataki, M., & Dorner, Z. (2025). Grassy and herbaceous interrow cover crops in European vineyards: A review of their short-term effects on water management and regulating ecosystem services. *Land (Basel)*, *14*(8), 1526. DOI: 10.3390/land14081526

Zampetti, F., Geremia, S., Bavota, G., & Di Penta, M. "CI/CD pipelines evolution and restructuring: A qualitative and quantitative study." In *2021 IEEE International Conference on Software Maintenance and Evolution (ICSME)*, pp. 471-482. IEEE, 2021.

Zanakis, S. H., Solomon, A., Wishart, N., & Dublish, S. (1998). Multi-attribute decision making: A simulation comparison of select methods. *European Journal of Operational Research*, *107*(3), 507–529. DOI: 10.1016/S0377-2217(98)00128-3

Zaremba, A., & Demir, E. (2023). ChatGPT: Unlocking the future of NLP in finance. *Modern Finance*, *1*(1), 93–98.

Zeng, R., Abate, M. C., Cai, B., Addis, A. K., & Derebe Dereso, Y. (2025). A systematic review of contemporary challenges and debates on Chinese food security: Integrating priorities, trade-offs, and policy pathways. *Foods*, *14*(6), 1057. DOI: 10.3390/foods14061057 PMID: 40232105

Zeng, Y., Mai, S., Yan, W., & Hu, H. (2023). Multimodal reaction: Information modulation for cross-modal representation learning. *IEEE Transactions on Multimedia*, *26*, 2178–2191.

Zhang, L., Bate, B., Cui, J., Feng, Y., Yu, J., Cui, Z., Wang, H., & Li, Q. (2025). Biochar input to saline-alkali farmland can improve soil health and crop yield: A meta-analysis. *Agriculture*, *15*(5), 561. DOI: 10.3390/agriculture15050561

Zhang, Y., & Tang, Q. (2025). Integrating AI-generated content tools in higher education: A comparative analysis of interdisciplinary learning outcomes. *Scientific Reports*, *15*(1), 25802. DOI: 10.1038/s41598-025-10941-y PMID: 40670544

Zhang, Z., Tyc, J., & Hensel, M. (2025). An ecogeomorphological approach to land-use planning and natural hazard risk mitigation: A literature review. *Land (Basel)*, *14*(9), 1911. DOI: 10.3390/land14091911

Zhao, Y., Yuan, X., Ran, W., Zhao, Z., Su, D., & Song, Y. (2025). The ecological restoration strategies in terrestrial ecosystems were reviewed: A new trend based on soil microbiomics. *Ecology and Evolution*, *15*(3), e70994. DOI: 10.1002/ece3.70994 PMID: 40060716

Zheng, A., & Casari, A. (2018). *Feature engineering for machine learning: Principles and techniques for data scientists*. O'Reilly Media.

Zhou, L., Pan, S., Wang, J., & Vasilakos, A. V. (2017). Machine learning on big data: Opportunities and challenges. *Neurocomputing*, *237*, 350–361. DOI: 10.1016/j.neucom.2017.01.026

Zhuchkova, S. M., Busalaeva, E. I., Markova, T. N., & Semenova, V. A. (2024). Obesity in medical students : What has changed in 18 years. *Therapy*, *10*(3), 146–152. DOI: 10.18565/therapy.2024.3.146-152

Zhu, J., Zhang, J., & Feng, Y. (2022). Hard budget constraints and artificial intelligence technology. *Technological Forecasting and Social Change*, *183*, 121889. DOI: 10.1016/j.techfore.2022.121889

Zhu, Z., Zhu, G., Zhang, Y., Shi, J., Huang, X., & Fang, Y. (2024). Eigenobfu: A novel network topology obfuscation defense method. *IEEE Transactions on Network Science and Engineering*.

Zilliz. (2024). How does deep learning scale to large datasets? Zilliz AI FAQ. https://zilliz.com/ai-faq/how-does-deep-learning-scale-to-large-datasets

Zivot, E., & Andrews, D. W. K. (2002). Further Evidence on the Great Crash, the Oil-Price Shock, and the Unit-Root Hypothesis. *Journal of Business & Economic Statistics*, *20*(1), 25–44. DOI: 10.1198/073500102753410372

Zuheros, C., Martínez-Cámara, E., Herrera-Viedma, E., Katib, I. A., & Herrera, F. (2023). Explainable Crowd Decision Making methodology guided by expert natural language opinions based on Sentiment Analysis with Attention-based Deep Learning and Subgroup Discovery. *Information Fusion*, *97*, 101821. DOI: 10.1016/j.inffus.2023.101821

About the Contributors

Serkan Derici has been working at Nevşehir Hacı Bektaş Veli University, Faculty of Economics and Administrative Sciences, Department of Business Administration since 2017. He is currently an Assistant Professor Dr. Derici received his bachelor's degree from Nevşehir University, Department of Business Administration in 2012. He studied at Erciyes University for one year with the Farabi student exchange programme. He completed his master's degree in Business Administration at the same university and then his doctorate. Derici started his academic life as a research assistant at Istanbul Gelisim University School of Applied Sciences, Department of Aviation Management in 2016. His current research mainly focuses on quantitative methods, machine learning, supply chain management, big data, digitalisation, artificial intelligence, fuzzy logic, operations management and multi-criteria decision making.

***

Wafaa Benkayba is a chemistry researcher specializing in electrochemistry and analytical chemistry. Her multidisciplinary work covers water treatment, pharmaceutical and cosmetic analysis, food authenticity, green nanomaterials, and the development of electrochemical sensors. She also contributes to research on antimicrobial resistance (AMR) and traditional medicine. Her research bridges analytical sciences, environmental sustainability, and public health, with a strong commitment to developing evidence-based, sustainable, and culturally informed solutions to address global health and environmental challenges.

Fatih Çemrek serves at Eskişehir Osmangazi University, Faculty of Science, Department of Statistics, Division of Applied Statistics. His primary research interests include quantitative decision methods, data analytics, business statistics, optimization, and efficiency analysis. Dr. Çemrek's academic studies mainly focus on energy economics, sustainable growth, time series forecasting models, artificial neural networks, and support vector machines. In recent years, his publications have examined the relationship between energy consumption and economic growth in Turkey and the European Union countries. He has also developed AI-based models for forecasting CO_2 emissions and pension fund returns. He frequently collaborates with scholars such as Hüseyin Naci Bayraç and Özer Demir, contributing to research that integrates statistical modeling with applications in economics, environment, and finance. His work has been published in both national and international journals, reflecting a strong interdisciplinary approach to applied statistical modeling and econometric analysis.

Susanta Das is a seasoned educator with a proven record of teaching and mentoring diverse students, working at multiple institutions, and providing service to the various initiatives of organizations. He received his Ph.D. and M.A. degrees from Western Michigan University (WMU), USA, and M.Sc. degree from Banaras Hindu University (BHU), India all in Physics. He continued his research as a Marie-Curie post-doctoral fellow at Stockholm University, Stockholm, Sweden on beam diagnostics for the DESIREE (Double ElectroStatic Ion Ring ExpEriment) in the project DITANET (Diagnostic Techniques for particle Accelerators – a Marie-Curie initial training NETwork), at the Indian Institute of Science Education and Research-Kolkata, India on high-pressure physics as a Project Scientist-B, and at the University of Electro-Communications, Tokyo, Japan on ion-surface interactions as a Post-doctoral fellow. Throughout his career, Dr. Das worked and collaborated with many researchers, Ph.D., master, and visiting students from different countries. He further visited many countries to discuss research and an international conference participant (UK, Italy, Germany, Greece, Belgium, Bulgaria, Romania, Brazil etc.). He worked at several institutions before assuming his current role at Ajeenkya D Y Patil University. He received the Marie-Curie post-doctoral fellowship in Sweden, Science Academies' Summer Research Fellowship in India, Gowen Frostic Doctoral Fellowship, Department Graduate Research and Creative Scholar Award by WMU, and Leo R. Parpart Doctoral Fellowship by Dept. of Physics, WMU, among many others, throughout his academic journey. He co-authored several research articles in WoS/Scopus indexed international journals, conference proceedings, and scholarly book chapters and co-edited several books published by renowned international publishing houses (IGI-Global, Springer Nature, Nova Science Publishers, Wiley, Scrivener-Wiley, Notion Press). At present, Dr. Das continues to delve into cutting-edge fields, with a keen interest in nanotechnology, UN SDGs, and data science. His multifaceted contributions, spanning teaching, research, and administrative leadership, showcase a dedicated professional who is instrumental in advancing the vision and mission of the institutions he serves.

Ebrucan İslamoğlu is a researcher at Nevşehir Hacı Bektaş Veli University, specializing in Islamic banking, financial consumer behavior, social welfare, and applied quantitative methods. Her work focuses on the impact of ethical and interest-free financial practices on decision-making and social outcomes. She applies multi-criteria decision-making techniques, hybrid forecasting models, and structural equation modeling to analyze investment behavior, organizational performance, and sustainability indicators. Her research has been published in national and international journals and presented at academic conferences, reflecting a strong interdisciplinary and applied research orientation.

Abdellah Lemfarrak PhD candidate in Environmental Sciences, I conduct research on sustainable water management, with a particular focus on treated wastewater reuse and urban ecological practices in arid and semi-arid regions. My work explores the intersection between water reuse technologies, environmental regulation, and green space irrigation. I am also involved in quality management systems (ISO 9001) and interdisciplinary approaches to ecosystem resilience and resource optimization.

Tanishka Prashant Madke is currently pursuing her B.Tech. in Computer Engineering at the School of Engineering, Ajeenkya D Y Patil University, Pune, Maharashtra, India. She has a strong passion for research and has authored multiple articles in NOVA Science Publishers and IGI-Global, USA. Her research interests encompass AI (Artificial Intelligence), Big Data, climate change, and environmental sustainability. She is particularly focused on contributing to the United Nations Sustainable Development Goals (SDGs), especially those related to climate action, innovation in technology, and sustainable cities. Through her work, she aims to explore how AI and emerging technologies can drive sustainable solutions, bridging the gap between innovation and global challenges.

Ghada Nakhla is the Undergraduate Mathematics Programme Leader in the Faculty of Arts and Sciences at Edge Hill University. She holds a PhD from Lancaster University and teaches across a wide range of mathematical disciplines, including calculus, linear algebra, statistical inference, statistical modelling, discrete mathematics, and data analytics. In addition to her teaching, she supervises MSc and PhD students, supporting the development of future researchers. Her research portfolio is extensive, and she has served as principal investigator on numerous projects. Her primary interests lie in advanced statistical methods, with a particular focus on the applications of the General Linear Model and Structural Equation Modelling in diverse fields. Her scholarship has been published in peer reviewed journals, featured as chapters in edited volumes, and presented at conferences.

Beyond statistical modelling, Ghada's work extends into educational psychology. She has established a strong research agenda examining the role of fear in academic settings, exploring how emotional and psychological dynamics shape student performance. Her studies reveal both the hindering and motivating effects of fear, providing nuanced insights into how emotions influence academic engagement. Through this interdisciplinary approach, Ghada contributes not only to the advancement of quantitative research methods but also to a deeper understanding of the psychological factors that drive teaching, learning, and student success.

Amisha Shaikh is currently doing her B.Tech. in Computer Engineering at the School of Engineering, Ajeenkya D Y Patil University, Pune, Maharashtra, India. She has broad research interests from understanding the properties of biogenic gold nanoparticles to their application in various fields. In addition to nanoparticles, her interests extend into AI (Artificial Intelligence), Biotechnology, and Cybersecurity. She strives to contribute significantly in the possibilities of nanoparticles and AI, focusing on their vast applications. She aims to conduct research in these domains and make them more applied.

Okan Tiring is a researcher specializing in psychology, e-sports, perfectionism, life satisfaction, motivation, and stress-related behaviors. His work examines the influence of self-concept, positive future expectations, parental attitudes, and cognitive flexibility on psychological outcomes. He has published studies in national and international journals and conference proceedings, applying quantitative methods, literature reviews, and applied psychological models to understand individual and social behavior. His research contributes to both educational and clinical psychology contexts.

Index

www.ingramcontent.com/pod-product-compliance
Lightning Source LLC
LaVergne TN
LVHW080911300926
849162LV00003B/636